Reviews of Environmental Contamination and Toxicology

VOLUME 187

Reviews of Environmental Contamination and Toxicology

Continuation of Residue Reviews

Editor
George W. Ware

Founding Editor
Francis A. Gunther

VOLUME 187

Springer

Springer
New York: 233 Spring Street, New York, NY 10013, USA
Heidelberg: Postfach 10 52 80, 69042 Heidelberg, Germany

ISSN 0179-5953

ISBN-10: 1-4899-9645-1
ISBN-13: 978-1-4899-9645-9
ISBN: 978-0-387-32885-0 (eBook)

springer.com

Foreword

International concern in scientific, industrial, and governmental communities over traces of xenobiotics in foods and in both abiotic and biotic environments has justified the present triumvirate of specialized publications in this field: comprehensive reviews, rapidly published research papers and progress reports, and archival documentations. These three international publications are integrated and scheduled to provide the coherency essential for nonduplicative and current progress in a field as dynamic and complex as environmental contamination and toxicology. This series is reserved exclusively for the diversified literature on "toxic" chemicals in our food, our feeds, our homes, recreational and working surroundings, our domestic animals, our wildlife and ourselves. Tremendous efforts worldwide have been mobilized to evaluate the nature, presence, magnitude, fate, and toxicology of the chemicals loosed upon the earth. Among the sequelae of this broad new emphasis is an undeniable need for an articulated set of authoritative publications, where one can find the latest important world literature produced by these emerging areas of science together with documentation of pertinent ancillary legislation.

Research directors and legislative or administrative advisers do not have the time to scan the escalating number of technical publications that may contain articles important to current responsibility. Rather, these individuals need the background provided by detailed reviews and the assurance that the latest information is made available to them, all with minimal literature searching. Similarly, the scientist assigned or attracted to a new problem is required to glean all literature pertinent to the task, to publish new developments or important new experimental details quickly, to inform others of findings that might alter their own efforts, and eventually to publish all his/her supporting data and conclusions for archival purposes.

In the fields of environmental contamination and toxicology, the sum of these concerns and responsibilities is decisively addressed by the uniform, encompassing, and timely publication format of the Springer triumvirate:

Reviews of Environmental Contamination and Toxicology [Vol. 1 through 97 (1962–1986) as Residue Reviews] for detailed review articles concerned with any aspects of chemical contaminants, including pesticides, in the total environment with toxicological considerations and consequences.

Bulletin of Environmental Contamination and Toxicology (Vol. 1 in 1966) for rapid publication of short reports of significant advances and discoveries in the fields of air, soil, water, and food contamination and pollution as well as methodology and other disciplines concerned with the introduction, presence, and effects of toxicants in the total environment.

Archives of Environmental Contamination and Toxicology (Vol. 1 in 1973) for important complete articles emphasizing and describing original experimental or theoretical research work pertaining to the scientific aspects of chemical contaminants in the environment.

Manuscripts for *Reviews* and the *Archives* are in identical formats and are peer reviewed by scientists in the field for adequacy and value; manuscripts for the *Bulletin* are also reviewed, but are published by photo-offset from camera-ready copy to provide the latest results with minimum delay. The individual editors of these three publications comprise the joint Coordinating Board of Editors with referral within the Board of manuscripts submitted to one publication but deemed by major emphasis or length more suitable for one of the others.

Coordinating Board of Editors

Preface

The role of *Reviews* is to publish detailed scientific review articles on all aspects of environmental contamination and associated toxicological consequences. Such articles facilitate the often-complex task of accessing and interpreting cogent scientific data within the confines of one or more closely related research fields.

In the nearly 50 years since *Reviews of Environmental Contamination and Toxicology* (formerly *Residue Reviews*) was first published, the number, scope and complexity of environmental pollution incidents have grown unabated. During this entire period, the emphasis has been on publishing articles that address the presence and toxicity of environmental contaminants. New research is published each year on a myriad of environmental pollution issues facing peoples worldwide. This fact, and the routine discovery and reporting of new environmental contamination cases, creates an increasingly important function for *Reviews*.

The staggering volume of scientific literature demands remedy by which data can be synthesized and made available to readers in an abridged form. *Reviews* addresses this need and provides detailed reviews worldwide to key scientists and science or policy administrators, whether employed by government, universities or the private sector.

There is a panoply of environmental issues and concerns on which many scientists have focused their research in past years. The scope of this list is quite broad, encompassing environmental events globally that affect marine and terrestrial ecosystems; biotic and abiotic environments; impacts on plants, humans and wildlife; and pollutants, both chemical and radioactive; as well as the ravages of environmental disease in virtually all environmental media (soil, water, air). New or enhanced safety and environmental concerns have emerged in the last decade to be added to incidents covered by the media, studied by scientists, and addressed by governmental and private institutions. Among these are events so striking that they are creating a paradigm shift. Two in particular are at the center of ever-increasing media as well as scientific attention: bioterrorism and global warming. Unfortunately, these very worrisome issues are now super-imposed on the already extensive list of ongoing environmental challenges.

The ultimate role of publishing scientific research is to enhance understanding of the environment in ways that allow the public to be better informed. The term "informed public" as used by Thomas Jefferson in the age of enlightenment conveyed the thought of soundness and good judgment. In the modern sense, being "well informed" has the narrower meaning of having access to sufficient information. Because the public still gets most of its information on science and technology from TV news and reports, the role for scientists as interpreters and brokers of scientific information to the public will grow rather than diminish.

Environmentalism is the newest global political force, resulting in the emergence of multi-national consortia to control pollution and the evolution of the environmental ethic. Will the new politics of the 21st century involve a consortium of technologists and environmentalists, or a progressive confrontation? These matters are of genuine concern to governmental agencies and legislative bodies around the world.

For those who make the decisions about how our planet is managed, there is an ongoing need for continual surveillance and intelligent controls, to avoid endangering the environment, public health, and wildlife. Ensuring safety-in-use of the many chemicals involved in our highly industrialized culture is a dynamic challenge, for the old, established materials are continually being displaced by newly developed molecules more acceptable to federal and state regulatory agencies, public health officials, and environmentalists.

Reviews publishes synoptic articles designed to treat the presence, fate, and, if possible, the safety of xenobiotics in any segment of the environment. These reviews can either be general or specific, but properly lie in the domains of analytical chemistry and its methodology, biochemistry, human and animal medicine, legislation, pharmacology, physiology, toxicology and regulation. Certain affairs in food technology concerned specifically with pesticide and other food-additive problems may also be appropriate.

Because manuscripts are published in the order in which they are received in final form, it may seem that some important aspects have been neglected at times. However, these apparent omissions are recognized, and pertinent manuscripts are likely in preparation or planned. The field is so very large and the interests in it are so varied that the Editor and the Editorial Board earnestly solicit authors and suggestions of underrepresented topics to make this international book series yet more useful and worthwhile.

Justification for the preparation of any review for this book series is that it deals with some aspect of the many real problems arising from the presence of foreign chemicals in our surroundings. Thus, manuscripts may encompass case studies from any country. Food additives, including pesticides, or their metabolites that may persist into human food and animal feeds are within this scope. Additionally, chemical contamination in any manner of air, water, soil, or plant or animal life is within these objectives and their purview.

Manuscripts are often contributed by invitation. However, nominations for new topics or topics in areas that are rapidly advancing are welcome. Preliminary communication with the Editor is recommended before volunteered review manuscripts are submitted.

Tucson, Arizona G.W.W.

Table of Contents

Rev Environ Contam Toxicol 187:1–65

Does Pesticide Risk Assessment in the European Union Assess Long-Term Effects?

Michael C. Newman, Mark Crane, and Graham Holloway

Contents

Communicated by George W. Ware.

M.C. Newman (✉)
College of William and Mary, Virginia Institute of Marine Science, P.O. Box 1346, Rt. 1208 Greate Rd., Gloucester Point, VA 23062 USA.

M. Crane
Watts & Crane Associates, 23 London Street, Faringdon, Oxfordshire SN7 7AG, U.K.

G. Holloway
School of Animal and Microbial Sciences, University of Reading, Whiteknights, Reading RG6 6AJ, U.K.

I. Introduction

> *A number of authors have argued that the application of more ecology to the science of toxicology would facilitate a more comprehensive evaluation of risk. . . . However, it is only recently that authors have proposed using ecological factors as predictors of recovery rather than simply as variables that may confound the effects of toxicity.*
>
> (Sherratt et al. 1999)

Arguably, current pesticide risk assessments only indirectly address long-term pesticide effects on nontarget species populations and assemblages (Cairns 1984, 1992; Pratt and Cairns 1996). Most environmental professionals would quickly agree if this statement were being made about past assessments of legacy pesticides, such as DDT and other organochlorines (Newman 2001; Clements and Newman 2002) and might also add that the so-called legacy pesticides are still used in many, especially developing, countries (McLaughlin and Mineau 1995; Castillo et al. 1997; Nakamaru et al. 2002). However, considerable disagreement would emerge if the statement were made about present-day assessments of widely used pesticides in the developed world.

Current assessments of long-term ecological effects are performed mostly, although not exclusively, with data from individual-based, short-term tests (Crommentuijn et al. 2000). Risks or hazards are then predicted for individuals and populations by extrapolation to natural settings and natural communities. Although some ecological issues are addressed as certain pesticide assessments progress to higher tier investigations, most current assessments of long-term pesticide effects rely heavily on individual-based effect metrics derived from relatively short duration studies. Mesocosms and enclosure studies that have more ecological realism than laboratory ecotoxicity tests are less common than perhaps warranted. For example, Annex II of European Directive 91/414/EEC describes Tier 1 assessments for terrestrial wildlife as employing individual-based, laboratory assays and progressing to field or enclosure studies only at higher assessment tiers. In the United States, the proposed ECOFRAM approach (ECOFRAM 1999) also relies on acute toxicity tests at the lowest tier. Spatially explicit population analyses and ecological effects detection in mesocosms occur only at the Tier 4 stage, with only a small number of pesticide assessments progressing to this tier. In European pesticide risk assessments, the Guidance Document on Risk Assessment for Birds and Mammals under Council Directive 91/414/EEC stipulates:

> *One aim of the ecological risk assessment is to predict effects on the population level, although this is difficult or impossible to measure directly. The usual approach is based on the consideration that effects on populations will not occur if the survival rate, reproduction rate and development of individuals are not affected.*

Some professionals endorse this approach as a pragmatic and sufficient one for predicting ecological effects based on well-founded autoecological principles. Others reject it as producing sciolistic predictions (i.e., prediction from indirect, dubious evidence) of long-term ecological effects from demonstrably inadequate, reductionist tests. For example, the continued and widespread decline of bird (Beaumont 1997; Krebs et al. 1999) and amphibian (Wake 1991) species has been invoked to support the argument that current methods are ineffectual in assessing the long-term effects of pesticide use on nontarget species persistence. The problem with overreliance on single species tests is summarized in the three quotes below:

> *In our opinion, single-species bioassays generally miss the essence of the [ecological risk assessment] problem, even though they are quick, rather inexpensive to obtain, and straightforward to interpret (as far as they go)."*
>
> (Joern and Hoagland 1996)

> *If pesticide registration is to become more responsive to ecological issues, the information and approach to determining potential effects must be made explicitly ecological. The current focus is foremost on chemistry and toxicology."*
>
> (Kapustka et al. 1996)

> *. . . many ecologists are convinced that ecological studies on the community level can make significant contributions to determining the relative risk of toxic materials that will be introduced into the environment, and that these should be part of the regulatory decision process.*
>
> (Taub 1997)

Given this stark divergence between established practice and the views of many ecotoxicologists, it becomes important to know precisely what is intended when a pesticide risk assessor derives an Ecologically Acceptable Concentration (EAC) to be used to support authorization of a pesticide. Directive 91/414/EEC states that an EAC is the concentration at or below which no *unacceptable* "influences on the environment in general . . ." are to be expected.[1] It is obvious from this definition that ample understanding of long-term exposure effects on nontarget species and species assemblages is essential for deriving a defensible EAC. Crucial at this point would be a clear answer to the simple question, "Do current methods provide enough insight into long-term ecological effects to define pesticide EACs?" Based on the information provided below, our answer to this important question is no.

Practicality must be part of any answer to this question. Risk assessment methods must be pragmatic so that the regulated community and regulators can effectively judge costs and benefits, and so that a rational decision can be made even when uncertainties remain. However, this pragmatism must be balanced against the requirement that procedures be consistent

[1]Environment is defined in the Directive as "the water, air, land and wild species of fauna and flora, and any interrelationship between them, as well as any relationship with living organisms."

with ecological knowledge and paradigms, as such a balance is essential to meeting society's needs and the requirement to move toward sustainable practices. Decision makers require insight from expedient, yet *defensible*, tools. The key to optimal decision making by managers is to focus on the most expedient methods while minimizing satisficing, which is the satisfaction of vested interests by abandoning the best decision.

In this report, the ecological and ecotoxicological literature is reviewed to highlight reports of ecological impacts of pesticides on nontarget individuals, species populations, and species assemblages. Insights are drawn from conservation biology as well. This information is then used to address the question of current method adequacy.

This review considers patterns of recovery and recolonization for species possessing different life history qualities and assesses the utility of population projection models in fostering ecologically sound management. Modeling software based on individuals, populations, and species assemblages that has potential use in pesticide regulation and management is also considered. In addition, emerging probability-based methods are reviewed relative to the central question of method adequacy.

II. Effects in Ecosystems from Pesticides Applied According to Good Agricultural Practice

A. General

> *In the framework of Directive 91/414/EEC risks arising from direct toxic effects are considered secondary ecological effects e.g. due to decline in food resources are currently not evaluated, although they are within the scope of the Directive*
>
> (Guidance Document on Risk Assessment for Birds and Mammals under Council Directive 91/414/EEC)

> *Pesticides are notorious for altering the interrelationships of species in ecosystems.*
>
> (Pimentel 1992)

Pesticide effects on ecosystem components and processes are extensively documented in the open literature. Some direct effects on individuals have clear consequences for species populations within communities. Also, many studies have developed quantitative models for predicting pesticide concentration effects on vital rates (Forbes and Calow 1999, 2003) that allow some extrapolation to population consequences. Some pesticide exposures indirectly affect an individual by harming other species populations with which it interacts, although such indirect effects are often erroneously considered as less important (e.g., "secondary ecological effects" in the above quote) than direct effects on individuals in the ecological risk assessment literature. This relegation is inconsistent with ecological theory. Indeed, one of the most commonly cited examples of indirect effects in ecology involves the World Health Organization spraying of DDT and dieldrin in Borneo to protect the Dayak people from malaria (discussed in Clements and Newman 2002). Spraying to eliminate mosquitoes indirectly caused a rapid increase in rat populations because the cats that normally controlled the rat populations were poisoned after these pesticides underwent biomagnification within the trophic sequence insects → geckos → cats. The DDT also killed a parasitic wasp that controlled caterpillars that fed on roof thatch, causing roofs to collapse. None of these indirect effects would have been predicted from individual-based toxicity testing of DDT.

The argument could be made that, although they exist, indirect effects are uncommon. However, Atchison et al. (1996) provide ample evidence that effects on species interactions are to be expected in many instances. Furthermore, species interactions are generally acknowledged as important in determining the structure of ecological communities and for proper ecosystem functioning. The conventional model for simple species competition can be used to illustrate this point. Species competition is relevant to and can be affected by pesticide applications, for example, the study of lindane by Blockwell et al. (1998). The basic Lotka–Volterra model for two competing species is the following:

$$\frac{dN_1}{dt} = r_1 N_1 \left[1 - \frac{N_1}{K_1} - \frac{\alpha_{12} N_2}{K_1}\right]$$

$$\frac{dN_2}{dt} = r_2 N_2 \left[1 - \frac{N_2}{K_2} - \frac{\alpha_{21} N_1}{K_2}\right]$$

where N_i = the number of individuals in the competing species populations 1 or 2, r_i = the intrinsic rate of increase, K_i = the population carrying capacities, and α_{ij} = the competition coefficient quantifying the effect of species j on species i. Stability analysis of this model shows that it is not necessary for the population growth rate (r_i) of a competing species population to drop below 0 as a result of the direct effect of a pesticide for that species population to go locally extinct. If a pesticide changes either of the following conditions, the competing species cannot coexist: $K_1 < K_2/\alpha_{21}$ or $K_2 < K_1/\alpha_{12}$ (Gilpin and Ayala 1973; Gilpin and Justice 1972). Changes in competition among many species, or changes in predator–prey interactions, can endanger a species population as much as direct effects on individuals. Consequently, pesticide effects on species interactions are highlighted in this report to illustrate their prevalence.

In contrast to the foregoing discussion, the argument could be made that, although common, indirect effects are less critical than direct effects. However, there is no reason to believe that a reduction in a population's carrying capacity or growth rate due to diminished habitat quality is any less detrimental than a reduction of similar magnitude due to direct effects. A plausible argument could even be made that direct effects on individuals are less important than indirect alterations of habitats or communities because a population will recover faster than will the community that comprises the altered habitat. Habitat alteration might involve direct removal of important plant species, or keystone or dominant species that determine the community structure. Ernst and Brown (2001) suggest that a significant delay can occur before a vacated keystone position is occupied by another species. The community structure might take a long time to return to its original state after pesticide concentrations decline to toxicologically insignificant levels. In addition, Matthews et al. (1996), Landis et al. (1996), and other ecologists argue that an altered community might remain so for a very long time. For these reasons, direct and indirect effects are discussed initially as being of equal importance.

B. Aquatic Ecosystems

General. Aquatic organisms interact in many ways and to varying degrees, ranging from those that are essential, e.g., dispersal of unionoid mussel glochidia only by a specific host fish species, to those that are relatively trivial, e.g., unionoid mussel and physid snail competition for food. Some communities have clear dominant or keystone species that influence structure and energy flow. Other species groups show considerable functional redundancy. Loss of a species from a guild displaying much redundancy might seem trivial, but loss of a keystone or dominant species could be catastrophic for a community in the near term.

The influence of keystone species varies in intensity among communities (Menge et al. 1994), and the retention of redundancy is important in ecosystems (Naeem 1998). Keystone species, being often large and mobile species, can be particularly sensitive to toxicants (Clements and Newman 2002). When examining the concept of ecological redundancy, it becomes apparent that redundancy does not provide a license to sacrifice species without incurring a cost. With the important exceptions of driver species such as dominant and keystone species, sets of species in communities display varying degrees of functional redundancy. Removal of one or a few of these redundant species might not influence ecosystem function noticeably (Pratt and Cairns 1996); however, the cumulative effect of incremental species loss can adversely impact ecosystem inertia, resilience, and elasticity. Therefore, decisions based on the assumption of ample, albeit unspecified, amounts of redundancy in a community are prone to nonconservative error.

It follows that the relative importance of direct and indirect pesticide effects depends on the specific risk assessment endpoint of concern. As important examples, the direct and indirect effects of pesticides on amphibian, fish, invertebrate, plant, and microbial endpoints in aquatic systems are briefly reviewed in this section. Such effects can be divided into potential effects (i.e., hazards) identified primarily from laboratory exposure experiments and those effects that are realized and have been measured in the field. It is important to distinguish between these because pesticides are designed to be hazardous to pests, so it is unsurprising that they are also hazardous to some nontarget organisms. It is the risk that this hazard will be realized in nontarget species that is the focus of most pesticide risk assessment, including that sanctioned under EU Directive 91/414.

Amphibians: Potential Effects. Amphibians, both anurans and caudates, are good assessment endpoints (Sparling et al. 2000; USEPA 2002). They are often and easily studied, and their habitat requirements, physiologies, and life histories make them sensitive to pollution's effects. Many are keystone species (e.g., Burton and Likens 1975), with some being context-dependent keystone species in that the importance of their roles changes with environmental conditions (Fauth 1999). A final factor that makes amphibians good assessment endpoints is the present global decline of amphibian species (Blaustein and Wake 1990, 1995; Halliday 1998), which has drawn the public's attention to their vulnerability and placed many on threatened or endangered species lists.

Both direct and indirect potential effects of pesticides on amphibians are well documented. Conventional lethality metrics for many amphibian species are archived in the USEPA ECOTOX (AQUIRE) database and in the open literature, especially for effects on larvae (e.g., the standard FETAX test; ASTM 1993; Bantle 1995). The Canadian government maintains a reptile and amphibian effects (RATL) web site (http://www.cws-scf.ec.gc.ca/nwrc-cnrf/ratl/about_e.cfm; see also Pauli et al. 2000) that currently references more than 3000 studies of diverse contaminant effects on amphibians. Some recent examples of direct effects are for degradation products of chloroacetanilide herbicides (alachlor and metolachlor) that caused teratogenic effects in *Xenopus laevis* (Osano et al. 2002).

Atrazine may delay growth and metamorphosis (Brown Sullivan and Spence 2003) and cause gonadal deformities (Carr et al. 2003) and emasculation (Withgott 2002) in frogs when exposed in the laboratory. However, the concentration at which this occurs is currently subject to debate, and the relevance of these laboratory effects to field populations is also controversial. Recently, Hayes et al. (2002a,b) reported effects of atrazine on male gonads at concentrations as low as 0.1 μg/L, which is substantially lower than the concentrations of atrazine found in many surface waters across the world, including the U.K. However, Carr et al. (2003) were unable to repeat this finding in similar experiments, which Hayes (2003) claims was because of experimental flaws in their study.

Evidence of indirect effects is also easily found in the literature. The simultaneous presence of carbaryl and a predator produced much higher mortality of *Hyla versicolor* tadpoles than predicted from laboratory exposure to carbaryl alone. Endosulfan also influenced tadpole (*Litoria citropa*) susceptibility to predation by odonates (Broomhall 2002). Pesticide exposure increased trematode infection of *Rana sylvatica* and, in turn, this caused an increase in the prevalence of frog limb malformations (Kiesecker

2002). Clearly, pesticide effects on amphibians range from direct effects on individuals, to changes in predator–prey dynamics, to impacts on symbiosis.

Amphibians: Realized Effects. It is difficult to document realized effects of pesticides on amphibians. There are specific indications that cholinesterase-inhibiting pesticides have played a part in the long-term anuran decline in regions of California (Sparling et al. 2001). However, other evidence is ambiguous at this time. Harris et al. (1998a,b) surveyed gross, genetic, physiological, and biochemical endpoints in frogs (*Rana pipiens* and *R. clamitans*) from eight Canadian wetland sites, four within apple orchards exposed to pesticides and four within relatively uncontaminated reference sites (although background atrazine contamination was ubiquitous at most sites). Concentrations of organochlorine and organophosphate pesticides in frog body fat, water, and sediments were also determined, and laboratory and *in situ* assays with frogs were used in complementary toxicity studies. The authors found insecticides in water, sediment, and frog fat sampled from the orchard sites, but there were no consistent effects on frog toxicological endpoints and there were no clear relationships between these endpoints and pesticide concentrations.

Hayes et al. (2002b) reported a connection between atrazine exposure and gonadal deformities in field populations of *R. pipiens*. However, Carr and Solomon (2003) point out that the conclusions of Hayes et al. (2002b) violate several widely accepted criteria of cause and effect, and that other researchers have been unable to find a relationship between atrazine usage and amphibian deformities, such as intersexuality (Reeder et al. 1998). Kiesecker (2002) did find an association in the field between limb deformities in *R. sylvatica*, trematode infection of the frogs, and exposure to agricultural runoff containing atrazine, esfenvalerate, and malathion. However, Carr et al. (2002) suggested that deficiencies in meeting accepted criteria for cause and effect mean that the value of this study is uncertain.

Fish: Potential Effects. Fish laboratory tests of pesticides constitute a major component of aquatic effects assessment; therefore, studies documenting direct effects of pesticides on fish are plentiful. Information on indirect effects is also easily extracted from the ecotoxicology literature, and Atchison et al. (1996) reviewed numerous studies describing toxicant effects on fish interspecific interactions. Many of these interactions would not be categorized as direct effects on the individual's survival or reproduction, but they have clear potential to adversely affect an individual's overall fitness. As an example, methyl parathion changed the prey preference of killifish (Farr 1978) and notionally affected optimal foraging. In another study, mirex exposure shifted pinfish predation efficiency (Tagatz 1976). In a third study, rockfish predation on Chinook salmon increased with antisapstain fungicide exposure (Kruznski and Birtwell 1994). A review of these and other studies led Clements and Newman (2002) to conclude that "...

the majority of studies attempting to measure the influence of contaminants on predator–prey interactions have shown significant effects." Conventional laboratory ecotoxicity tests for pesticide effects do not quantify these pervasive, indirect effects. However, with the appropriate design, laboratory, mesocosm, and field studies can potentially detect and quantify such effects.

Fish: Realized Effects. There is evidence that pesticides have adversely affected fish species in the field. For example, Baughman et al. (1989) state that of 126 fish kills documented for coastal counties in South Carolina from 1977 to 1984, 56% were "related to" runoff of various insecticides, including methyl parathion, toxaphene, and endosulfan. Fulton and Key (2001) summarize information from a field study in which caged mummichogs (*Fundulus heteroclitus*) were exposed at U.S. tidal creek sites during two growing seasons (1988–1989). The sites were located next to tomato fields to which the organochlorine insecticide endosulfan, the synthetic pyrethroid fenvalerate, and the organophosphate azinphos-methyl were applied. Brain acetylcholinesterase (AChE) activity was inhibited after five of seven of the field deployments and was associated with measured concentrations of azinphos-methyl in water similar to those causing AChE inhibition in laboratory tests.

Invertebrates: Potential Effects. Potential direct and indirect effects of pesticides on invertebrates, including nontarget aquatic insects, are also well documented. Standard toxicity tests for zooplankton (e.g., *Daphnia* sp.) and benthic invertebrate species (e.g., chironomids) have produced abundant evidence of potential direct effects on individuals. Indirect effects are also likely for invertebrate populations and species assemblages as shown by numerous studies, some typical examples of which are provided here.

Evidence of invertebrate population effects is easy to find. In the 1980s, Daniels and Allan (1981) and Day and Kaushik (1987) clearly demonstrated the population consequences for zooplankton of pesticide exposure. Such effects on vital rates vary with population density and life history in complex ways (Petersen and Petersen 1988; Forbes and Calow 2003).

Pesticides also impact invertebrate species assemblages indirectly. Chlorpyrifos spraying of ditch mesocosms reduced insect and crustacean populations (Van den Brink et al. 1996), reducing competition and allowing gastropod and oligochaete populations to increase many fold. Similarly, Woin (1998) reported increased oligochaete population sizes after pond mesocosm application of fenvalerate, notionally caused by decreased predation and competition for food. During recovery, these oligochaete species were replaced by ostracods that compete for similar food sources but are less prone to predation. Although fenvalerate in water rapidly fell to nondetectable concentrations within 40 d of application, insect species assemblages in the highest dosed mesocosms remained distinct from those of control mesocosms more than 2 yr after spiking.

Invertebrates: Realized Effects (Individuals). Matthiessen et al. (1995) showed through use of an *in situ* bioassay that carbofuran runoff could be responsible for toxicity in an English receiving stream. They performed a study at the Rosemaund experimental farm in which carbofuran was applied to oilseed rape at 3 kg a.i./ha. One month later rainfall caused runoff into drains, which then led to a transient 24-hr peak concentration of carbofuran in adjacent stream water of 26 µg/L. Caged *Gammarus pulex* placed in the stream ceased feeding and then died after this event, and subsequent laboratory studies showed that exposure to carbofuran at this level was a likely cause. In a later trial at the same site, there was also significant mortality of *G. pulex* (Williams et al. 1996). On this occasion, the cause was most likely to have been chlorpyrifos, which was found in a drain 100 m upstream at a concentration of 4.3 µg/L. Unfortunately, no measurements were made nearer to the caged *G. pulex*. Benthic fauna sampled from the Rosemaund stream were impoverished when surveyed in 1994, with a low Biological Monitoring Working Party (BMWP) score, and *G. pulex* were either absent or sparse at all sampling stations. However, this could not be attributed unequivocally to pesticide impact, as there were no unimpacted upstream sites for comparison, and a control selected in a nearby stream showed equally poor faunal richness (Williams et al. 1996).

In the U.S., the grass shrimp (*Palaemonetes pugio*) was used by Baughman et al. (1989) to examine the toxicity of fenvalerate in a South Carolina estuary. Shrimp were placed in cages for 96 hr at a potentially contaminated site and a reference site, and monitored daily for survival. Daily grab samples of water were taken for pesticide analysis instead of more preferable composite samples. Mortality of shrimp was associated with measurable concentrations of endosulfan and fenvalerate, with the latter the most likely cause, based upon information from laboratory toxicity tests.

Germinomas and teratoid siphon anomalies found in clams (*Mya arenaria*) were associated with applications of herbicides (picloram, 2,4-D, and 2,4,5-T) in blueberry cultivation and silviculture near coastal areas (Gardner et al. 1991). However, chemical analysis of the clams did not provide evidence of elevated herbicide concentrations, so demonstration of cause and effect remained circumstantial.

Invertebrates: Realized Effects (U.K. Assemblages). Benthic invertebrates are commonly monitored in the U.K. to assess the effects of contaminants in streams, and several studies are available that have tried to link pesticide exposure to adverse effects on benthic invertebrate assemblages.

Eleven years ago, Ashby-Crane et al. (1994) reviewed the current level of knowledge on the extent and nature of pesticide impacts in the U.K. and concluded that there were few relevant studies. A study by Morrison and Wells (1981) was one of these few. They found that invertebrate drift in a Scottish stream increased after aerial spraying of a forest with fenitrothion.

Concentrations in the stream rose to 18μg/L after spraying, but declined to background levels within 48hr. Caged insects survived in the stream, so the major effect appeared to be displacement through drift.

In another earlier study, Crossland et al. (1982) examined the effects on invertebrates in three ponds in Suffolk of cypermethrin applications at 70g a.i./ha in adjacent fields. After 5hr, water samples taken from the surface of the ponds contained up to 26μg cypermethrin/L, and samples taken at a depth of 20–30cm contained only up to 0.07μg/L. Sampling of zooplankton and benthic macroinvertebrates did not reveal any marked effects of pesticide exposure on organism abundance when compared with natural fluctuations. However, no statistical analyses were performed on the invertebrate data, and there was evidence of immobilized pond skaters (*Gerris* sp.), rat-tailed maggots (*Eristalis* sp.), and a water boatman (*Notonecta* sp.) in one sprayed pond, so invertebrate exposure and effects did occur. The authors reported that these surface-dwelling invertebrates recovered after 1d in uncontaminated water.

Shires and Bennett (1985) extended this investigation of cypermethrin and studied its impact when sprayed aerially at 25g a.i./ha onto a field in Kent surrounded by ditches on three sides. Cypermethrin concentrations were measured in samples taken at 30–50cm below the water surface, caged fish were analyzed for residues, water samples were returned to the laboratory for bioassays with *G. pulex*, and zooplankton and benthic macroinvertebrates were sampled from the ditches. The highest concentration of cypermethrin was 0.03μg/L, measured at one sampling station after 2d, but most measurements were below the detection limit of 0.01–0.02μg/L. The highest concentration of cypermethrin in fish tissue was 2μg/kg wet weight. Substantial mortality of *G. pulex* of up to 82% occurred in some samples taken from the ditches up to 2d after spraying. Figures in their paper suggest that there may also have been large effects on several zooplankton and benthic invertebrate species for several weeks, although this was not the conclusion reached by the authors, except for water mites and corixids. Natural fluctuations in organism abundance and the lack of a control site make the data difficult to interpret.

Furse et al. (1995) investigated the impacts of agricultural activity on U.K. headwater streams. They examined samples of invertebrates from 131 sites in four separate river systems, i.e., the Derwent (Yorkshire), Lugg (Herefordshire), Cam (Cambridgeshire), and Stour (Dorset). This investigation did not focus on pesticide impacts and concentrations of chemicals were not measured, with the exception of alkalinity and nitrate. Instead, associations were examined between different types of land cover adjacent to headwater streams, stream environmental variables, and macroinvertebrate taxa present within the streams. Furse et al. (1995) found evidence that headwater stream quality, based upon the type of macroinvertebrate assemblage present, was poorer than expected. Streams in arable landscapes had the poorest quality, but the precise cause of this was unclear.

The authors therefore speculated that nutrient enrichment, channelization, drought, and agricultural runoff might all have been important.

Sheahan et al. (1999) expanded upon the earlier work of Furse et al. (1995) and Matthiessen et al. (1995) by investigating the link between pesticide contamination and biological effects at four headwater stream study sites. They used several laboratory-based and *in situ* bioassays during this study, including the following: (1) the effect of field-sampled water on *Daphnia magna* survival, *Raphidocellis subcapitata* (algal) growth, and *G. pulex* acetylcholinesterase activity in the laboratory; (2) the effect of automatically collected samples concentrated by solid-phase extraction on *D. magna* survival, *R. subcapitata* growth, *Lemna minor* electrical conductivity, and *Chironomus riparius* acetylcholinesterase activity in the laboratory; (3) the effect of field-collected sediments on *C. riparius* survival, growth, and emergence, and (4) *G. pulex in situ* survival and feeding rate. Natural macroinvertebrate populations were also collected by kick sampling and drift netting in the Sheahan et al. (1999) study, and natural populations of epilithic algae were collected by placing glass plates in the stream and removing them at weekly intervals. Stage-triggered automatic samplers were used to take water samples for analysis at each site. The main pesticides detected were herbicides, although low concentrations of the fungicides fenpropimorph and propiconazole, were also found early in the season. The compound detected at the highest concentration in any of the samples was the herbicide diuron which was found at a concentration of 27 μg/L. However, throughout the 3-yr study, no pesticides were found at any of the four sites at a concentration and frequency predicted to be lethal to aquatic animals and plants. There was also very limited evidence of specific linkages between biological effects and pesticide applications, although generally impoverished macroinvertebrate assemblages were present at some sampling sites. Sheahan et al. (1999) concluded that:

> *It is apparent from this study and a considerable number of previous studies that the major difficulty faced in field monitoring programmes is the discrimination of low-level effects of contaminants on ecosystems from natural variation.*

ADAS (2000) performed a series of studies to assess the efficacy of no-spray zones for protecting aquatic systems. Four study sites were chosen: three with slow-flowing or static ditches at the field margin and one with a faster-flowing stream. Chlorpyrifos was the main pesticide applied during this study, with applications up to the stream/ditch edge, or with a buffer zone of 6 m for arable crops and 18 m at an orchard site. A no-spray control was also included during each pesticide application. Caged organisms (*G. pulex*, *C. riparius*, and *D. magna*) were placed in the surface waters and also in what the authors referred to as "standard ditch tanks" suspended over the surface waters. These tanks, which measured 1 m wide and 30 cm deep, were used to represent the standard ditch used in risk assessments under Directive 91/414/EEC. The results showed that caged organisms in both natural systems and standard ditch tanks were affected on several occasions after pesticide application up to the edge of the watercourse, particularly in the tanks and in the ditches where dilution was relatively low. Organisms were also affected on a smaller number of occasions when a no-spray zone of 6 m or 18 m was observed. Invertebrate assemblages were also monitored at the flowing stream site, using Surber samplers and drift nets, so that effects on naturally occurring populations could be compared with effects on caged organisms. Application of chlorpyrifos to the edge of the stream on two occasions was associated with reduced AChE activity and feeding in caged *G. pulex*. However, there was no strong evidence for either immediate or long-term effects on the abundance of naturally occurring *G. pulex* populations or on the overall structure of macroinvertebrate assemblages.

The evidence in the published literature for a link between pesticide exposure and adverse effects on invertebrates is therefore sparse for the U.K. environment. There is some evidence from the extensive survey by Furse et al. (1995) that aquatic macroinvertebrate assemblages in headwaters flowing through arable farmland areas are not as diverse as would be expected, but it is not possible to link this specifically to pesticides rather than other covarying factors. The Matthiessen et al. (1995) and ADAS (2000) studies show that there is the potential for lethal and sublethal effects from runoff and spray drift. However, the severity of these for the structure and function of aquatic assemblages either remains unknown (Matthiessen et al. 1995), or appears to be small (ADAS 2000). The Sheahan et al. (1999) study, designed specifically to examine the link between pesticide exposure and aquatic effects, was unable to find convincing evidence of such a link, but only four sites were investigated. These may not have been good representatives of the most vulnerable sites, or might not have been studied at times appropriate to the timing of pesticide application. We are therefore left wondering whether the apparent absence of a link between pesticide exposure and effects reflects the true situation, or whether the rather limited research conducted to date has simply missed the evidence.

Invertebrates: Realized Effects (Non-U.K. Assemblages). Heckman (1981) carried out a yearlong survey of the ecology of orchard drainage ditches near Hamburg, Germany. He was able to compare the results from his survey with those found in a similar survey from the 1950s before widespread pesticide use in the orchards. This study showed that there had been some large changes in species composition over the period of 25 yrs between the two surveys. However, the author acknowledged that attribution of clear causes to these effects was not possible in a study of this type.

One of the few studies that demonstrates a link between pesticide contamination from runoff and effects on natural populations of organisms in Europe is reported in several papers by Schulz and Liess (Liess 1998; Schulz 1998; Liess and Schulz 1999; Schulz and Liess 1999). The study site in

Germany is an agricultural area where sugar beet, winter barley, and winter wheat were grown with the assistance of ethyl parathion, fenvalerate, and deltamethrin applications. Spray drift apparently does not affect the study stream and impacts are from runoff alone, the source of which was identified by the presence of bankside erosion rills. Water in the study stream was sampled with an automatic sampler triggered by increased specific conductivity and set to sample over a 1-hr period when pesticide concentrations were assumed to peak. Sediment was sampled every 2 wk, using a passive sampler described in Liess et al. (1999). Macroinvertebrates were sampled nine times between March 1994 and April 1995 using Surber and emergence samplers.

The study was designed to exclude hydraulic stress as a stressor by using two streamside microcosms, one of which was shut off during precipitation to prevent runoff components from entering. Caged caddis flies (*Limnephilus lunatus*) and amphipods (*G. pulex*) were suspended in these microcosms and counted every 10 d. There were 11 runoff events during the study, 3 of which were associated with concentrations of ethyl parathion considered to be higher than background levels. These events alone were associated with a significant reduction in species number with disappearance of 8 of the 11 abundant species and a reduction in abundance of the remaining 3 species. Four species recovered within 6 mon, 9 species recovered within 11 mon, and 2 species remained at a low density for about a year. Hydraulic stress and turbidity associated with precipitation and runoff were excluded as a primary cause of these effects, because some high hydraulic stress and turbidity events did not lead to these effects on invertebrates in the main stream and because caged invertebrates survived better in the streamside microcosm without runoff water than in the microcosm with runoff water.

Other studies in the same German catchment measured invertebrate drift (Liess and Schulz 1999) and used *in situ* bioassays with *L. lunatus* and *G. pulex* placed in the natural stream and scored weekly (Schulz and Liess 1999). Lower survival in the bioassays was associated with pesticide runoff events and reductions in the abundance of natural macroinvertebrate populations. However, there were some discrepancies in the concentrations of ethyl parathion and fenvalerate that were associated with biological effects: an event with lower concentrations was associated with greater effects, and all concentrations associated with effects measured in the field were lower than those causing effects in the laboratory (Liess 1994). The authors suggest that this could have been the result of chemical sampling error (not adequately capturing either the intensity or duration of exposure), or because of differences in the sensitivity of organisms exposed in the laboratory and field. These discrepancies do not match some of the criteria used to establish cause and effect in ecoepidemiological studies (Suter 1993) and thus slightly weaken the case for a linkage between pesticide exposure and biological effects in these studies.

Schulz and Liess (1999) also acknowledge as problematic the lack of an upstream control station at their study site. Such a station was impossible to establish because the headwater stream being studied was contaminated with pesticides at its source. However, they believe that it is justified to infer pesticide effects because of the temporal association of pesticide contamination and effects on macroinvertebrates and the lack of a similar association between biological effects and other potentially stressful environmental factors.

Schulz and Liess took the approaches they developed in northern Germany to an orchard study site principally contaminated with azinphos-methyl on the Lourens River in South Africa. Their methods for sampling suspended sediment and runoff also worked well in this setting, and there was evidence for reduced macroinvertebrate species richness and abundance at the contaminated site (Dabrowski and Schulz 2003; Schulz et al. 2002). A direct link between pesticide contamination and effects on natural assemblages of macroinvertebrates has not been established to date. However, the high concentrations of insecticides measured in the Lourens River make such a link very plausible. Supporting evidence for this is that mortality occurred among chironomid larvae exposed to spray drift-contaminated samples of river and tributary water and in bowls of water placed at different distances from sprayed orchards (Schulz et al. 2001).

Crossland et al. (1982) describe a study in France in which cypermethrin was applied to grapevines with mist blowers at 30 or 45 g a.i./ha and effects were measured in two streams and a drainage ditch. Initial concentrations of cypermethrin at the water surface were 140–1010 μg/L, declining to 20 μg/L or less after 3 hrs. Subsurface water samples contained 0.4–1.7 μg/L cypermethrin immediately after spraying, generally declining to <1 μg/L within a few hours. Collection of invertebrates in drift nets showed a great increase in numbers of drifting Gammaridae, Ephemeroptera, Coleoptera, and Hemiptera about 2 hr after spraying. However, sampling of invertebrates remaining in the substrate did not suggest that population abundance had been adversely affected.

Davies and Cook (1993) found very large effects on the drift of invertebrates after aerial spraying of cypermethrin in Tasmania. Concentrations of cypermethrin in an exposed stream were 0.1–0.5 μg/L and led to a 200-fold increase in drifting invertebrates, particularly true flies, mayflies, stoneflies, caddisflies, and beetles. At one contaminated site, most of the stoneflies, mayflies, and amphipods caught in drift nets were dead. Some invertebrates, such as mayfly and stonefly species, were completely absent from substrate samples after spraying, but population abundances of most taxa recovered over a period of up to 6 mon.

Leonard et al. (1999, 2000) studied the impact of endosulfan use in cotton on macroinvertebrates in the Namoi River, New South Wales. Initially (Leonard et al. 1999), they concentrated effort on several dominant species of mayfly nymph and caddisfly larvae and Surber sampled each month at eight sites along the river: two upstream reference sites, two low-exposure pesticide sites, and four high-pesticide exposure sites. This approach was expanded in a subsequent study (Leonard et al. 2000) to consideration of the entire macroinvertebrate assemblage at 17 sites. Pesticide exposure at each site was characterized by taking sediment samples and through use of *in situ* passive samplers containing trimethylpentane, which were replaced monthly. The multivariate field data were analyzed using CANOCO (ter Braak and Smilauer 1998) and principal response curves to determine links between biological and physicochemical data, and field results were supported by laboratory toxicity testing of indigenous mayfly and caddisfly species to confirm their sensitivity to endosulfan.

These studies showed that passive samplers produced less variable and more interpretable results than sediment samples, although there was still a positive correlation between the two. Endosulfan was the main insecticide detected along with the herbicide prometryn, and concentrations of this insecticide were negatively correlated with the abundance of most of the invertebrate taxa that were studied. The authors acknowledged that several other factors associated with distance downstream covaried with endosulfan concentrations on some sampling dates, but the weight of evidence suggested that endosulfan had an important impact on invertebrate densities.

Hatakeyama et al. (1990) reported the effects on macroinvertebrate drift of aerial spraying of fenitrothion at 1500 g/ha in Japan. Water samples were collected from an exposed stream every 2 hr on the day of spraying. Fenitrothion concentrations reached a peak of around 20 μg/L, decreasing exponentially to below 1 μg/L within a day. The high fenitrothion concentrations were associated with a large increase in daytime invertebrate drift, particularly of mayflies and a caddisfly species, with consequent adverse effects on invertebrate abundance in the stream substrate. Recovery of invertebrate assemblages occurred within about 3 mon, largely because of recolonization from upstream.

Hatakeyama and Yokoyama (1997) investigated the relationship between benthic macroinvertebrate faunal and shrimp bioassay responses at several stations along a river surrounded by rice paddies in its lower reaches. These paddies were sprayed with a variety of insecticides and herbicides. Increases in shrimp mortality (to 100% after some insecticide spraying events) were associated with higher concentrations of fenthion (transient concentrations up to 50 μg/L) and lower species diversity at pesticide-contaminated stations.

Many North American studies that have investigated a direct link between pesticide use and adverse effects on naturally occurring aquatic organisms have focused on aerial spraying of pesticides in forest or coastal areas (Eidt 1975; Hurd et al. 1996). However, Clarke and Ainsworth (1993) also review work by Kirby-Smith et al. (1992) and Scott et al. (1990, 1992) into effects on estuarine ecosystems of pesticide runoff from arable and vegetable crops. Kirby-Smith et al. (1992) found that storm-induced runoff led to peak concentrations of up to 100 μg/L alochlor, 0.3 μg/L terbufos, and 1.6 μg/L permethrin. No biological effects were detected, which the authors attributed to low concentrations of pesticides, brevity of exposure, localization of contamination, and partitioning of pesticides to suspended material, which rendered them unavailable. In contrast, Scott et al. (1990, 1992) were able to demonstrate an association between pesticide exposure and effects on natural fauna. Measured peak concentrations of pesticides were 1.0 μg/L endosulfan, 0.9 μg/L fenvalerate, and 7.0 μg/L azinphosmethyl, and mortality among fish and crustacean populations was observed during runoff events. Subsequent reductions in population densities of invertebrates and fishes were also observed, with recovery taking several weeks to months.

Examples of studies on aerial pesticide applications include that by Kreutzweiser et al. (1989), who found that aerial application of the herbicide glyphosate was not associated with generally increased invertebrate drift in a Canadian stream at measured concentrations up to 162 μg/L. However, there was weak evidence that a transient increase in the drift of *Gammarus* sp. and mayflies (*Paraleptophlebia* sp.) may have been associated with herbicide application.

Ernst et al. (1991) found that the fungicide chlorothalonil caused mortalities in caged fish and invertebrates at measured concentrations of 0.17–0.88 mg/L in a Canadian pond. However, effects on naturally occurring benthic invertebrates were difficult to interpret, as the single control and exposed ponds differed substantially in species abundance before spraying occurred, and there were no clear and consistent trends in abundance after spraying.

Beyers et al. (1995) studied the impact on invertebrate drift of aerial application of carbaryl in North Dakota. Carbaryl concentrations in the Little Missouri River reached a maximum of 85.1 μg/L 1 hr after one application, declining to 12.3 μg/L after 4 hr. Mayflies, particularly Heptageniidae, entered the drift immediately after spraying, but this effect was transient.

The number of studies in which the effect of pesticides on field communities has been demonstrated is therefore small. Several of these studies are associated with aerial application of pesticides, which is a rare practice in the European Union (EU), or involve crops such as rice or cotton that cannot be grown in Northern Europe. In a recent review of insecticide effects on aquatic communities, Schulz (2004) also concluded that knowledge about pesticide effects on communities of organisms in the field was sparse.

Plants: Potential Effects. Aquatic plants are taxonomically diverse (Philbrick and Les 1996) and fill crucial roles in aquatic ecosystems. They provide animals with food, habitat, and solid substrate and foster essential biogeochemical transformations. They are used extensively in the lowest pesticide risk assessment tiers, with heavy reliance on unicellular algae and duckweed (*Lemna gibba*) testing. Consequently, much laboratory information exists for quantifying direct effects on aquatic plants, including abundant information on microalgal population responses.

Plants: Realized Effects. Field monitoring programs for algae and higher plants are common in the literature (Stewart et al. 1999) but are not usually linked explicitly to pesticide exposure. The effect of aerially and manually sprayed glyphosate on diatom populations colonizing glass slides, and found in sediments, in two Canadian streams and a pond was investigated by Sullivan et al. (1981). Concentrations of glyphosate were not measured in either the streams or the pond. They concluded that habitat and seasonal difference were more influential than possible glyphosate exposure, as the population abundance of several species was higher in the exposed streams and ponds when compared with controls.

Although there have been some studies of pesticide effects on macrophytes, these are usually reports in which macrophytes are the target species or the pesticide application was experimental (Gardner and Grue 1996). Lewis and Wang (1999) lamented that "reported attempts to field validate the available laboratory-derived data reported in the scientific literature are almost non-existent." Van Geest et al. (1999) compared laboratory and mesocosm-derived metrics of effect of the herbicide linuron and found that laboratory-estimated maximum permissible concentration, using an algal species, were protective of macrophyte-dominated mesocosm communities.

Microbes. Aquatic microbes and microbial community structure and function are influenced by pesticides, (Cuppen et al. 2000). Community structure is often assumed to be more sensitive than community function and, perhaps, less resilient to stressor effects. For example, periphyton community structure was more sensitive to the biocide tri-*n*-butyltin than periphyton photosynthesis and net production (Dahl and Blanck 1996). The underlying explanation is that functional redundancy exists and, within limits, functions can remain unchanged if structure changes; this was the case for chlorpyrifos-exposed ditch mesocosms in which changes in community structure resulted in minor effects on ecosystem metabolism (Kersting and Van den Brink 1997). A similar study focusing on dissolved oxygen and pH dynamics in response to linuron spikes indicated that ecosystem adaptation occurred and that lower-tier studies would not have detected this adaptation (Kersting and Van Wijngaarden 1999).

C. Terrestrial and Wetland Ecosystems

Mammals: Potential Effects. Information exists to show the potential for direct pesticide effects on mammals, with much of it from laboratory testing. Hart (2003) notes that the EU Guidance Document on Risk Assessment for Birds and Mammals recommends that risk assessment effects levels (NOEL intake values) be derived from "toxicity studies conducted routinely for human health assessment." The effect metric value for the most sensitive species is then used in an ecological assessment. Alternatively, a species sensitivity distribution method can be used.

Mammals: Realized Effects. Nearly half the 60 terrestrial mammals extant in Britain are associated with agricultural habitat in one form or another, and pesticides are listed as a potential threat to approximately half the 40 tabulated species in a report by Harrington and MacDonald (2002). However, they concluded from the sparse information available that direct pesticide effects are probably not contributing significantly to mammal declines in the U.K.

Incidental reports provide additional field data about direct pesticide effects (Barnett et al. 2002), in which poisonings of such diverse U.K. mammals as badgers, foxes, rabbits, and bats were investigated. The report of Barnett et al. (2002) reflects the conclusions of an earlier one covering 1990–1994 in which approved uses of pesticides in Europe accounted for only a small fraction of reported wildlife poisonings (de Snoo et al. 1999). Of the 384 incidents reported in the U.K. (1990–1994), more than a quarter (103) involved poisoning of a mammalian species. Many poisonings were eventually linked to intentional misuse and so do not contribute useful information on direct effects from approved pesticide use. Information from incident reports is also difficult to assess due to the *ad hoc* manner in which such information is compiled in different EU countries.

There is less information on the indirect or important sublethal effects of pesticides on mammals, but consistent themes do emerge from what information does exist. Long-term risks to mammals have been documented in some cases. For example, there was evidence of risk to fruit-eating bats 1 yr after dieldrin spraying to control tsetse flies in Cameroon (Muller et al. 1981). The mechanisms for indirect or sublethal effects are as diverse as are those of direct, acute effects measured in the laboratory.

Pesticides that act as endocrine modifiers disrupt sexual differentiation and potentially change the demographics of exposed rodent populations (Gray et al. 1998). Chlordane, dieldrin, metabolites of vinclozolin, procymidone, phthalates, and other pesticides are thought to be endocrine modifiers. It is challenging to predict population effects from endocrine disruption in field-exposed individuals. *In utero* exposure to the fungicide vinclozolin did change the anatomical and hormonal status of male gray-tailed voles (*Microtus canicaudus*) in field enclosures, but this resulted in minimal change in population growth rate (Caslin and Wolfe 1999).

One of the major challenges in detecting effects on mammals is summarized succinctly by Pimentel (1992): "If pesticide effects on bird and mammal populations are to be measured, then extremely large ecosystems have to be treated." Although acknowledged as important, the effects of pesticides on mammal populations within a heterogeneous spatial context are difficult to assess. Edge and Schauber (2000) studied gray-tailed voles (*Microtus canicaudus*) and deer mice (*Peromyscus maniculatus*) and concluded that vegetation structure, via its effect on pesticide vertical distribution after spraying, influences effects on rodent populations. In another nonpesticide study, reserpine administered to a field population of pika (*Ochotona pallasi*) changed the aggressive defense of spatial territories and, in doing so, disrupted effective resource partitioning (Shilova 1990). It is plausible that endocrine modifiers have similar effects on other territorial mammals.

Even wider scales than those described above need to be considered for an effective risk assessment for some mammal populations. Maurer and Holt (1996) argued that mammal metapopulation dynamics could determine the consequences of pesticide application. Metapopulation-based assessments provide insight into species persistence within a complex landscape, not simply in an isolated patch. Exploring life history traits, Fagan et al. (2001) found that a large number of populations have such locally variable population dynamics that dispersal, habitat patches, and corridors were essential to consider in any predictions of population persistence. Knowledge of basic autecological qualities of a mammal species was found to be enough to place them into one of three groups differing in extinction risk under pressure. Such important and tractable qualities are not commonly considered in pesticide assessments.

Pesticides can adversely impact the interaction of mammalian species with other species. Diazinon, a relatively persistent insecticide that can be ingested while feeding on dead insects or grooming, influenced the normally strong competition between cotton rats (*Sigmodon hispidus*) and prairie voles (*Microtus ochrogaster*) in field enclosures. Although no outright mortality was observed at either of two application rates, reproduction was impacted more for cotton rats than for prairie voles. The prairie vole's higher tolerance to diazinon tipped the competition in enclosures in its favor. Sheffield and Lochmiller (2001) further concluded: "This suggests that negative impacts on populations and community structure and function may persist longer than diazinon persists in the environment." They observed that the noted effects would not have been apparent from conventional laboratory testing of the two species separately.

Other species interactions can be influenced by pesticides. Pimentel (1992) suggests that predators and parasites can be affected more by pesticides than are their herbivorous prey or hosts. Chemical agents can adversely impact natural mammal population interactions by altering the structure of the community within which they exist. Rodenticide applied to kill black-tailed prairie dogs resulted in increased vegetation densities that, in turn, decreased the population density of deer mice (*Peromyscus maniculatus*), a species adapted to open habitat (Deisch et al. 1990). Freemark and Boutin (1994) suggested that herbicide impacts on mammal populations are more commonly associated with habitat modification than direct mortality. Plant morphology, species abundances, species composition, diversity, and spatial interspersion can be changed by herbicides. Freemark and Boutin (1995) list (in their table 5) indirect herbicide effects on wild mammal species as involving a change in diet related to decreased survival and reproduction, decreased diversity, reduced dispersal, and lowered persistence. Many of these indirect or subtle effects would not be addressed normally in guidance supporting Directive 91/414/EEC, although they are acknowledged as being relevant to the Directive's mandate.

Birds: Potential Effects. Considerable avian effects data have been generated in support of environmental legislation. Directive 91/414/EEC requires testing of acute, short-term, and long-term effects on birds. Acute and short-term effects are derived from LD_{50} and LC_{50} tests, respectively, and long-term effects are expressed as NOEC values. Acute testing involves oral ingestions, and short-term testing involves a 5-d dietary exposure. Long-term testing focuses on reproductive effects.

Birds: Realized Effects.

> *Although there are limitations with field investigations, particularly uncontrollable variables that must be addressed, the value of a well-designed field study far outweighs its shortcomings.*
>
> (Blus and Henny 1997)

Pesticides have had demonstrable adverse effects on field populations of birds. Unambiguously, a single DDE application to Clear Lake (California) reduced the number of western grebe (*Aechmophorus occidentalis*) from 1000 to 0 breeding pairs (Blus and Henny 1997). Equally clear in retrospect was the influence of DDT on raptors and piscivorous bird species reproduction and, consequently, population declines (Hickey and Anderson 1968). Several legacy pesticides still influence bird species through more subtle changes to reproduction (Hunt and Hunt 1977; Fry and Toone 1981; McLaughlin and Mineau 1995; Bishop et al. 2000; Nakamaru et al. 2002). Blus and Henny (1997) make the important point that early laboratory toxicity testing of DDT provided no evidence to suggest that bird populations would experience adverse effects upon field exposure.[2]

Outright poisonings occur today, although many in the U.K. and the rest of Europe have been linked to illegal setting of poisoned baits (de Snoo et al. 1999; Barnett et al. 2002). de Snoo et al. (1999) suggest that only a small percentage of reported bird poisonings are associated with approved pes-

[2]This concern is not restricted to legacy pesticides. Mineau et al. (1994) assessed the hypothesis that conventional avian reproduction testing with quail and mallards allow identification of pesticide effects to field bird populations. They concluded: "[Our analyses] cast doubt on the ability to extend the results of avian reproduction studies to any potentially affected bird species." Their comments were echoed by Matz et al. (1998).

ticide use. In contrast, an avian field study in Canada suggested to Mineau (2002) that avian mortality occurs "regularly and frequently" in agricultural fields. Differences in opinions on the prevalence of poisonings appear to arise from the difficulty in detecting and enumerating field kills. Pimentel (1992) estimates that 80% of bird carcasses disappear within a day of death and only 1 in 100 deaths of forest or grassland species is noted in surveys because it is so difficult to find carcasses in such habitats. McLaughlin and Mineau (1995) suggest that insecticides such as carbofuran that are applied in granular form kill large numbers of songbirds on field edges and that the cumulative effect is a regional decline in songbird populations.

Reports of pesticide field poisonings vary in key details and complexity. A chlordane poisoning of roosting birds in New Jersey was linked to emergence of tainted beetles and beetle grubs, suggesting to Stansley and Roscoe (1999) that chlordane poisoning of birds might be more common than previously thought. Poisoned common grackles (*Quiscalus quiscula*), European starling (*Sturnus vulgaris*), and American robins (*Turdus migratorius*) were also consumed by Cooper's hawks (*Accipiter cooperi*), leading to a secondary poisoning of an endangered species. Famphur, an organophosphorus pesticide applied topically to cattle, resulted in poisoning of magpies (*Pica pica*) that commonly gather in association with cattle and ingest their hair. Consumption of dead magpies led to secondary poisoning of hawks (*Buteo jamaicensis*) in Washington State (Henny et al. 1985; Blus and Henny 1997).

Effects of pesticides on avian reproductive fitness are also evident. Azinphosmethyl and carbaryl spraying of apple orchards increased nestling hunger signaling of tree swallows (*Tachycineta bicolor*) (Bishop et al. 2000). However, fledgling weight, a good predictor of first-year survival, was not affected for these swallows.

Some information suggests an influence of pesticides in combination with other factors, but provides no definitive conclusions. In the U.K., 13 farmland bird species have declined an average of 30% from 1968 to 1995; however, populations of 29 habitat generalist bird species increased an average of 25% during the same time (Krebs et al. 1999). Similar reports of decline are present in the scientific literature, including those concerned with North American, British, and other European avifauna (Sotherton and Holland 2003). Pesticides are thought to contribute to these trends by altering arthropod prey species abundances and distributions and by changing the composition of the plant community that provides avian habitat and sustenance. Furadan, a carbamate, and Decis, a pyrethroid, decreased grasshopper numbers by 90% in one grassland study. Correlated with this decrease were a drop in Baird's sparrow (*Ammodramus baidrii*) nest productivity and an increase in abandoned Baird's sparrow territories (Martin et al. 2000). In this same study, the foraging distances of chestnut-collared longspur (*Calcarius ornatus*) increased, but nesting success did not appear to be affected. Malathion eliminated a key prey species of blue tit (*Parus caeruleus*) from a sprayed site in Spain but the tit population was unaffected because another prey species increased in numbers to replace the first (Blus and Henny 1997).

Some long-term trends seen in the field are less ambiguous due to considerable effort expended in gathering data. Savidge (1978) found that herbicide effects in a California Jeffery pine plantation reduced bird species abundance and number for 6 yr after application. Schroeder and Sturges (1975) reported that Brewer's sparrow (*Spizella breweri*) densities declined by 99% 2 yr after herbicide application to sagebrush habitat. However, the clearest field evidence for indirect effects of pesticides on birds is for the grey partridge (*Perdix perdix*) in the U.K. (Potts 1986). Dimethoate and other herbicides indirectly affect grey partridge by reducing the plant habitat of their invertebrate prey (Blus and Henny 1997) and by reducing nesting cover (Freemark and Boutin 1995). Insecticides exacerbate this condition, so the required high-protein diet of young partridge is difficult to maintain in treated areas. Young birds must forage over wider areas and are subject to high mortality rates as a result (Blus and Henny 1997) (Fig. 1). As a specific regional example of the long-term consequences of these effects, grey partridge abundances determined from 1970 to 1995 in Sussex were negatively correlated with herbicide use and positively correlated with the densities of dicotyledonous weeds (Ewald and Aebischer 1999). Work such as this supports the conclusions below:

> *Strong evidence exists for adverse effects of changes in habitat pattern on beneficial insects and arthropods in the United Kingdom, and on birds in North America and Western Europe.*
>
> (Freemark and Boutin 1995)

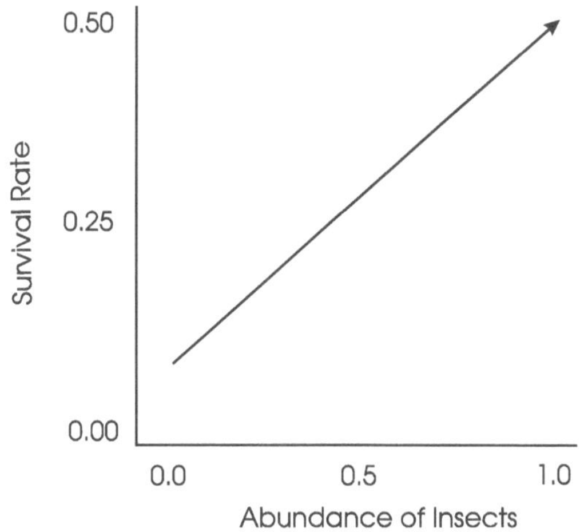

Fig. 1. Grey partridge chick survival is related to insect abundance on cereals. (Modified with permission from Routledge/Taylor & Francis, from Sotherton and Holland 2003.)

> *Many species of farmland birds are in decline in the United Kingdom, and there is considerable evidence that the indirect effects of pesticides are the cause.*
>
> (Sotherton and Holland 2003)

Herpetofauna:[3] *Potential Effects.* Reptiles are used in laboratory and field effects assessments only occasionally, although attempts are now being made to partially remedy this deficiency. As an example, recognizing the need for a standard species, Talent et al. (2002) recently determined that fence lizards (*Sceloporus occidentalis* and *S. undulates*) are excellent subjects for laboratory testing.

In a survey of the journal *Environmental Toxicology and Chemistry,* Hopkins (2000) found that only 1.4% of published articles concerning vertebrates focused on reptiles. Articles addressing amphibians were five times more abundant than those addressing reptiles. The Canadian reptile and amphibian effects web site and report (Pauli et al. 2000) contain much more data for amphibians than reptiles. These observations are incongruous with conservation biologists estimates that more reptile species exist today than amphibian species, 4680 amphibian and 7150 reptile species (Gibbons et al. 2000). It is also inconsistent with observations of the European Committee for the Conservation of Nature and Natural Resources that "at least 30% of Europe's amphibian and 45% of its reptile species were in danger of extinction" (Hall and Henry 1992).

An explanation for the differences in numbers of reptile and amphibian pesticide studies could be the perception of their relative susceptibilities. Amphibians, including their eggs and larvae, are linked to aquatic systems in such a way as to create a high potential for exposure, and amphibians are widely known to be in global decline. However, reptile populations are also in global decline (Gibbons et al. 2000), and many have plausible pesticide exposure pathways. The relatively long time to reach reproductive maturity, their mechanisms of sex determination, and the longevity of many reptiles make them particularly sensitive to reproductive effects of contaminants, including pesticides. Increasing their risk of high exposure, many are carnivores occupying relatively high trophic positions in communities (Hopkins 2000). As telling, albeit obscure, evidence of reptile sensitivity to pesticides, Hall (1980) reported that the U.S. National Pest Control Association had at one time recommended DDT to control snakes.

Herpetofauna: Realized Effects. Lizards have been the subjects of field studies of pesticide effects at locations ranging from the Canary Islands

[3]Although amphibians and reptiles constitute what is normally considered the herpetofauna, they are treated separately here. Most relevant amphibians are so intimately linked to aquatic environments that their discussion seemed most appropriate in the aquatic systems section.

(Fossi et al. 1995; Sanchez-Hernandez 2003), to NW Zimbabwe (Lambert 1993, 1994), to Somaliland (Lambert 1997), and to South Africa (Alexander et al. 2002), but the best known reptile studies are those documenting changes in sexual characteristics of Florida alligators. Guillette and coworkers showed that endocrine modifiers influence penis size and hormone levels of wild alligators (Crain et al. 1998; Guillette et al. 1996, 1999; Milnes et al. 2002). Such changes in alligators in Florida's Lake Apopka resulted in high adult mortality, low clutch viability, and low juvenile densities (Crain and Guillette 1998). These effects are notionally linked to endocrine-modifying pesticides.

Nontarget Terrestrial Invertebrates: Potential Effects. Concern about nontarget, often beneficial, invertebrate species has prompted the issue of guidance by various groups. Although US EPA information needs as described by Touart and Maciorowski (1997) for pesticide testing did not include consideration of nontarget terrestrial invertebrate data,[4] the Guidance Document of the European Standard Characteristics of Non-target Arthropod Regulatory Testing (ESCORT 2) Working Group provides details for preliminary and higher tier assessment of pesticide effects on terrestrial invertebrates (Candolfi et al. 2001). Field testing is presented as a potential tool and associated methods are described in Candolfi et al. (2000); however, all Tier I and most higher tier information is derived from laboratory tests. Beneficial species [the cereal aphid parasitoid *Aphidius rhopalosiphi*, the predatory mite *Typhlodromus pyri*, the thrip predators *Orius laevigatus* and *Chrysoperla carnea* (green lacewing) and *Coccinella septempunctata* (ladybeetle), and the root maggot parasite *Aleochara bilineata* (rove beetle)] are the species of choice for these studies. Annex II of Directive 91/414/EEC requires that two sensitive nontarget species be routinely used. The very sensitive *A. rhopalosiphi* and *T. pyri* (Candolfi et al. 1999) were recommended as Tier I test species so that a "reasonable worst case" was represented at that step (Tones et al. 2001; Defra Project No. PN0937). Details for nontarget arthropods in *The Guidance Document on Terrestrial Ecotoxicology Under Council Directive 91/414/EEC* (European Commission 2002a) were derived from and remain consistent with the ESCORT 2 conference tiered methods.

Effects on pollinators are a particular concern to ecological risk assessors, with the honeybee *Apis mellifera* receiving the most attention. *The Guidance Document on Terrestrial Ecotoxicology Under Council Directive 91/414/EEC* (European Commission 2002a) provides routine testing

[4]However, US EPA pesticide risk assessments for outdoor use products normally include honeybee toxicity data. This information is used to produce label precautionary language. (Les Touart, U.S. EPA OPPTS, personal communication.)

descriptions for honeybee acute toxicity, metabolite testing, and other metrics of exposure or effect. Thompson (2002) provides details about possible sublethal effects of pesticides on honeybees, which include effects on longevity, division of labor, foraging, colony development, larval development, colony mate recognition, and repellency. The first few effects are associated with the age-determined division of labor in honeybee hives and the expected length of a bee's life that it spends performing different activities for the colony. Pesticide effects on foraging and age-specific division of labor can influence hive survival.

Nontarget Terrestrial Invertebrates: Realized Effects. Field surveys in England and Wales (1981–1991) (Greig-Smith et al. 1994) indicated that honeybee poisonings in the 1980s occurred primarily because of misapplication of triazophos on rapeseed. Dimethoate application to several crop types also produced a number of poisonings.

Thompson and Hunt (1999) point out that bumblebees (*Bombus* spp.) are also important pollinators that have been declining in abundance for several decades. Pesticides are implicated in this decline. They warn that 19 bumblebee species have ecological factors distinct from those of the honeybee and that risk assessments based on honeybees are insufficient to protect bumblebees. Bumblebees can be more vulnerable because only the queen survives overwinter to establish a new colony the next year and the number of individuals in a hive is much smaller than that in honeybee hives. Underreporting of mortality is more likely for bumblebees than for honeybees because honeybees often are attended by beekeepers.

Nontarget Terrestrial Plants: Potential Effects. Standard toxicity testing methods for quantifying pesticide effects on nontarget plants have been established (Pfleeger et al. 1991; Boutin et al. 1993). Current methods are thought to be conservative due to application differences between test and field exposures (American Crop Protection Association and the Canadian Crop Protection Association 2001). Despite the perceived conservative nature of the established tests, considerable variability exists in species sensitivity (Fletcher et al. 1990), suggesting to Boutin and Rogers (2000) the need for testing more species than currently required by the EPA.

Nontarget Terrestrial Plants: Realized Effects. Direct effects of pesticides, particularly herbicides, on nontarget plant species are the focus of much deserved attention (Boutin et al. 1993; MAFF Pesticides Safety Directorate 1999). Off-field impacts can involve harm to specific rare or protected species (Catling and Porebski 1998) or to overall plant diversity (McLaughlin and Mineau 1995). Figure 2 (top panel) shows the influence of British agriculture on dicotyledonous seed densities in soils (Marshall et al. 2001). Nontarget species provide habitat and food for many animal species (Boutin et al. 1995). As an example, Fig. 2 (bottom panel) also shows

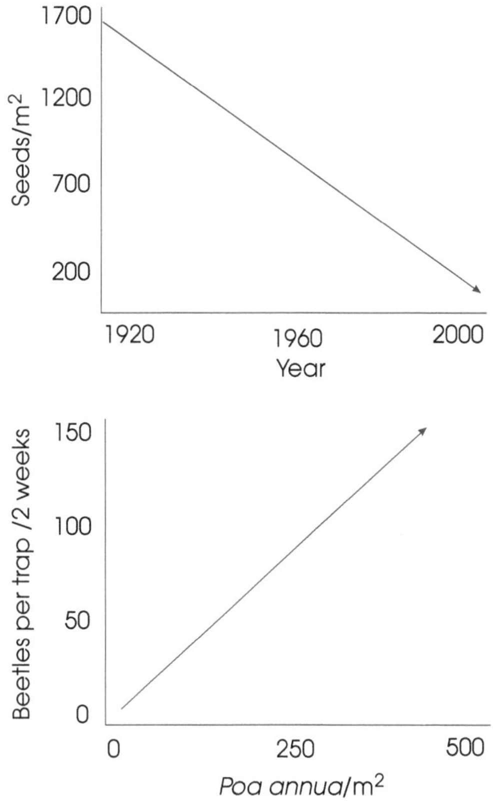

Fig. 2. The cumulative influence of U.K. farming on soil dicotyledonous seed densities (*top panel*) (modified with permission from Blackwell Publishing, from Robinsion and Sutherland 2002) and the relationship between an annual grass species (*Poa annua*) densities and those of ground beetles (*bottom panel*) (modified with permission from the USDA, from Norris and Kogan 2000).

the influence of densities of an annual grass species on the densities of ground beetles (Norris and Kogan 2000).

The impact of pesticides on noncrop plants is difficult to determine in terms of diminished utility to humans. As we have already seen, animals such as birds that depend on these plants are impacted (McLaughlin and Mineau 1995). However, wild plants species also have direct uses by humans. A Canadian plant priority list for protection was recently developed based on current economic value and plant potential to supply important genes during genetic manipulation of crop species (Catling and Porebski 1998).

Drift onto nearby fields can damage nontarget crops, and the increasing use of potent herbicides creates particularly high risk to nontarget crops (Fletcher et al. 1993, 1996). As an example, the sulfonylurea herbicide, chlorsulfuron, has a high potential for such damage (Fletcher et al. 1995). Also, chlorsulfuron drift after spraying of wheat fields impacted the yield of cherries (*Prunus avium*) in an adjacent orchard (Bhatti et al. 1995).

III. Long-Term Implications of Pesticide Exposure for Ecosystems

> *Almost half of the land surface of Britain is enclosed farming and pesticides, at some time, are applied over most of this area. The agricultural environment, therefore, has huge relevance to wildlife diversity.*
>
> (Harrington and MacDonald 2002)

A. Habitat and Resource Modification

Can plausible inferences be made about long-term habitat or resource modifications caused by the approved use of pesticides, based on the evidence discussed so far? Without doubt, agricultural activities have produced long-lasting changes on the British landscape and associated biota. The specific suggestion that pesticides play a significant role in habitat and ecological resource change is supported by the following theory and evidence.

1. The community conditioning hypothesis suggests that community structure may remain in an altered state for a long time after pesticide residues have dropped to toxicologically insignificant levels. Sheffield and Lochmiller (2001) provide an example of modified species interactions that persisted for longer than pesticide residues.
2. The removal of a keystone or dominant species alters habitat qualities (Power et al. 1996). Although it is important to prevent this (Mills et al. 1993), current toxicity tests performed during the authorization of pesticides cannot ensure that keystone species populations remain extant. Therefore, habitat or resource alteration as a result of unintentional removal of keystone species could occur.
3. Although commonly invoked, the general application of the ecosystem redundancy hypothesis has not been adequately addressed, and an equally plausible hypothesis, the rivet popper hypothesis, according to which the loss of any species is of concern, might be a more accurate and conservative depiction of ecosystem consequences of pesticide exposures (Pratt and Cairns 1996).
4. British studies of grey partridge provide strong evidence that herbicide use has diminished habitat quality and, consequently, decreased partridge population densities (Blus and Henny 1997; Ewald and Aebischer 1999).
5. Population densities of 13 farmland bird species have fallen 30% on average in the U.K. from 1968 to 1995, but 29 generalist species have increased (Krebs et al. 1999), suggesting a decrease in habitat or resource quality for bird species that prefer farmlands.

6. Other instances of medium-term adverse effects of agricultural practice on aquatic species (Furse et al. 1995) and terrestrial species (Martin et al. 2000) have been reported, suggesting that longer-term effects are plausible.

In contrast to the foregoing, the following theory and evidence detract from any definitive statement that pesticides play an important role in general changes to the British landscape and associated biota.

1. Most communities display a degree of resistance to change and an ability to recover after stress-induced change (Cairns and Niederlehner 1993; Pratt and Cairns 1996).
2. Although current toxicity tests in support of pesticide authorizations generate imprecise predictions of population or community vitality or viability, there are indications that effects observed from these tests are, on average, conservative relative to population effects (Caslin and Wolfe 1999; Bishop et al. 2000; Forbes and Calow 2003).
3. Other plausible causal agents co-occur with pesticides, such as physical habitat modification, introduced species, and increased fertilizer use. The presence of these other potential causes confounds the identification of pesticides as a major factor contributing to long-term habitat or resource modifications.

B. Local Extinction and Reduced Biodiversity

Evidence can also be compiled from the literature covered so far about the risk of local extinction and reduced biodiversity. Support for the view that pesticides contribute to local extinctions or reduced biodiversity includes the following.

1. Laboratory studies have identified mechanisms capable of disrupting species interactions and, consequently, species persistence and community structure.
2. Publications addressing widespread declines in bird (Beaumont 1997; Krebs et al. 1999) and amphibian (Wake 1991) species suggest that pesticides might play a role. Sparling et al. (2001) provide one example of amphibian decline that is directly linked to pesticide use.
3. Mineau (2002) suggests that bird mortality likely occurs frequently in Canadian agricultural fields.
4. Pesticides can modify endocrine function in ways that impact population viability (Gray et al. 1998). For example, changes in alligator sexual characteristics and reproductive viability in Florida lakes are notionally linked to endocrine-modifying pesticides (Crain et al. 1998; Guillette et al. 1996, 1999; Milnes et al. 2002).
5. Pesticides can adversely influence vital rates, and changes in these vary in complex ways among species (Forbes and Calow 2003).

6. Many indirect effects, not quantified in conventional tests, are documented for diverse taxa exposed to pesticides. These effects could lead to local extinctions and, gradually, to reduced diversity.
7. Although commonly invoked, the general applicability of the ecosystem redundancy hypothesis has not been adequately ascertained. The equally plausible, and more conservative, rivet popper hypothesis might provide a more appropriate framework for protecting ecosystems (Pratt and Cairns 1996).
8. Long-term risk to mammalian species has been documented (Muller et al. 1981).
9. Some mesocosm studies (Van den Brink 1996; Woin 1998) document species abundance shifts as a consequence of pesticide exposure.
10. Bird populations are impacted by legacy (Hickey and Anderson 1968) and modern (Schroeder and Sturges 1975; Savidge 1978; Blus and Henny 1997; Martin et al. 2000; Mineau 2002; Sotherton and Holland 2003) pesticides. Some impacts have long-term consequences for bird populations.

In contrast, theory and evidence detracting from any definitive conclusion that pesticides influence local extinction and reduced biodiversity are the three provided above for habitat and resource modification, and the following.

1. Phenotypic plasticity and life history strategies are important features of most populations (Stearns 1992) and could lessen the effects of pesticides on populations (Stearns and Crandall 1984; Sibly and Calow 1989).
2. Harrington and MacDonald (2002) saw no indication in the sparse data available that the decline in U.K. mammalian species is linked to pesticides.
3. Some bird studies suggest that sublethal effects on nestling (Bishop et al. 2000) and adult birds (Blus and Henny 1997) occur, but compensation results in these effects having no apparent adverse consequences on overall fitness.

C. Recovery and Recolonization

Evidence supporting a negative view about community recovery and recolonization after pesticide exposure includes the following.

1. There can be a significant delay between the removal of a keystone species and its replacement in a community (Ernst and Brown 2001).
2. Negative impacts on mammalian populations can persist longer than pesticide residues do in treated environments (Sheffield and Lochmiller 2001).

3. Although life history traits can help identify mammal species that rely on dispersal to remain viable (Fagan et al. 2001), such traits are not formally considered in present risk assessment methods.
4. The community conditioning hypothesis suggests that community structure can remain altered after pesticide residues have dropped to toxicologically insignificant levels.

In contrast, evidence detracting from any conclusion about the adverse effects of pesticides on community recovery and recolonization include all of those provided above for habitat and resource modification and also local extinction and reduced biodiversity.

IV. An Ecological Vantage on Pesticide Risk Assessment

A. Organismal

Most of the lower-tier information applied in pesticide risk assessment is interpreted from an autecological vantage. Such a vantage is directly useful for endangered or threatened species because the taking of even one individual is illegal for these species. For such species, information about direct effects on individuals is essential. Such information might also be relevant for charismatic species for which the taking of an individual would be very undesirable. For example, a 1984 poisoning of Atlantic brant (*Branta bernicla*) led to the U.S. regulatory authorities discontinuing diazinon use on turf grasses despite the lack of any evidence of risk to populations of this unlisted species (Bacietto 1998). Application of diazinon to golf courses and sod farms was regarded as constituting an unreasonable risk to individuals of a charismatic species.

The autecological vantage is also useful, but insufficient, for predicting effects on species populations. This approach has a long standing in ecology where it was used to examine the relationship between individual organisms or species and their physical, chemical, and biological environment. Liebig's law of the minimum (Liebig 1840), Shelford's law of tolerances (Shelford 1911, 1913), and the concept of a fundamental niche grow out of this premise that knowledge of the tolerances or requirements of individuals can be used to predict species distributions and abundances. However, this vantage is insufficient for explaining or predicting all important aspects of population or community ecology, and the synecological vantage is an essential one for general prediction or description of ecological systems (Preston 2002). This view can perhaps be summed up by a quote from 20 years ago:

> *Although this discussion may appear hostile to single species toxicity testing efforts, it is not intended to be. Single species tests are exceedingly useful and are presently the major and only reliable means of estimating probable damage from anthropogenic stress. Furthermore, a substantial majority, perhaps everyone in this meeting is certainly aware of the need for com-*

munity and system level toxicity testing. How then does one account for the difference between awareness and performance?

(Cairns 1984)

B. Population

The EC Guidance Document on Aquatic Ecotoxicology (European Commission 2000b) and EC Guidance Document on Terrestrial Ecotoxicology (European Commission 2000a) explicitly states that species populations are to be protected by the EC risk assessment process for pesticides. Although the issue is gradually being remedied, the tools most often used in pesticide risk assessment provide imperfect, sciolistic insights into population level effects. For example, Grant (1998) indicates that substantial changes in some vital rates such as those often measured in toxicity tests can have little impact on populations because density-dependent population dynamics shift to compensate for the toxicant's effects.

Newman (2001) identified four reasons why current toxicity tests might not provide sufficiently accurate predictions of consequences for field populations. First, toxicity testing often focuses on the most sensitive stage in an organism's life cycle based on the assumption that protection of individuals in this most sensitive stage will ensure protection of the associated species population. However, the most sensitive stage of a life history may not be the most important one relative to maintaining a viable population, as illustrated by Kammenga et al. (1996). Second, metrics such as the LC_{50}, LOEC, or NOEC cannot be incorporated directly into demographic models used to project population change through time. Third, postexposure mortality is ignored in most laboratory-derived metrics but field populations can experience significant mortality after exposure to a toxicant ends (Newman and McCloskey 2000; Zhao and Newman 2004). Fourth and finally, the underlying assumptions for some conventional concentration–effect models have not been resolved. As an example, the probit model is based on the concept of the individual effective dose or concentration. It has been hypothesized that each individual has a unique concentration above which it will die and below which it will survive exposure (Bliss and Cattell 1943; Finney 1947). The distribution of tolerances in any population is assumed to be lognormal. An alternate hypothesis is that each individual has the same chance of dying as any other and whether it dies depends on chance alone. Which of the hypotheses dominates in a particular population exposure scenario currently remains ambiguous. The consequences of this ambiguity are significant because the two hypotheses predict very different population consequences with repeated exposures (Newman and McCloskey 2000).

A risk assessment for atrazine by Solomon et al. (1996) provides another illustration of subtle but crucial population-related issues left unexplored in current ecological risk assessment (ERA) approaches. Although not considered in the very thorough ERA by Solomon et al. (1996), Hayes et al.

(2002a,b; Hayes 2003) and Withgott (2002) indicate that atrazine acts as an endocrine modifier and is suspected to affect the reproductive fitness of amphibians at environmentally realistic concentrations. Dodson et al. (1999) found that atrazine also influenced the production of males in *Daphnia pulicaria* cultures, and that *D. pulicaria* fecundity and survival were much less sensitive than male production. Both these effects could significantly change population vitality but were not considered in the original ERA for atrazine. Solomon et al. (1996) focused on the effects of atrazine on primary production, maintenance of macrophyte community structure, and long-term viability of fish populations in their higher-tier ERA. In their effects characterization, they commented on the lack of information for amphibians but concluded that, "the limited data suggest that amphibians are tolerant of atrazine."

Appropriately designed laboratory (Van der Hoeven and Gerritsen 1997; Snell and Serra 2000), mesocosm (Van den Brink 1996; Sherratt et al. 1999), enclosure (Caslin and Wolfe 1999; Wang et al. 2001), and field (Schroeder and Sturges 1975; Savidge 1978) studies can provide valuable insights into the population effects of pesticides, and more studies of these types are included in both predictive and retrodictive pesticide risk assessments each year. As more such tests are applied to ERA, the strength of associated inferences about long-term effects of pesticides to populations will improve. As is occurring now in conservation biology, more and more emphasis is slowly being placed in ERA on the risk of local extinction under specified exposure conditions (Tanaka 2003).

> *Practical problems emerge due to our preoccupation with measuring effects in a way more appropriate for predicting fate of exposed individuals instead of exposed populations. In my opinion, current tests to predict population-level consequences are no less peculiar than one described in the poem Science by Alison Hawthorne Deming in which the mass of the soul is estimated by weighing mice before and after they are chloroformed to death. The incongruity of the test is more fascinating than its predictive power.*
>
> (Newman 2001)

C. Community and Species Assemblage

As discussed already, interactions among species populations are also essential in any ERA to understanding the long-term consequences of pesticide use. The value of individual-based effect metrics from the laboratory is ambiguous for this purpose and, in some cases, demonstrably inadequate. Laboratory-based designs for quantifying community effects exist (Cairns et al. 1986) but lack the realism of field or mesocosm studies. Field studies have tremendous value for this purpose, and methods exist for extracting insights about effects on species assemblages or communities (Savidge 1978). Studies involving mesocosms or enclosures (Liber et al. 1992; Sheffield and Lochmiller 2001) may also provide valuable information about these effects, but the practical temporal scale is shorter for most mesocosm and enclosure studies than for field studies so that, although such studies are more realistic than laboratory studies, they still may not reflect the true field situation with complete accuracy or precision (Crane 1997).

> *. . . in many situations species interactions are important and indirect effects complicate ecological assessments. Just as laboratory toxicologists recognize the influences of certain abiotic factors . . . on chemical effects, community ecotoxicologists understand that responses of individual populations cannot be measured in isolation from other populations.*
>
> (Clements and Newman 2002)

V. Conservation Biology Vantage on Pesticide Risk Assessment

A. Pesticides and Nontarget Species of Conservation Concern

Realized Effects. Changes in the way that farmland is managed has unquestionably reduced insect abundance over the last 30yr in the U.K. and other countries (Wilson et al. 1999; Benton et al. 2002). Pesticides could have contributed to this decline both through direct poisoning by insecticides (Dover et al. 1990; Cardwell et al. 1994) and the removal of food plants by herbicide application (Marshall et al. 2001). Many bird species declining on farmland are reliant on insects and other invertebrates during the breeding season as staple nestling food (Cramp 1988, 1992; Cramp and Perrins 1993, 1994a,b; Benton et al. 2002; Gruar et al. 2003). Furthermore, there is a correlation between the declining abundance of insects on farmland and the abundance of several bird species (Benton et al. 2002). Such a correlation, however, is not definitive proof that pesticide usage is the principal cause of the decline. There is ample evidence that agricultural intensification is responsible for reducing bird populations on farmland (Chamberlain et al. 1999; Henderson et al. 2000) and, while pesticide usage has evolved during the intensification process, it is not the only practice that has changed. For example, the U.K. corncrake (*Crex crex*) population crashed during the 20th century through the intensification of farmland practices (Green and Gibbons 2000), but here the driving factor was the shift from hay cutting to silage production (Green and Stowe 1993; Stowe et al. 1993). Silage is cut much earlier in the year than hay with the consequence that corncrakes, which nest in long grass, are unable to produce fledglings before the nests are destroyed during mechanical cutting of grass (Broyer 1994; Green 1996). The skylark *Alauda arvensis* and the lapwing *Vanellus vanellus* provide further examples of species that have declined on farmland through reasons other than pesticide application (Chamberlain et al. 2000; Wilson et al. 2001).

There are few examples where pesticides are known to have affected the population sizes of nontarget species in the long term. Exceptions include the bioaccumulation of organochlorine (OC) pesticides through food chains into top avian predators, such as the peregrine falcon *Falco peregri-*

nus (Ratcliffe 1980) and the sparrowhawk *Accipiter nisus* (Newton 1986), and the effect of the loss of insects in field margins on the grey partridge (Chiverton 1999), as described earlier. Once the effect of OC pesticides was realized, appropriate legislation was introduced in developed countries and the bird of prey populations recovered to earlier levels (Millsap et al. 1998; Wingfield Gibbons et al. 1993; Horne and Fielding 2002). For the grey partridge, the introduction of conservation headlands conclusively demonstrated a link between pesticide application, both herbicide and insecticide, and a reduction in insect food required by the precocious grey partridge chicks (Green 1984; Borg and Toft 2000; Southwood and Cross 2002). The headland experiments further showed that pesticide application could be sympathetically designed to achieve both acceptable agricultural and conservation targets. Where implemented, headlands have resulted in a dramatic increase in the local abundance of grey partridges (Rands 1985, 1986; Chiverton 1999). In addition, a remarkable increase has also been recorded in the numbers of certain very rare species of farmland flowers, such as pheasant's-eye (*Adonis annua*), shepherds needle (*Scandix pectinveneris*), and cornflower (*Centaurea cyanus*). Headlands are also more heavily used by butterflies (Dover et al. 1990) and small mammals (Tew et al. 1992).

Perceived Effects. Despite a general lack of direct evidence, pesticides and veterinary medicines are frequently cited as possible factors causing population declines in currently scarce species. For example, the hornet robberfly *Asilus crabroniformis* breeds on farmland and is associated with large grazing animals (Holloway et al. 2003a). In Britain, the species is Biodiversity Action Plan listed and scarce, and exists as a series of fragmented populations (Smith 2000; Clements and Skidmore 2002). Ivermectin, a veterinary medicine, is cited as a factor contributing to the decline of hornet robberflies (Smith 2000). Although research has shown that ivermectin can influence the numbers of coprophagous flies emerging from dung (McCracken and Foster 1993; Wardhaugh et al. 2001), the effect is considerably less dramatic for dung beetles (Kadiri et al. 1999). By extension, it is therefore assumed that ivermectin can negatively influence the hornet robberfly (Clements and Skidmore 1998; Smith 2000, 2001), even though neither a direct nor indirect effect has ever been shown. It is quite common for pesticides in general to be blamed for problems experienced by rare species without supporting data.

There is a general assumption within the conservation community that pesticides are detrimental to wildlife. Concern about pesticide application extends mainly to the effects of spray drift. For example, where pesticide application is occurring close to a Site of Special Scientific Interest, a barrier such as scrub, is often allowed to develop between the two sites to intercept any drift (D. Sheppard, English Nature, personal communication) as a precaution against any possible adverse effects.

Potential Effects. Many species have become rare on farmland and beyond as a result of habitat loss and degradation. Consequently, many populations are highly fragmented across the landscape, substantially altering plant (Jacquemyn et al. 2003; Luoto et al. 2003) and animal (Sherman and Runge 2002; Schmiegelow and Monkkonen 2002; Ryttman 2003; Bellamy et al. 2003; Watson et al 2003) population dynamics. Some populations, particularly of insect species such as silver spotted skipper (*Hesperia comma*) (Hill et al. 1999), Glanville fritillary (*Melitaea cinxia*) (Saccheri et al. 1998) and bog fritillary (*Proclossiana eunomia*) (Sawchik et al. 2002), but also some vertebrate species such as red squirrels (*Sciurus vulgaris*) (van Apeldoorm et al. 1994) and nuthatches (*Sitta europaea*) (van Langevelde 2000), function as metapopulations, i.e., a series of populations only loosely connected through occasional dispersion of individuals among populations (Hanski 1999). A situation that is also common in the wild occurs when local populations have gone extinct and the distance separating populations is too great to allow exchange of individuals through dispersion (Hill et al. 1996; Epperson 2000; Kudoh 2001; Doebeli and Killinback 2003). When this happens, genetic variation in an isolated population begins to erode at a rate negatively correlated with the size of the population (Frankham et al. 2002).

Small populations go through population bottlenecks and, particularly if the population remains small for several generations (inevitable if a species is rare), a substantial amount of genetic variation can be lost (Frankham et al. 2002). *k*-strategy species may be at more risk from extended bottlenecks than *r*-strategists because of their slower rates of recovery (Begon et al. 1986). Notionally, lower genetic variation results in a decline in adaptability, an increased probability that an entire population will be lost as a result of a single event, and a general reduction in fitness (Ralls and Ballou 1983; Frankham et al. 2002). There are examples of species that survive in the wild with apparently very little genetic variation, for example, the northern elephant seal *Mirounga angustirostrus* (Hoelzel et al. 1993; Hoelzel 1999) and the fallow deer *Dama dama* (Pemberton and Smith 1985) in Britain but, of course those populations that have disappeared as a result of reduced genetic variation no longer exist to allow their levels of variation to be examined. Working with adders (*Vipera berus*), Madsen et al. (1999) unequivocally demonstrated that low genetic variation is associated with low fitness for this species.

With rare, fragmented populations in agricultural landscapes, pesticides could play a role in the long-term reduction in fitness of some populations by repeatedly driving them through bottlenecks. Each time this happens, the genetic variation would be reduced more and more until the population enters an extinction vortex (Tanaka 2000; Lane and Alonso 2001; Rowe and Beebee 2003; McGinnity et al. 2003) from which there is no prospect of recovery. In addition to this process, if pesticide application results in the loss of one or more populations within a metapopulation, increasing the distance among populations to beyond the maximum for successful emigration, populations could become isolated and enter a process of genetic erosion.

Pesticide Risk Assessment and Conservation. The risk of pesticide application to vertebrate wildlife is currently assessed using small numbers of bird (usually Japanese quail, *Coturnix coturnix,* or Northern bobwhite quail, *Colinus virginianus,* and mallard duck, *Anas platyrhynchos*) or mammal (rat, *Rattus norwegicus* or mouse, *Mus* spp.) species under laboratory conditions. It is fair to say that pesticide risk assessment requires a leap of faith to extend the results from the small number of species tested under artificial conditions to a natural and complex field scenario. Furthermore, proposals for the safe use of pesticides under field conditions often focus on species of economic value. For example, to protect honeybees (*Apis melifera*), it is recommended that spraying be carried out early in the morning or late in the evening to coincide with the cooler temperatures before and after flight. However, bumblebees (*Bombus* sp.) and many species of solitary bees are able to fly under cooler conditions than honeybees and become nontarget casualties of pesticide application. It is difficult to integrate pesticide considerations into conservation because of the multitude of possible effects. Consequently, there is a general precautionary assumption made that pesticides applied for agricultural purposes are detrimental to wildlife and that sites valuable to conservation, for whatever reason, should be protected from pesticides [although note that pesticides are sometimes applied for conservation purposes, for example, to control invasive weed species (http://www.english-nature.org.uk/pubs/Handbooks/default.asp)].

B. Future Considerations

It is possible that the role that pesticides might play in the reduction of animal and plant populations, through whatever routes, may come under closer scrutiny in the future for several reasons. First, the increased awareness of long-term genetic consequences of short-term poisoning events may focus more attention on the distribution of pesticides across farmland habitats. Another issue is global warming (Hulme and Viner 1998), resulting in weather pattern changes in many countries. It is quite possible that the growth rate of arable weeds could increase considerably in temperate regions as a result of warmer, and probably wetter, conditions at particular times of the year (Hossel 2001). Pest insects might also survive in higher numbers. For example, the winter of 1989–1990 was exceptionally mild in the U.K., which triggered a big increase in the number of aphids spreading disease to winter cereal crops. Crop yields fell by as much as 1 t/ha (Hossel 2001). Global warming is likely to result in increased pesticide application. However, the general public have become more aware of the amount of toxic substances used in food production and an increase in the quantities applied is likely to meet opposition, so outcomes are difficult to predict.

Another factor that may result in increased scrutiny of pesticides is governmental activity such as the new biodiversity strategy for England entitled "Working with the Grain of Nature" (http://www.defra.gov.uk/wildlife-countryside/ewd/ biostrat/). The target is to ensure that species are part of "healthy functioning ecosystems" and to ensure that "biodiversity considerations become embedded in all the main sectors of economic activity." With a focus on ecosystems comes an acceptance that species exist within complex food webs and communities (Giler and O'Donovan 2002). A change of practice affecting one component of such a community could have a cascade effect all the way up to a species of conservation concern. Agricultural practices are considered to be very important within the context of the U.K. government's overall vision for the country's biodiversity.

C. Long-Term Effect Data Required by the Conservation Community

There is a paucity of data that unequivocally demonstrate a link between pesticide use and long-term population reductions in farmland environments. For example, most of the species of conservation concern in the U.K. are not associated with farmland (www.ukbap.org.uk), and it is generally accepted that, for the majority of species, habitat loss or alteration is clearly the most significant factor threatening species extinction at the moment. Consequently, effects of pesticides are rarely considered. Part of the reason for this is that following a poisoning event, long-term monitoring is infrequently carried out and the potential ways in which pesticides could alter community structure are not studied. The investigation of both these issues is limited by the amount of available resource and, with so many immediate problems, long-term or more-complex considerations fall low on the agenda. Furthermore, with the political desire in many countries to see wildlife conservation largely directed by local communities, the scientific expertise to appreciate and investigate the more-insidious, but nevertheless potentially damaging, changes is not usually available. This type of research is the realm of a few academics within universities or overstretched non-government organizations.

Management of farmland to consider biodiversity is generally voluntary and often entails joining an incentive scheme, such as the Countryside Stewardship Scheme in the U.K. (http://www.defra.gov.uk/erdp/schemes/css/default.htm). The way that farmland within the scheme is managed is dependent on the conservation objectives of that part of the countryside. Within such schemes, particular pesticide application regimens need to be approved, and here knowledge of the longer-term effects of pesticide application would be very useful. For example, spraying with a given active substance may result in the decline of a suite of butterfly species, but how long would it take for the butterfly community to recover? Armed with this information, regulatory authorities may be able to recommend a suitable period of time between subsequent sprays both to maintain the butterflies and to achieve the pest control objective. Alternatively, a pesticide may be particularly detrimental to a small number of species that could affect community structure and ultimately a species of conservation concern. For example, a proposed herbicide application could eliminate food for meadow grasshoppers (*Chorthippus parallelus*), in field margins in an area. This species is not rare, but its powers of dispersal are very limited (Chinery 1993), so it could take some years for the area to be recolonized. Meadow grasshoppers are the staple food source of growing cirl bunting (*Emberiza cirlus*) nestlings, so the proposed application regimen could ultimately eradicate a rare, target species through an indirect route.

In conclusion, there are very few examples known to conservationists in which pesticides have been shown to have direct or long-term damaging effects on wildlife. However, potential effects of pesticides are rarely considered in conservation and pesticide issues, for example, in connection with nature reserves, are examined on a case-by-case basis. The conservation community would welcome a general reduction in the use of pesticides on farmland, often as a result of the perceived impact of their use rather than direct evidence. Reductions are most likely to be achieved through government or consumer pressure, e.g., the move toward organically grown foods (de Boer 2003). An integration of pesticide application and effects into conservation strategy would be desirable but can only be realized if appropriate data are collected. Currently, these data are generally unavailable.

VI. Quantitative Tools for Population and Community Analysis

A. Population Analyses

Population effects can be quantified in a variety of ways. Application of demographic methods to chronic *Daphnia* test data is possible with only minimal changes to current methods (Newman and McCloskey 2002; Section 4.3 of ECOFRAM 1999). Nacci et al. (2002) applied a demographic approach to analyzing toxicant exposed field populations. Caswell (1996) describes a straightforward approach to performing demographic analyses for toxicant-exposed populations and, in his book (Caswell 2001), provides many details about the matrix approach to demographic analysis. ECOFRAM (1999; Section 4.4 of the ECOFRAM Aquatic Report) provides general details for implementing population models in studies of population effects. With knowledge of the relevant exposure duration, a stochastic projection of population dynamics over the exposure period can be used to estimate the risk of local extinction at different exposure concentrations, i.e., a statement of population risk can be made with these methods

(Newman 2001). If spatially explicit modeling is required, such an approach can be taken with a metapopulation model (O'Connor 1996; Newman 2001). Mackay et al. (2002) provide an example of a spatially explicit, population-based ERA.

Several software packages are available for analysis of demographic information. Population viability analysis (PVA) has been available as a tool in wildlife conservation for over a decade (Soulé 1987) and involves the estimation of extinction probabilities through analyses that incorporate identifiable threats to population survival into models of the extinction process. Models such as VORTEX are available to carry out PVA (Lacy 1993), and there are many examples where PVA scenarios have been modeled and tested (Ball et al. 2003; Kaye and Pyke 2003). Studies have also been carried out using PVA to provide recommendations for the survival of rare vertebrates, such as the eastern barred bandicoot *Perameles gunnii* (Lacy and Clarke 1990), and invertebrates, such as Fender's blue butterfly (*Icaicia icaroides fenderi*). Although PVA has undoubtedly great potential use in conservation, it has yet to be widely adopted as a tool by practitioners. The RAMAS program (Ferson and Akçakaya 1990) is another inexpensive program useful for deterministic and stochastic modeling of populations and PVA and is available in a version that explicitly considers ecotoxicological information. A further shareware example of a population modelling tool is the PopTools (http://www.cse.csiro.au/poptools) add-in to Excel that implements demographic methods to fit population data and to make deterministic or stochastic population projections.

B. Community and Assemblage Analyses

The conventional approach to effects characterization produces a series of relevant effect metrics such as LC_{50} values for test species for comparison with expected exposure concentrations. Recently, a probabilistic approach has emerged for many risk characterizations beyond Tier I.

There are considerable differences in sensitivities among species, and a focus on the most sensitive of a small number of tested species does not take full advantage of the available data. Species sensitivity distributions (SSDs) make fuller use of the available effects data. An SSD is a distribution of effect metrics for individual species thought to represent collectively the species of concern. Effect metrics for the test species are ranked from lowest to highest and their ranks converted to approximate proportions. The paired proportions and effect metric concentrations are then fitted to one of several models (Posthuma et al. 2002).[5] Some SSD models include

[5]Grist et al. (2002) describe a nonparametric method that circumvents the need to identify a well-fitting distributional model.

all species; however, some require separate models for taxonomic subsets of species. For example, Solomon et al. (1996) performed a retrodictive ERA for atrazine in the North American cornbelt by exploring SSD models for logical species groupings (see fig. 21 in Solomon et al. 1996).

In many recent "probabilistic" risk assessments, the creation of a SSD relies heavily on one of several statistical distributions such as the log logistic, lognormal, or triangular distribution. For aquatic risk assessments of pesticides, data points in these distributions are taken from endpoints in acute or chronic toxicity tests. In acute distributions, data points are normally taken from tests used to derive LC_{50} or EC_{50} values. For chronic toxicity distributions, no-observed-effect-concentrations (NOEC values) are commonly used. In acute toxicity tests, exposures to a contaminant are generally of short duration (24–96 hr), and chronic toxicity tests are conducted over a full life cycle or an early life stage of an organism.

For a pesticide ERA, it is important to identify a threshold hazard concentration above which ecological effects are likely to occur. With SSDs, this is often approximated by selecting a low centile of the distribution. The resulting metric is commonly called a hazardous concentration (HC_p). Normally, the 5th or 10th centiles (HC_5 and HC_{10}) have been arbitrarily used in ERA (Aldenberg and Slob 1993; Wagner and Løkke 1991). These lower centiles of an effect concentration distribution have been applied historically in deriving U.S. water quality standards and, in the case of the HC_5, have more recently been recommended in the EU Technical Guidance Document. Software available for fitting SSDs and estimating HC_5 values has been developed by Van Vlaardingen et al. (2003).

As discussed by the Aquatic Dialogue Group (1994), a risk assessment that relies solely on the protection of a certain proportion of exposed species might not be protective if keystone, dominant, or legally protected species are ranked below the specified proportion on a SSD. In choosing a proportion from a distribution of acute or chronic effects, one makes the assumption that "protecting" a certain proportion of species will be protective of the structure and function of an ecosystem and that the available single-species toxicity tests are representative of the ecosystem to be protected or the universe of species in the environment. In reality, it would be remarkably fortuitous if the issue of protecting crucial species were adequately addressed in laboratory testing of small numbers of conventional species. Also, the argument could readily be made that acute LC_{50} or chronic NOEC information is not adequate for predicting a concentration to protect a species population existing in a natural community (Hopkins 1993; Jagoe and Newman 1997; Newman and McCloskey 2002; Newman and Unger 2003). A certain amount of sciomancy exists in the SSD approach as applied today. Regardless, it is now a common risk assessment practice, and Maund et al. (2001) introduced some supporting evidence for using the 10th centile of acute distributions based upon ecologically significant effects observed at higher concentrations in field studies.

Effect quantification for population and species assemblages is less common than that for individuals yet, based on the materials discussed to this point, the need for such metrics is high in pesticide risk assessment. Fortunately, relevant methods are being applied more and more frequently, and higher-tier methods such as mesocosm experiments, enclosure studies, and field surveys are amenable to their use.

Community effects can also be extracted from laboratory, enclosure, mesocosm, and field studies using conventional ecological methods. Newman (1995), Matthews et al. (1998), and Clements and Newman (2002) provide information specific to their application for the risk assessment of chemicals such as pesticides. The influence of simple community interactions such as predator–prey interactions can be quantified (Tagatz 1976[6]) in laboratory assays, but this is not often done to support ERA activities. More commonly, mesocosm community structure metrics are used in predictive pesticide risk assessment activities and field community structure metrics are used in retrodictive risk assessment activities. The most common indices are species richness, diversity, and equitability indices. Communities or species assemblages might be compared for different exposures using distance metrics. Multivariate methods such as ordination or clustering methods can assess differences and similarities in species assemblages that have different exposure histories. All these methods rely on community structure information such as species abundances or presence–absence data.

Multimetric methods can incorporate structural and functional qualities of species assemblages during assessments of effect (Clarke 1999; Clements and Newman 2002). The most common multimetric index is the index of biological integrity (IBI) (Karr et al. 1986; Karr 1991, 1993). Karr's IBI attempts to quantify the integrity of a system of concern relative to an undisturbed or intact system of the same type in the same geographical region. As such, the IBI score for a system has quantitative meaning only in comparison to that of an undisturbed system. The IBI concept has been applied successfully to summarize the integrity of numerous sets of mesocosm or field data.

Numerous software packages implement conventional ecological metrics and multivariate methods. One example of such shareware is the BioDiversity package available from Neil McAleece (biodiversity@nhm.ac.uk) of The Natural History Museum and Scottish Association of Marine Science. Other frequently used software for multivariate analysis of ecological community data includes PRIMER (Plymouth Routines In Multivariate Research; see http://www.primer-e.com/in) and CANOCO (see: http://www.plant.dlo.nl/

[6]Table 2.6 in Clements and Newman (2002) gives details for several predator–prey experiments conducted to quantify the effects of toxicants on species interactions. Six involve pesticides as the stressor.

default.asp?section=products&page=/products/canoco/right.htm). Pastorok et al. (2002) provide a comprehensive review of available modeling tools and software for analysis of chemical effects on populations, ecosystems, and landscapes.

A development in wildlife conservation that followed the Bern Convention (1979) was the realization that, to conserve species, there is a need to address the state of the habitats in which they exist (http://www.english-nature.org.uk/baps/habitats/), because the two are inextricably linked. Attention to habitat and landscape issues has provided a platform for the use of Geographic Information System (GIS) software in wildlife conservation. GISs have been used in geographical studies for some time, but possible applications in conservation have only recently been appreciated. As with PVA, there are examples of potential uses of GIS in conservation (Markus et al. 2003; Holloway et al. 2003b), but its application has yet to be fully embraced by practitioners.

VII. Discussion and Conclusions

Despite recent advances, both in the acquisition of data and in its analysis, I doubt that any multispecies community is sufficiently well understood for us to make confident predictions about its response to particular disturbances, especially those caused by man.

(May 1984)

The question was asked at the onset of this review, "Do current methods provide enough insight about long-term ecological effects to define pesticide Ecologically Acceptable Concentrations?" One detailed answer is presented now.

It is difficult to envisage a risk assessment process capable of consistently predicting long-term pesticide effects founded primarily on individual-based effect metrics. A risk assessment process that includes more population or community effects metrics would substantially reduce uncertainty in predicting long-term ecological effects of pesticide use. The best illustration of this point is the current assumption that conservative calculations using individual-based effect metrics allow conservative expression of protection of communities from unreasonable risk. In many cases, a predictive risk assessment progressing to Tier 2 would not include any information on indirect effects that, as has been discussed, might be common and important in ecological systems. Furthermore, laboratory test species tend to be *r*-strategy species. This condition creates high uncertainty about predictions of field population persistence for *k*-strategy species.

Can the current ERA process form the foundation for estimating long-term ecological risks of pesticide use? Current tests do provide useful information about the direct effects of pesticides. It would be unwise to abandon these tests completely and to require only complex and expensive ecosystem studies during the predictive assessment of pesticides. However, more laboratory tests focused on population level effect metrics can be performed and would generate more insight than currently possible. For example, risk predictions from the species sensitivity distribution (SSD) approach would be greatly improved by using risk of local population extinction instead of risk of exceeding an LC_{50} value. It is currently impractical to attempt this with most published information from laboratory toxicity tests performed to date, either because information on time-dependent survival and fecundity is not collected or because it is not reported in sufficient detail.

Another improvement would be if strict adherence to the requirements of legislation such as Directive 91/414, with listing of the results of sometimes inappropriate toxicity tests, were replaced by more formal adoption of the principles of ecological risk assessment: this would include formal construction of a conceptual model for each active substance of exposure sources and pathways, and identification of potentially sensitive receptors in Tier 1, rather than only at higher tiers, if at all. In drafting the risk assessment, as much effort should be spent on integrating known ecological relationships into the assessment as is presently being spent on the compiling of concentration and individual-based effects data. A rich literature on ecology remains grossly underexploited in pesticide risk assessment activities.

The continuing use of mesocosm or enclosure studies might also result in more ecologically relevant information for conducting predictive risk assessments because of an increased likelihood of seeing an indirect effect of a pesticide before it is authorized and enters into wide use. However, most such studies currently performed in Europe omit vertebrate species because their extensive feeding can confound measurements of invertebrate and plant populations in mesocosms. Thus, indirect effects on vertebrates remain unmeasured, and those that are observed in mesocosms may be due to the absence of top predators. The realism of mesocosms when predicting the effects of pesticides therefore remains uncertain.

Finally, and of great importance, more postauthorization monitoring could serve as a safety net for the predictive risk assessment of pesticides, which cannot assess all plausible direct and indirect effects for all systems or pesticide mixtures. Pesticides are currently authorized for use in most countries by combining toxicity data with modeled predictions of exposure concentrations. There is some evidence that model scenarios combine conservative assumptions that do not occur widely in the environment (Hendley et al. 2001). Kapustka et al. (1996) describe the "ecological disconnect" that results from the gulf between the current simple pesticide authorization procedures and the complex ecological protection goals of these procedures, and suggest that:

> *"It is remarkable that with such imperfect information, so few reports of pesticide incidents occur. Alternatively, one could argue that the registration*

> *process is exceptionally conservative to the extent that some beneficial uses are excluded without cause."*

Their answer to the problem of ecological disconnect is to treat risk estimates made during pesticide registration as working hypotheses. These measures would then require postauthorization testing, through field monitoring of concentrations and biological effects in the environment during a probationary use phase. This idea seems to be a sensible and logical approach, and it is surprising that it has not been widely adopted by regulatory authorities.

Field studies to examine the biological effects of pesticides must be able to distinguish between effects caused by natural stressors (e.g., extreme weather events), other anthropogenic stressors (e.g., habitat modification or exposure to nonpesticide chemical pollution), and natural covariates (e.g., the prevailing climate, geology, and geography), as these may all influence the structure of organism assemblages (Wickham et al. 1997). In addition to this, the effects of stressors can only be determined in relation to a reference condition (Rykiel 1985), which is the condition that would occur in the absence of stress.

Unfortunately, many reported field studies do not meet the underlying assumptions of hypothesis testing statistics because they are unreplicated or pseudoreplicated and unable to assign treatments at random (Beyers et al. 1995). For example, comparison of upstream and downstream sites will likely include factors that covary with pesticide contamination, such as other agricultural impacts (e.g., nutrients, sediments, or physical disturbance; Sallenave and Day 1991). Investigations performed in this way almost inevitably attract criticism, especially when the politically charged subject of pesticide use and effects is under study. To address such defects, Suter (1993) emphasizes the need for rigorous logic when attempting to unravel the ecological epidemiology of pollution effects in field studies. Both he and researchers with experience in field studies of pesticides (Beyers et al. 1995) identify Koch's postulates and Hill's factors (Hill 1965), both appropriated from medical epidemiology, as useful for focusing attention on the criteria needed to demonstrate causality in field studies.

Koch's postulates, modified for use in environmental toxicology, are as follows.

1. The effects of a toxicant must be regularly associated with exposure to the toxicant and any contributory causal factors. According to Suter (1993), such a regular association should normally consist of Kant's criteria for causation (law of succession and concept of action). These are that cause and effect must always occur together, and that the effect must follow, not precede, the cause.
2. Indicators of exposure to the toxicant must be found in the affected organisms. This could be established by measuring either the toxicant in

the organism or measuring a relevant biomarker induced by the toxicant.
3. The toxic effects must be observed when normal organisms or assemblages are exposed to the toxicant under controlled conditions, and any contributory factors should contribute in the same way during the controlled exposures. This criterion is best met through use of laboratory toxicity tests or mesocosms (Schulz et al. 2002) to confirm that organisms are affected to the same degree at measured toxicant concentrations.
4. The same indicators of exposure and effects must be identified in the controlled exposures as in the field. Again, this is best established through laboratory or mesocosm experiments to confirm that similar concentrations of toxicant within organisms lead to similar effects.

Koch's modified postulates are augmented by Hill's criteria, which were used by Gilbertson (1997) and coworkers to establish that organochlorine pollutants had adversely affected fish, other wildlife and humans in the Great Lakes basin. The main elements of Hill's criteria are as follows:

1. Specificity: Does only the potential cause lead to the effect, and does the potential cause lead only to the effect? Meeting this criterion would be fortunate in most environmental investigations, as it is often the case that multiple causes have multiple potential effects. However, a very high degree of cholinesterase inhibition, outside the normal range of natural variability, would be an example of a specific effect that is very likely to be caused only by exposure to organophosphorus or carbamate pesticides.
2. Strength of association: How precise is the relationship between the potential cause and the observed effect?
3. Time order: Does the effect follow the cause temporally? (Kant's law of succession).
4. Consistency on replication: Is the association repeatedly observed at different times and places by different investigators?
5. Coherence: Does a cause–effect interpretation of the data seriously conflict with generally known facts? Are there plausible mechanisms of toxic action?

Design of studies around Hill's and Koch's criteria is a logical approach to answering questions about the link between pesticides and effects in nature, as found by researchers studying organochlorine effects in the Great Lakes (Gilbertson 1997). We believe that this requires five main elements:

1. Studies should be extensive, rather than intensive, and include measurements at as many sites as possible, so that the consistency of association between pesticides and effects can be assessed.
2. Pesticide concentrations in surface waters should be measured rather than inferred through modeling, so that the presence of a causal agent can be demonstrated.

3. Ideally, researchers should possess detailed knowledge of pesticide application times, rates, and locations near particular aquatic sampling sites so that the time order of possible pesticide cause and effect can be examined more accurately. This provisio may not be necessary if organism exposure concentrations are well characterized in the field, although it would almost certainly be of great use when identifying potential study sites.
4. Concentrations of pesticides in organisms should also be measured as a direct indicator of exposure.
5. The plausibility of potential pesticide cause and effect relationships in the field should be demonstrated by laboratory studies to confirm that concentrations of pesticides found in the field can cause the observed effects on particular taxa.

Summary

Current methods would allow reasonable predictions of long-term effects of pesticide application if three changes were instituted. First, more population-based laboratory studies should be applied in predictive pesticide risk assessment. Second, ERA should include as much effort on collating and integrating ecological knowledge into the assessment in Tier 1 as is currently expended on gathering chemical and toxicological information on exposure and effects. Production of a formal conceptual ecological risk assessment model for each product or active substance for which authorization is sought would provide an appropriate framework for integrating and applying such knowledge. Third, in acknowledgment of the uncertainties in the predictive risk assessment process, more postauthorization monitoring should be done.

The application of Occam's razor to pesticide risk assessment makes good sense, as it does in any other field of science. However, we must take care that simplicity in risk assessment process does not lead to oversimplification:

> *Essentially all science is the study of either very small bits of reality or simplified surrogates for complex whole systems. How we simplify can be critical. Careless simplification leads to misleading simplistic conclusions.*
> (Slobodkin 1994)

Acknowledgments

This work was funded by the U.K. Department for Environment Food and Rural Affairs (Defra) under R&D Contract PS2302, and by the Environment Agency of England and Wales, but the views expressed are those of the authors.

References

ADAS (2000) The Efficacy of No-Spray Buffer Zones in Protecting Field Boundary Watercourses from Pesticide Spray Drift. Report on Project PS0417 to the Ministry of Agriculture, Fisheries and Food, London, UK.

Aldenberg T, Slob W (1993) Confidence limits for hazardous concentrations based on logistically distributed NOEC toxicity data. Ecotoxicol Environ Saf 25:48–63.

Alexander GJ, Horne D, Hanrahan SA (2002) An evaluation of the effects of deltmethrin on two non-target lizard species in the Karoo, South Africa. J Arid Environ 50:121–133.

American Crop Protection Association and Canadian Crop Protection Institute (2001) Joint comments of the American Crop Protection Association (ACPA) and the Canadian Crop Protection Institute (CPI) regarding the NAFTA proposal for the assessment of chemicals on non-target plants (NTP), p. 68.

Aquatic Dialogue Group (1994) Aquatic Dialogue Group: Pesticide Risk Assessment and Mitigation. SETAC Press, Pensacola, FL.

Ashby-Crane RE, Clarke S, Mainstone CP (1994) Impact of Pesticides on River Ecology. R&D Note 269, National Rivers Authority, Bristol, UK.

ASTM (1993) Standard guide for conducting the frog embryo teratogenesis assay: *Xenopus* (FETAX). In: Annual Book of ASTM Standards. American Society for Testing and Materials, Philadelphia.

Atchison GJ, Sandheinrich MB, Bryan MD (1996) Effects of environmental stressors on interspecific interactions of aquatic animals. In: Newman MC, Jagoe CH (eds) Ecotoxicology. A Hierarchical Treatment. CRC/Lewis Press, Boca Raton, pp 319–345.

Bacietto JJ (1998) A framework for ecological risk assessment: beyond the quotient method. In: Newman MC, Strojan CL (eds) Risk Assessment: Logic and Measurement. CRC/Lewis Press, Boca Raton, pp 11–22.

Ball SJ, Lindenmayer DB, Possingham HP (2003) The predictive accuracy of population viability analysis: a test using data from two small mammal species in a fragmented landscape. Biodivers Conserv 12:2393–2413.

Bantle JA (1995) FETAX: a developmental toxicity assay using frog embryos. In: Rand GM (ed) Fundamentals of Aquatic Toxicology: Effects, Environmental Fate, and Risk Assessment, 2nd Ed. Taylor & Francis, Washington, DC, pp 207–230.

Barnett EA, Fletcher MR, Hunter K, Sharp EA (2002) Draft. Pesticide poisoning of animals 2001: Investigations of suspected incidents in the United Kingdom. A report of the Environmental Panel of the Advisory Committee on Pesticides, pp 43.

Baughman DS, Moore DW, Scott GI (1989) A comparison and evaluation of field and laboratory toxicity tests with fenvalerate on an estuarine crustacean. Environ Toxicol Chem 8:417–429.

Beaumont P (1997) Where have all the birds gone? Pestic News 30:3.

Begon M, Harper JL, Townsend CR (1986) Ecology: Individuals, Populations and Communities. Blackwell, London.

Bellamy PE, Rothery P, Hinsley SA (2003) Synchrony of woodland bird populations: the effect of landscape structure. Ecography 26:338–348.

Benton TG, Bryant DM, Cole L, Crick HQP (2002) Linking agricultural practice to insect and bird populations: a historical study over three decades. J Appl Ecol 39:673–687.

Beyers DW, Farmer MS, Sikoski PJ (1995) Effects of rangeland aerial application of Sevin-4-Oil® on fish and aquatic invertebrate drift in the little Missouri River, North Dakota. Arch Environ Contam Toxicol 28:27–34.

Bhatti MA, Al-Khatib K, Felsot AS, Parker R, Kadir S (1995) Effects of simulated chlorsulfuran drift on fruit yield and quality of sweet cherries (*Prunuus avium* L.). Environ Toxicol Chem 14:537–544.

Bishop CA, Ng P, Mineau P, Quinn JS, Struger J (2000) Effects of pesticide spraying on chick growth, behaviour, and parental care in tree swallows (*Tachycineta bicolor*) nesting in an apple orchard in Ontario, Canada. Environ Toxicol Chem 19:2286–2297.

Blaustein AR, Wake DB (1990) Declining amphibian populations: a global phenomenon. Trends Ecol Evol 5:203–204.

Blaustein AR, Wake DB (1995) The puzzle of declining amphibian populations. Sci Am 272:52–57.

Bliss CI, Cattell M (1943) Biological assay. Annu Rev Physiol 5:479–539.

Blockwell SJ, Taylor EJ, Jones I, Pascoe D (1998) The influence of fresh water pollutants and interaction with *Asellus aquaticus* (L.) on the feeding activity of *Gammarus pulex* (L.). Arch Environ Contam Toxicol 34:41–47.

Blus LJ, Henny CJ (1997) Field studies of pesticides and birds: unexpected and unique relations. Ecol Appl 7:1125–1132.

Borg C, Toft S (2000) Importance of insect prey quality for grey partridge chicks *Perdix perdix*: a self-selection experiment. J Appl Ecol 37:557–563.

Boutin C, Rogers CA (2000) Pattern of sensitivity of plant species to various herbicides: an analysis with two variables. Ecotoxicology 9:255–271.

Boutin C, Freemark KE, Keddy CJ (1993) Proposed guidelines for registration of chemical pesticides: nontarget plant testing and evaluation. Tech Rep Series No. 145. Environment Canada, Canadian Wildlife Service, Ottawa.

Boutin C, Freemark KE, Keddy CJ (1995) Overview and rationale for developing regulatory guidelines for nontarget plant testing with chemical pesticides. Environ Toxicol Chem 14:1465–1475.

Broomhall S (2002) The effects of endosulfan and variable water temperature on survivorship and subsequent vulnerability to predation in *Litoria citropa* tadpoles. Aquat Toxicol 61:243–250.

Brown Sullivan K, Spence KM (2003) Effects of sublethal concentrations of atrazine and nitrate on metamorphosis of the African clawed frog. Environ Toxicol Chem 22:627–635.

Broyer J (1994) The decline of the corncrake in France and the management of meadow habitats. Alauda 62:1–7.

Burton TM, Likens GE (1975) Energy flow and nutrient cycling in salamander populations in the Hubbard Brook Experimental Forest, New Hampshire. Ecology 56:1068–1080.

Cairns J Jr (1984) Are single species toxicity tests alone adequate for estimating environmental hazard? Environ Monit Assess 4:259–273.

Cairns J Jr (1992) Paradigms flossed: the coming of age of environmental toxicology. Environ Toxicol Chem 11:285–287.

Cairns J Jr, Niederlehner BR (1993) Ecological function and resiliency: neglected criteria for environmental impact assessment and ecological risk analysis. Environ Prof 15:116–124.

Cairns J Jr, Pratt JR, Niederlehner BR, McCormick PV (1986) A simple cost-effective multispecies toxicity test using organisms with a cosmopolitan distribution. Environ Monit Assess 6:207–220.
Campbell KR, Bartell SM, Shaw JL (2000) Characterizing aquatic ecological risks from pesticides using a diquat dibromide case study. II. Approaches using quotients and distributions. Environ Toxicol Chem 19:760–774.
Candolfi MP, Bakker F, Cañez V, Miles M, Neumann C, Pilling E, Primiani M, Roman K, Schmuck R, Storck-Weyhermüller S, Ufer A, Walttersdorfer A (1999) Sensitivity of non-target arthropods to plant protection products: could *Typhlodromus pyri* and A*phidius* spp. be used as indicator species? Chemosphere 39:1357–1370.
Candolfi MP, Bigler F, Campbell P, Heimbach U, Schmuck R, Angeli G, Bakker F, Brown K, Carli G, Dinter A, Forti D, Forster R, Gothmann A, Hassan S, Mead-Briggs M, Melandri M, Neumann P, Pasqualini E, Powell W, Reboulet J-N, Romijn K, Sechser B, Thieme T, Ufer A, Vergnet C, Vogt H (2000) Principles for regulatory testing and interpretation of semi-field and field studies with non-target arthropods. J Pestic Sci 73:141–147.
Candolfi MP, Barrett KL, Campbell PJ, Forster R, Grady N, Huet M-C, Lewis G, Oomen PA, Schmuck R, Vogt H (2001) Guidance Document on Regulatory Testing and Risk Assessment Procedures for Plant Protection Products to Non-target Arthropods. SETAC, Pensacola, FL.
Cardwell C, Hassall M, White P (1994) Effects of headland management on carabid beetle communities in Breckland cereal fields. Pedobiologia 38:50–62.
Carr JA, Solomon KR (2003) Is atrazine causing frog deformities at very low doses? SETAC Globe 4(1):30–32.
Carr JA, Du Preez LH, Giesy JP, Gross TS, Kendall RJ, Smith EE, Solomon KR, van der Kraak G (2002) Critique of Kiesecker JM. 2002. Synergism between trematode infection and pesticide exposure: a link to amphibian limb deformities in nature? In: Proceedings of the National Academy of Science of the United States of America. Prepared by the Atrazine Endocrine Ecological Risk Assessment Panel, ECORISK, Inc., Ferndale, WA.
Carr JA, Gentles A, Smith EE, Goleman WL, Urquidi LJ, Thuett K, Kendall RJ, Giesy JP, Gross TS, Solomon KR, van der Kraak G (2003) Response of larval *Xenopus laevis* to atrazine: assessment of growth, metamorphosis, and gonadal and laryngeal morphology. Environ Toxicol Chem 22:396–405.
Caslin TM, Wolfe JO (1999) Individual and demographic responses of the gray-tailed vole to vinclozolin. Environ Toxicol Chem 18:1529–1533.
Castillo LE, de La Cruz E, Ruepert C (1997) Ecotoxicology and pesticides in tropical aquatic ecosystems of Central America. Environ Toxicol Chem 16:41–51.
Caswell H (1996) Demography meets ecotoxicology: untangling the population level effects of toxic substances. In: Newman MC, Jagoe CH (eds) Ecotoxicology. A Hierarchical Treatment. CRC/Lewis, Boca Raton, pp 255–292.
Caswell H (2001) Matrix Population Models, 2nd Ed. Sinauer, Sunderland.
Catling PM, Porebski S (1998) Rare wild plants of potential or current economic importance in Canada: a list of priorities. Can J Plant Sci 78:653–658.
Chamberlain DE, Wilson AM, Browne SJ, Vickery JA (1999) Effects of habitat type and management on the abundance of skylarks in the breeding season. J Appl Ecol 36:856–870.

Chamberlain DE, Vickery JA, Gough S (2000) Spatial and temporal distribution of breeding skylarks *Alauda arvensis* in relation to crop type in periods of population increase and decrease. Ardea 88:61–73.
Chinery M (1993) Insects of Britain and Northern Europe, 3rd Ed. Harper Collins, London.
Chiverton PA (1999) The benefits of unsprayed cereal crop margins to grey partridges *Perdix perdix* and pheasants *Phasianus colchicus* in Sweden. Wildl Biol 5:83–92.
Clarke KR (1999) Nonmetric multivariate analysis in community-level ecotoxicology. Environ Toxicol Chem 18:118–127.
Clarke KR, Ainsworth M (1993) A method of linking multivariate community structure to environmental variables. Mar Ecol Prog Ser 92:205–219.
Clements WH, Newman MC (2002) Community Ecotoxicology. Wiley, Chichester.
Clements DK, Skidmore P (1998) The autecology of the hornet robberfly *Asilus crabroniformis* L. in Wales, 1997. Countryside Council for Wales Contract Science Rep No. 263.
Clements DK, Skidmore P (2002) The autecology of the hornet robberfly *Asilus crabroniformis* L. in Wales, 1997–99. Countryside Council for Wales Contract Science Rep No. 525.
Crain DA, Guillette LJ Jr (1998) Reptiles as models of contaminant-induced endocrine disruption. Anim Reprod Sci 53:77–86.
Crain DA, Guillette LJ Jr, Pickford DB, Percival HF, Woodward AR (1998) Sex-steroid and thyroid hormone concentrations in juvenile alligators (*Alligator mississippiensis*) from contaminated and reference lakes in Florida, USA. Environ Toxicol Chem 17:446–452.
Cramp S (1988) Handbook of the Birds of Europe, the Middle East and North Africa: The Birds of the Western Palearctic, vol V. Tyrant Flycatchers to Thrushes. Oxford University Press, Oxford.
Cramp S (1992) Handbook of the Birds of Europe, the Middle East and North Africa: The Birds of the Western Palearctic, vol VI. Warblers. Oxford University Press, Oxford.
Cramp S, Perrins CM (1993) Handbook of the Birds of Europe, the Middle East and North Africa: The Birds of the Western Palearctic, vol VII. Flycatchers to Shrikes. Oxford University Press, Oxford.
Cramp S, Perrins CM (1994a) Handbook of the Birds of Europe, the Middle East and North Africa: The Birds of the Western Palearctic, vol VIII. Crows to Finches. Oxford University Press, Oxford.
Cramp S, Perrins CM (1994b) Handbook of the Birds of Europe, the Middle East and North Africa: The Birds of the Western Palearctic, vol IX. Bunting and New World Warblers. Oxford University Press, Oxford.
Crane M (1997) Research needs for predictive multispecies tests in aquatic toxicology. Hydrobiologia 346:149–155.
Crommentuijn T, Sijm D, de Bruijn J, van Leeuwen K, van de Plassche E (2000) Maximum permissible and negligible concentrations for some organic substances and pesticides. J Environ Manag 58:297–312.
Crossland NO, Shires SW, Bennett D (1982) Aquatic toxicology of cypermethrin. III. Fate and biological effects of spray drift deposits in fresh water adjacent to agricultural land. Aquat Toxicol 2:253–270.

Cuppen JGM, Van den Brink PJ, Camps E, Uil KF, Brock TCM (2000) Impact of fungicide carbendazin in freshwater microcosms. I. Water quality, breakdown of particulate organic matter and responses of macroinvertebrates. Aquat Toxicol 48:233–250.
Dabrowski JM, Schulz R (2003) Predicted and measured levels of azinphos-methyl in the Lourens River, South Africa: comparison of runoff and spray drift. Environ Toxicol Chem 22:494–500.
Dahl B, Blanck H (1996) Pollution-induced community tolerance (PICT) in periphyton communities established under tri-*n*-butyltin (TBT) stress in marine microcosms. Aquat Toxicol 34:305–325.
Daniels RE, Allan JD (1981) Life table evaluation of chronic exposure to a pesticide. Can J Fish Aquat Sci 38:485–494.
Davies PE, Cook LSJ (1993) Catastrophic macroinvertebrate drift and sublethal effects on brown trout, *Salmo trutta*, caused by cypermethrin spraying on a Tasmanian stream. Aquat Toxicol 27:201–224.
Day K, Kaushik NK (1987) An assessment of the chronic toxicity of the synthetic pyrethroid, fenvalerate, to *Daphnia galeata mendota*, using life tables. Environ Toxicol Chem 44:13–26.
Deisch MS, Uresk DW, Linder RL (1990) Effects of rodenticides on deer mice in Western South Dakota. Great Basin Nat 50:347–353.
de Boer IJM (2003) Environmental impact assessment of conventional and organic milk production. Livestock Prod Sci 80:69–77.
de Snoo GR, Scheidegger NMI, de Jong FMW (1999) Vertebrate wildlife incidents with pesticides: a European survey. Pestic Sci 55:47–54.
Dodson SI, Merritt CM, Shannahan J-P, Shults CM (1999) Low exposure concentrations of atrazine increase male production in *Daphnia pulicaria*. Environ Toxicol Chem 18:1568–1573.
Doebeli M, Killingback T (2003) Metapopulation dynamics with quasi-local competition. Theor Pop Biol 64:397–416.
Dover J, Sotherton N, Gobbett K (1990) Reduced pesticide inputs on cereal field margins: the effects on butterfly abundance. Ecol Entomol 15:17–24.
ECOFRAM (1999) Draft Guidance, Ecological Committee on FIFRA Risk Assessment Methods (ECOFRAM) Aquatic Report, http://www.epa.gov/oppefed1/ecorisk/.
Edge WD, Schauber EM (2000) Factors affecting risk assessment of small mammals to pesticides. Environ Toxicol Chem 19:2735–2741.
Eidt DC (1975) The effect of fenitrothion from large-scale forest spraying on benthos in New Brunswick headwaters streams. Can Entomol 107:743–760.
Epperson BK (2000) Spatial genetic structure and non-equilibrium demographics within plant populations. Plant Species Biol 15:269–279.
Ernst SKM, Brown JH (2001) Delayed compensation for missing keystone species by colonization. Science 292:101–102.
Ernst W, Doe K, Jonah P, Young J, Julien G, Hennigar P (1991) The toxicity of chlorothalonil to aquatic fauna and the impact of its operational use on a pond ecosystem. Arch Environ Contamin Toxicol 21:1–9.
European Commission (Health & Consumer Protection Directorate – General). (2002a) Guidance Document on Terrestrial Ecotoxicology Under Council Directive 91/414/EEC. Draft Working Document, 17 October 2002. SANCO/10329/2002 rev 2 final.

European Commission (Health & Consumer Protection Directorate – General). (2002b) Guidance Document on Aquatic Ecotoxicology Working Document, 17 October 2002. SANCO/3268/2001, rev. 4 final.
Ewald JA, Aebischer NJ (1999) Pesticide use, avian food resources and bird densities in Sussex. Report No. 296. Joint Nature Conservation Committee, Peterborough, UK.
Fagan WF, Meir E, Prendergast J, Folarin A, Karieva P (2001) Characterizing population vulnerability for 758 species. Ecol Lett 4:132–138.
Farr JA (1978) The effect of methyl parathion on predator choice of two estuarine prey species. Trans Am Fish Soc 107:87–91.
Fauth JE (1999) Identifying potential keystone species from field data: an example from temporary ponds. Ecol Lett 2:36–43.
Ferson S, AkçHakaya HR (1990) Modeling Structure in Age-structured Populations. RAMAS/Age User Manual. Applied Biomathematics, Setauket, NY.
Finney DJ (1947) Probit Analysis. A Statistical Treatment of the Sigmoidal Response Curve. Cambridge University Press, Cambridge.
Fletcher JS, Johnson FL, McFarlane JC (1990) Influence of greenhouse versus field testing and taxonomic differences on plant sensitivity to chemical treatment. Environ Toxicol Chem 9:769–776.
Fletcher JS, Pfleeger TG, Ratsch HC (1993) Potential environmental risks associated with new sulfonylurea herbicides. Environ Sci Technol 27:2250–2252.
Fletcher JS, Pfleeger TG, Ratsch HC (1995) Chlorsulfuron influence on garden pea reproduction. Physiol Plant 94:261–267.
Fletcher JS, Pfleeger TG, Ratsch HC, Hayes R (1996) Potential impact of low levels of chlorsulfuron and other herbicides on growth and yield of nontarget plants. Environ Toxicol Chem 15:1189–1196.
Forbes VE, Calow P (1999) Is the per capita rate of increase a good measure of population-level effects in ecotoxicology? Environ Toxicol Chem 18:1544–1556.
Forbes VE, Calow P (2003) Contaminant effects on population demographics. In: Newman MC, Unger MA (eds) Fundamentals of Ecotoxicology, 2nd Ed. CRC/Lewis, Boca Raton, pp 221–224.
Fossi MC, Sanchez-Hernandez JC, Diaz-Diaz R, Lari L, Garcia-Hernandez JE, Gaggi C (1995) The lizard *Gallotia galloti* as a bioindicator of organophosphorus contamination in the Canary Islands. Environ Pollut 87:289–294.
Frankham R, Ballou JD, Briscoe DA (2002) Introduction to Conservation Genetics. Cambridge University Press, Cambridge.
Freemark KE, Boutin C (1994) Impacts of agricultural herbicide use on terrestrial wildlife: a review with special reference to Canada. Tech Rep No. 196. Canadian Wildlife Service, Headquarters, p. 53.
Freemark K, Boutin C (1995) Impacts of agricultural herbicide use on terrestrial wildlife in temperate landscapes: a review with special reference to North America. Agric Ecosyst Environ 52:67–91.
Fry DM, Toone CK (1981) DDT-induced feminization of gull embryos. Science 213:922–924.
Fulton MH, Key PB (2001) Acetylcholinesterase inhibition in estuarine fish and invertebrates as an indicator of organophosphorus insecticide exposure and effects. Environ Toxicol Chem 20:37–45.

Furse MT, Symes KL, Winder JM, Clarke RT, Blackburn JH, Gunn RJM, Grieve NJ, Hurley M (1995) The Faunal Richness of Headwater Streams: Stage 3. Impact of Agricultural Activity. Volume 1. Main Report. Project 242. National Rivers Authority R&D, Bristol, UK.

Gardner GR, Yevich PP, Hurst J, Thayer P, Benyi S, Harshbarger JC, Pruell RJ (1991) Germinomas and teratoid siphon anomalies in softshell clams, *Mya arenaria*, environmentally exposed to herbicides. Environ Health Perspect 90:43–51.

Gardner SC, Grue CE (1996) Effects of Rodeo® and Garlon® 3A on nontarget wetland species in central Washington. Environ Toxicol Chem 15:441–451.

Gibbons JW, Scott DE, Ryan TJ, Buhlmann KA, Tuberville TD, Metts BS, Greene JL, Mills T, Leiden Y, Poppy S, Winne CT (2000) The global decline of reptiles, déjá vu amphibians. Bioscience 50:653–666.

Gilbertson M (1997) Advances in forensic toxicology for establishing causality between Great Lakes epizootics and specific persistent toxic chemicals. Environ Toxicol Chem 16:1771–1778.

Giler, PS, O'Donovan G (2002) Biodiversity and ecosystem function: do species matter? Biol Environ 102B:129–139.

Gilpin ME, Ayala FJ (1973) Global models of growth and competition. Proc Natl Acad Sci USA 70:3590–3593.

Gilpin ME, Justice KE (1972) Reinterpretation of the invalidation of the Principle of Competitive Exclusion. Nature (Lond) 236:273–301.

Grant A (1998) Population consequences of chronic toxicity: incorporating density dependence into the analysis of life table response experiments. Ecol Model 105:325–335.

Gray LE Jr, Ostby J, Wolf C, Lambright C, Kelce W (1998) The value of mechanistic studies in laboratory animals for the prediction of reproductive effects in wildlife: endocrine effects on mammalian sexual differentiation. Environ Toxicol Chem 17:109–118.

Green RE (1984) The feeding ecology and survival of partridge chicks *Alectoris rufa* and *Perdix perdix* on arable farmland in East Anglia. J Appl Ecol 21: 817–830.

Green RE (1996) Factors affecting the population density of the corncrake *Crex crex* in Britain and Ireland. J Appl Ecol 33:237–248.

Green RE, Gibbons DW (2000) The status of the corncrake *Crex crex* in Britain in 1998. Bird Study 47:129–137.

Green RE, Stowe TJ (1993) The decline of the corncrake *Crex crex* in Britain and Ireland in relation to habitat change. J Appl Ecol 30:689–695.

Greig-Smith PW, Thompson HM, Hardy AR, Bew MH, Findlay E, Stevenson JH (1994) Incidents of poisoning of honeybees (Apis mellitera) by agricultural pesticides in Great Britain (1981–1991). Crop Protection 13:567–581.

Grist EPM, Leung KMY, Wheeler JR, Crane M (2002) Better bootstrap estimation of hazardous concentration thresholds for aquatic assemblages. Environ Toxicol Chem 21:1515–1524.

Gruar D, Peach W, Taylor R (2003) Summer diet and body condition of song thrushes *Turdus philomelos* in stable and declining farmland bird populations. Ibis 145:637–649.

Guillette LJ Jr, Pickford DB, Crain DA, Rooney AA, Percival HF (1996) Reduction in penis size and plasma testosterone concentrations in juvenile alligators living in a contaminated environment. Gen Comp Endocrinol 101:32–42.

Guillette LJ Jr, Brock JW, Rooney AA, Woodward AR (1999) Serum concentrations of various environmental contaminants and their relationship to sex steroid concentrations and phallus size in juvenile American alligators. Arch Environ Contam Toxicol 36:447–455.

Hall RJ (1980) Effects of Environmental Contaminants on Reptiles: A Review. Special Scientific Report – Wildlife 228. U.S. Fish and Wildlife, Washington, DC.

Hall RJ, Henry PFP (1992) Assessing effects of pesticides on amphibians and reptiles: status and needs. Herpetol J 2:65–71.

Halliday T (1998) A declining amphibian conundrum. Nature (Lond) 394:418–419.

Hanski I (1999) Metapopulation Ecology. Oxford University Press, Oxford.

Harrington LA, MacDonald DW (2002) A Review of the Effects of Pesticides on Wild Terrestrial Mammals in Britain. Wildlife Conservation Research Unit, University of Oxford, Oxford, UK.

Harris ML, Bishop CA, Struger J, Ripley B, Bogart J (1998a) The functional integrity of northern leopard frog (*Rana pipiens*) and green frog (*Rana clamitans*) populations in orchard wetlands. II. Effects of pesticides and eutrophic conditions on early life stage development. Environ Toxicol Chem 17:1351–1363.

Harris ML, Bishop CA, Struger J, van den Heuvel MR, van der Kraak GJ, Dixon DG, Ripley B, Bogart JP (1998b) The functional integrity of northern leopard frog (*Rana pipiens*) and green frog (*Rana clamitans*) populations in orchard wetlands. I. Genetics, physiology, and biochemistry of breeding adults and young-of-the-year. Environ Toxicol Chem 17:1338–1350.

Hart A (2003) Draft briefing paper. Scope and scale for improved assessments of long term risks to birds and mammals for regulatory purposes in the UK.

Hatakeyama S, Yokoyama N (1997) Correlation between overall pesticide effects monitored by shrimp mortality test and change in macrobenthic fauna in a river. Ecotoxicol Environ Saf 36:148–161.

Hatakeyama S, Shiraishi H, Kobayashi N (1990) Effects of aerial spraying on insecticides on nontarget macrobenthos in a mountain stream. Ecotoxicol Environ Saf 19:254–270.

Hayes T (2003) Effects of atrazine on sex differentiation of *Xenopus laevis*. SETAC Globe 4(1):30.

Hayes TB, Collins A, Lee M, Mendoza M, Noriega N, Stuart AA, Vonk A (2002a) Hermaphroditic, demasculinised frogs after exposure to the herbicide atrazine at low ecologically relevant doses. Proc Natl Acad Sci USA 99:5476–5480.

Hayes TB, Haston K, Tsui M, Hoang A, Haeffele C, Vonk A (2002b) Feminization of male frogs in the wild. Nature (Lond) 419:895–896.

Heckman CW (1981) Long-term effects of intensive pesticide applications on the aquatic community in orchard drainage ditches near Hamburg, Germany. Arch Environ Contamin Toxicol 10:393–426.

Henderson IG, Cooper J, Fuller RJ, Vickery J (2000) The relative abundance of birds on set-aside and neighbouring field in summer. J Appl Ecol 37:335–347.

Hendley P, Holmes C, Kay S, Maund SJ, Travis KZ, Zhang M (2001) Probabilistic risk assessment of cotton pyrethroids: III. A spatial analysis of the Mississippi, USA, cotton landscape. Environ Toxicol Chem 20:669–678.

Henny CL, Blus LJ, Kolbe EH, Fitzner RE (1985) Organophosphate insecticide (Famphur) topically applied to cattle kills magpies and hawks. J Wildl Manag 49:648–658.

Hickey JJ, Anderson DW (1968) Chlorinated hydrocarbons and eggshell changes in raptorial and fish-eating birds. Science 162:271–273.
Hill AB (1965) The environment and disease: association or causation? Proc R Soc Med 58:295–300.
Hill JK, Thomas CD, Lewis OT (1996) Effects of habitat patch size and isolation on dispersal by *Hesperia comma* butterflies: implications for metapopulation structure. J Anim Ecol 65:725–735.
Hill JK, Thomas CD, Lewis OT (1999) Flight morphology in fragmented populations of a rare British butterfly, *Hesperia comma*. Biol Conserv 87:277–283.
Hoelzel AR (1999) Impact of population bottlenecks on genetic variation and the importance of life-history: a case study of the northern elephant seal. Biol J Linn Soc 68:23–39.
Hoelzel AR, Halley J, O'Brien SJ, Campagna C, Arnborn T, le Boeuf B, Ralls K, Dover GA (1993) Elephant seal genetic variation and use of simulation models to investigate historical population bottlenecks. J Hered 84:443–449.
Holloway GJ, Dickson JD, Harris PW, Smith J (2003a) Dynamics and foraging behaviour of adult hornet robberflies, *Asilus crabroniformis*: implications for conservation management. J Insect Conserv 7:127–135.
Holloway GJ, Griffiths GH, Richardson P (2003b) Conservation strategy maps: a tool to facilitate biodiversity action planning illustrated using the heath fritillary butterfly. J Appl Ecol 40:413–421.
Hopkins SP (1993) Ecological implications of "95% protection levels" for metals in soil. Oikos 66:137–141.
Hopkins WA (2000) Reptile toxicology: challenges and opportunities on the last frontier in vertebrate ecotoxicology. Environ Toxicol Chem 19:2391–2393.
Horne G, Fielding AH (2002) Recovery of the Peregrine Falcon *Falco peregrinus* in Cumbria, UK, 1966–99. Bird Study 49:229–236.
Hossel JE (2001) Climate change and UK farmland birds: a review of implications. RSPB report no. 5060. p. 76.
Hulme M, Viner D (1998) A climate change scenario. Clim Change 39:145–176.
Hunt GL Jr, Hunt MW (1977) Female-female pairing in Western gulls (*Larus occidentalis*) in Southern California. Science 196:1466–1467.
Hurd MK, Perry SA, Perry WB (1996) Nontarget effects of a test application of diflubenzuron to the forest canopy on stream macroinvertebrates. Environ Toxicol Chem 15:1344–1351.
Jacquemyn H, van Rossum F, Brys R, Endels P, Hermy M, Triest L, de Blust G (2003) Effects of agricultural land use and fragmentation on genetics, demography and population persistence of the rare *Primula vulgaris*, and its implications for conservation. Belg J Bot 136:5–22.
Jagoe R, Newman MC (1997) Bootstrap estimation of community NOEC values. Ecotoxicology 6: 293–306.
Joern A, Hoagland KD (1996) In defense of whole-community bioassays for risk assessment. Environ Toxicol Chem 15:407–409.
Kadiri N, Lumaret J-P, Janati-Idrissi A (1999) Macrocyclic lactones: impact on non-target fauna in pastures. Ann Soc Entomol France 35:222–229.
Kammenga JE, Busschers M, Van Straalen NM, Jepson PC, Baker J (1996) Stress induced fitness is not determined by the most sensitive life-cycle trait. Funct Ecol 10:106–111.

Kapustka LA, Williams BA, Fairbrother A (1996) Evaluating risk predictions at population and community levels in pesticide registration: hypothesis to be tested. Environ Toxicol Chem 15:427–431.
Karr JR (1991) Biological integrity: a long-neglected aspect of water resource management. Ecol Appl 1:66–84.
Karr JR (1993) Defining and assessing ecological integrity: beyond water quality. Environ Toxicol Chem 12:1521–1531.
Karr JR, Fausch KD, Angermeier PL, Yant PR, Schlosser IJ (1986) Assessing Biological Integrity in Running Waters. A Method and Its Rationale. Illinois Natural History Survey Publication 5. Illinois Natural History Survey, Champagne, IL.
Kaye TN, Pyke DA (2003) The effect of stochastic technique on estimates of population viability from transition matrix models. Ecology 84:1464–1476.
Kersting K, Van den Brink PJ (1997) Effects of the insecticide Dursban® 4E (active ingredient chlorpyrifos) in outdoor experimental ditches: responses of ecosystem metabolism. Environ Toxicol Chem 16:251–259.
Kersting K, Van Wijngaarden RPA (1999) Effects of a pulsed treatment with the herbicide Afalon (active ingredient linuron) on macrophyte-dominated mesocosms. I. Responses of ecosystem metabolism. Environ Toxicol Chem 18:2859–2865.
Kiesecker JM (2002) Synergism between trematode infection and pesticide exposure: a link to amphibian limb deformities in nature? Proc Natl Acad Sci USA 99(15):9900–9904.
Kirby-Smith WW, Eisenreich SJ, Howe JT, Luettich RA Jr (1992) The Effects in Estuaries of Pesticide Runoff from Adjacent Farmlands. CR813415. Final Project Report. US Environmental Protection Agency, Gulf Breeze, FL.
Krebs JR, Wilson JD, Bradbury RB, Siriwardena GM (1999) The second Silent Spring? Nature (Lond) 400:611–612.
Kreutzweiser DP, Kingsbury PD, Feng JC (1989) Drift response of stream invertebrates to aerial applications of glyphosate. Bull Environ Contam Toxicol 42:331–338.
Kruzynski GM, Birtwell IK (1994) A predation bioassay to quantify the ecological significance of sublethal responses of juvenile chinook salmon (Oncorhynchus tshawytscha) to the antisapstain fungicide TCMTB. Can J Fish Aquat Sci 51:1780–1790.
Kudoh H (2001) Gene flow among plant populations in an ecological landscape. Jpn J Ecol 51:193–201.
Lacy RC (1993) VORTEX: a computer simulation model for Population Viability Analysis. Wildl Res 20:45–65.
Lacy RC, Clarke TW (1990) Population viability assessment of the eastern barred bandicoot in Victoria. In: Clarke TW, Seebeck JH (eds) The Management and Conservation of Small Populations. Chicago Zoological Society, Chicago, pp 131–145.
Lambert MRK (1993) Effects of DDT ground-spraying against tsetse flies on lizards in NW Zimbabwe. Environ Pollut 82:231–237.
Lambert MRK (1994) Ground-spray treatment with deltamethrin against tsetse flies in NW Zimbabwe has little short term effect on lizards. Bull Environ Contam Toxicol 53:555–561.
Lambert MRK (1997) Environmental effects of heavy spillage from a destroyed pesticide store near Hargeisa (Somaliland) assessed during the dry season, using reptiles and amphibians as bioindicators. Arch Environ Contam Toxicol 32:80–93.

Landis WG, Matthews RA, Matthews GB (1996) The layered and historical nature of ecological systems and the risk assessment of pesticides. Environ Toxicol Chem 15:432–440.
Lane SJ, Alonso JC (2001) Status and extinction probabilities of great bustard (*Otis tarda*) leks in Andalucia, southern Spain. Biodivers Conserv 10:893–910.
Leonard AW, Hyne RV, Lim RP, Chapman JC (1999) Effect of endosulfan runoff from cotton fields on macroinvertebrates in the Namoi River. Ecotoxicol Environ Saf 42:125–134.
Leonard AW, Hyne RV, Lim RP, Pablo F, van den Brink PJ (2000) Riverine endosulfan concentrations in the Namoi River, Australia: link to cotton field runoff and macroinvertebrate population densities. Environ Toxicol Chem 19:1540–1551.
Lewis MA, Wang W (1999) Biomonitoring using aquatic vegetation. Environ Sci Forum 96:243–274.
Liber K, Kaushik NK, Solomon KR, Carey JH (1992) Experimental designs for aquatic mesocosm studies: comparison of the "ANOVA" and "regression" design for assessing the impact of tetrachlorophenol in zooplankton populations in limnocorrals. Environ Toxicol Chem 11:61–77.
Liebig J (1840) Chemistry in Its Application to Agriculture and Physiology. Taylor & Walton, London.
Liess M (1994) Pesticide impact on macroinvertebrate communities of running waters in agricultural ecosystems. Verh Int Verein Limnol 25:2060–2062.
Liess M (1998) Significance of agricultural pesticides on stream macroinvertebrate communities. Verh Int Verein Limnol 26:1245–1249.
Liess M, Schulz R (1999) Linking insecticide contamination and population response in an agricultural stream. Environ Toxicol Chem 18:1948–1955.
Liess M, Schulz R, Liess MH-D, Rother B, Kreuzig R (1999) Determination of insecticide contamination in agricultural headwater streams. Water Res 33:239–247.
Luoto M, Rekolainen S, Aakkula J, Pykala J (2003) Loss of plant species richness and habitat connectivity in grasslands associated with agricultural change in Finland. Ambio 32:447–452.
Mackay CE, Colton JA, Bigham G (2002) Structuring population-based ecological risk assessments in a dynamic landscape. In: Newman MC, Roberts Jr MH, Hale RC (eds) Coastal and Estuarine Risk Assessment. CRC/Lewis, Boca Raton, pp 273–296.
Madsen T, Shine R, Olsson M, Wittzell H (1999) Restoration of an inbred adder population. Nature (Lond) 402:34–35.
MAFF Pesticides Safety Directorate (1999) Assessing pesticide risks to non-target terrestrial plants: a desk study. Commission No. PN0923.
Markus P, Timo P, Juha T (2003) Habitat preferences of the skylark *Alauda arvensis* in Southern Sweden. Ornis Fenn 80:97–110.
Marshall J, Brown V, Boatman N, Lutman P, Squire G (2001) The impact of herbicides on weed abundance and biodiversity, PN0940. A Report for the UK Pesticides Safety Directorate. IACR-Long Ashton Research Station.
Martin PA, Johnson DL, Forsyth DJ, Hill BD (2000) Effects of two grasshopper control insecticides on food resources and reproductive success of two species of grassland songbirds. Environ Toxicol Chem 19:2987–2996.
Matthews RA, Landis WG, Matthews GB (1996) The community conditioning hypothesis and its application to environmental toxicology. Environ Toxicol Chem 15:597–603.

Matthews RA, Matthews GB, Landis WG (1998) Application of community level toxicity testing to environmental risk assessment. In: Newman MC, Strojan CL (eds) Risk Assessment. Logic and Measurement, CRC/Lewis Press, Boca Raton, pp 225–253.

Matthiessen P, Sheahan D, Harrison R, Kirby M, Rycroft R, Turnbull A, Volkner C, Williams R (1995) Use of a *Gammarus pulex* bioassay to measure the effects of transient carbofuran runoff from farmland. Ecotoxicol Environ Saf 30:111–119.

Matz AC, Bennett RS, Landis WG (1998) Effects of azinphos-methyl on northern bobwhite: a comparison of laboratory and field results. Environ Toxicol Chem 17:1364–1370.

Maund SJ, Travis KZ, Hendley P, Giddings JM, Solomon KR (2001) Probabilistic risk assessment of cotton pyrethroids: V. Combining landscape-level exposure and ecotoxicological effects data to characterize risks. Environ Toxicol Chem 20:687–692.

Maurer BA, Holt RD (1996) Effects of chronic pesticide stress on wildlife populations in complex landscapes: processes at multiple scales. Environ Toxicol Chem 15:420–426.

May RM (1984) An overview: real and apparent patterns in community structure. In: Strong DR, Simberloff D, Abele LG, Thistle AB (eds) Ecological Communities: Conceptual Issues and the Evidence. Princeton University Press, Princeton, NJ, pp 3–18.

McCracken DI, Foster GN (1993) The effects of ivermectin on the invertebrate fauna associated with cow dung. Environ Toxicol Chem 12:73–84.

McGinnity P, Prodohl P, Ferguson A, Hynes R, Maoileidigh NO, Baker N, Cotter D, O'Hea B, Cooke D, Rogan G, Taggart J, Cross T (2003) Fitness reduction and potential extinction of wild populations of Altlantic salmon, *Salmo salar*, as a result of interactions with escaped farm salmon. Proc R Soc Lond B 270: 2443–2450.

McLaughlin A, Mineau P (1995) The impact of agricultural practices on biodiversity. Agric Ecosyst Environ 55:201–212.

Menge BA, Berlow EL, Blanchette CA, Navarrete SA, Yamada SB (1994) The keystone species concept: variation in interaction strength in a rocky intertidal habitat. Ecol Monogr 64:249–286.

Mills LS, Soule ME, Doak DF (1993) The keystone-species concept in ecology and conservation. BioScience 43:219–223.

Millsap BA, Kennedy PL, Mitchell A, Court G, Enderson JH, Rosenfield RN (1998) Review of the proposal to de-list the American peregrine falcon. Wildl Soc Bull 36:522–538.

Milnes MR, Woodward AR, Rooney AA, Guillette LJ (2002) Plasma steroid concentrations in relation to size and age in juvenile alligators from two Florida lakes. Comp Biochem Physiol (Part A) 131:923–930.

Mineau P (1994) An analysis of avian reproduction studies submitted for pesticide registration. Ecotox Env Safety 29:304–329.

Mineau P (2002) Estimating the probability of bird mortality from pesticide sprays on the basis of the field study record. Environ Toxicol Chem 21:1497–1506.

Morrison B, Wells D (1981) The fate of fenitrothion in a stream environment and its effect on the fauna, following aerial spraying of a Scottish forest. Sci Total Environ 19:233–253.

Muller P, Nagel P, Flacke W (1981) Ecological side effects of dieldrin against tsetse flies in Adamaoua, Cameroon. Oecologia (Berl) 50:187–194.
Nacci DE, Gleason TR, Gutjahr-Gobell R, Huber M, Munns Jr WR (2002) Effects of chronic stress on wildlife populations: a population modeling approach and case study. In: Newman MC, Roberts Jr MH, Hale RC (eds) Coastal and Estuarine Risk Assessment. CRC/Lewis, Boca Raton, pp 247–272.
Naeem S (1998) Species redundancy and ecosystem reliability. Conserv Biol 12:39–45.
Nakamaru M, Iwasa Y, Nakanishi J (2002) Extinction risk to herring gull populations from DDT exposure. Environ Toxicol Chem 21:195–202.
Newman MC (1995) Quantitative Methods in Aquatic Ecotoxicology. CRC/Lewis, Boca Raton.
Newman MC (2001) Population Ecotoxicology. Wiley, Chichester.
Newman MC, McCloskey JT (2000) The individual tolerance concept is not the sole explanation for the probit dose-effect model. Environ Toxicol Chem 19:520–526.
Newman MC, McCloskey JT (2002) Applying time to event methods to assess pollutant effects on populations. In: Crane M, Newman MC, Chapman PF, Fenlon J (eds) Risk Assessment with Time to Event Models. CRC/Lewis, Boca Raton, pp 23–38.
Newman MC, Unger MA (2003) Fundamentals of Ecotoxicology, 2nd Ed. CRC/Lewis, Boca Raton.
Newton I (1986) The Sparrowhawk. T & AD Poyser, Calton, Staffs, UK.
Norris RF, Kogan M (2000) Interactions between weeds, arthropod pests, and their natural enemies in managed ecosystems. Weed Sci 48:94–158.
O'Connor RJ (1996) Toward the incorporation of spatial temporal dynamics into ecotoxicology. In: Rhodes OE Jr, Chesser RK, Smith MH (eds) Population Dynamics in Ecological Space and Time. University of Chicago Press, Chicago.
Osano O, Admiraal W, Otieno D (2002) Developmental disorders in embryos of the frog *Xenopus laevis* induced by chloroacetanilide herbicides and their degradation products. Environ Toxicol Chem 21:375–379.
Pastorok RA, Bartell SM, Ferson S, Ginzburg LR (2002) Ecological Modeling in Risk Assessment: Chemical Effects on Populations, Ecosystems and Landscapes. Lewis, Boca Raton.
Pauli BD, Perrault JA, Money SL (2000) RATL: A Database of Reptile and Amphibian Toxicology Literature. Technical Report Series No. 357. Canadian Wildlife Service, Headquarters, Hull, Québec, Canada.
Pemberton JM, Smith RH (1985) Lack of biochemical polymorphism in British UK fallow deer *Dama dama*. Heredity 55:199–208.
Petersen RC Jr, Petersen LB-M (1988) Compensatory mortality in aquatic populations: its importance for interpretation of toxicant effects. Ambio 17:381–386.
Pfleeger T, McFarlane C, Sherman R, Volk G (1991) A short-term bioassay for whole plant toxicity. In: Gorsuch JW, Lower WR, Wang W, Lewis MA (eds) Plants for Toxicity Assessment, vol 2. American Society for Testing and Materials, Philadelphia, pp 355–364.
Philbrick CT, Les DH (1996) Evolution of aquatic angiosperm reproductive systems. Bioscience 46:813–826.
Pimentel D (1992) Ecological effects of pesticides on non-target species in terrestrial ecosystems. In: Tardiff RG (ed) Methods to Assess Effects of Pesticides on Non-target Organisms. Wiley, New York, pp 171–190.

Posthuma L, Suter GW II, Traas TP (eds) (2002) Species Sensitivity Distributions in Ecotoxicology. CRC/Lewis, Boca Raton.
Potts GR (1986) The Partridge. Pesticides, Predation and Conservation. Collin, London.
Power ME, Tilman D, Estes JA, Menge BA, Bond WJ, Mills SJ, Daily G, Castilla, JC, Lubchenco J, Paine RT (1996) Challenges in the quest for keystones. BioScience 46:609–620.
Pratt JR, Cairns J Jr (1996) Ecotoxicology and the redundancy problem: understanding effects on community structure and function. In: Newman MC, Jagoe CH (eds) Ecotoxicology. A Hierarchical Treatment. CRC/Lewis Press, Boca Raton, pp 347–370.
Preston BL (2002) Indirect effects in aquatic ecotoxicology: implications for ecological risk assessment, Environ Manag 29:311–323.
Ralls K, Ballou J (1983) Extinction: lessons from zoos. In: Schonewald-Cox CM, Chambers SM, MacBryde B, Thomas L (eds) Genetics and Conservation: a Reference for Managing Wild Animal and Plant Populations. Benjamin/Cummings, Menlo Park, pp 164–184.
Rands MRW (1985) Pesticide use on cereals and the survival of grey partridge *Perdix perdix* chicks in a field experiment. J Appl Ecol 22:49–54.
Rands MRW (1986) The survival of gamebird Galliformes chicks in relation to pesticide use in cereals. Ibis 128:57–64.
Ratcliffe D (1980) The Peregrine Falcon. T & AD Poyser, Calton, Staffs, UK.
Reeder AL, Foley GL, Nichols DK, Hansen LG, Wikoff B, Faeh S, Eisold J, Wheeler MB, Warner R, Murphy JE, Beasley VR (1998) Forms and prevalence of intersexuality and effects of environmental contaminants in sexuality in cricket frogs (*Acris crepitans*). Environ Health Perspect 106:261–266.
Robinson RA, Sutherland WJ (2002) Post-war changes in arable farming and biodiversity in Great Britain. J Appl Ecol 39:157–176.
Rowe G, Beebee TJC (2003) Population on the verge of a mutational meltdown? Fitness costs of genetic load for an amphibian in the wild. Evolution 57:177–181.
Rykiel EJ Jr (1985) Toward a definition of ecological disturbance. Aust J Ecol 10:361–365.
Ryttman H (2003) Breeding success of Wryneck *Jynx torquiila* during the last 40 years in Sweden. Ornis Svec 13:25–28.
Saccheri I, Kuussaari M, Kankare M, Fortelius W, Hanski I (1998) Inbreeding and extinction in a butterfly metapopulation. Nature (Lond) 392:491–494.
Sallenave RM, Day KE (1991) Secondary production of benthic stream invertebrates in agricultural watersheds with different land management practices. Chemosphere 23:57–76.
Sanchez-Hernandez JC (2003) Evaluating reptile exposure to cholinesterase-inhibiting agrochemicals by serum butyrylcholinesterase activity. Environ Toxicol Chem 22:296–905.
Savidge JA (1978) Wildlife in herbicide-treated Jeffery pine plantation in eastern California. J For 76:476–478.
Sawchik J, Dufrene M, Schickzelle N, Baguette M (2002) Metapopulation dynamics of the bog fritillary butterfly: modelling the effect of habitat fragmentation. Acta Oecol 23:287–296.
Schmiegelow FKA, Monkkonen M (2002) Habitat loss and fragmentation in dynamic landscapes: avian perspectives from the boreal forest. Ecol Appl 12:375–389.

Schroeder MH, Sturges DL (1975) The effect on the Brewer's sparrow of spraying big sagebrush. J Range Manag 28:294–297.
Schulz R (1998) Macroinvertebrate dynamics in a stream receiving insecticide-contaminated runoff. Verh Int Verein Limnol 26:1271–1276.
Schulz R (2004) Field studies on exposure, effects, and risk mitigation of aquatic nonpoint-source insecticide pollution: a review. J Environ Qual 33:419–448.
Schulz R, Liess M (1999) Validity and ecological relevance of an active in situ bioassay using *Gammarus pulex* and *Limnephilus lunatus*. EnvironToxicol Chem 18:2243–2250.
Schulz R, Peall SKC, Dabrowski JM, Reinecke AJ (2001) Spray distribution of two insecticides into surface waters in a South African orchard area. J Environ Qual 30:814–822.
Schulz R, Thiere G, Dabrowski JM (2002) A combined microcosm and field approach to evaluate the aquatic toxicity of azinphosmethyl to stream communities. Environ Toxicol Chem 21:2172–2178.
Scott GI, Fulton MH, Moore DW, Chandler GT, Key PB, Hampton TW, Marcus JM, Kackson KL, Bauchman SS, Trim AH, Williams L, Louden CJ, Patterson ER (1990) Agricultural Insecticide Runoff Effects on Estuarine Organisms: Correlating Laboratory and Field Toxicity Testing with Ecotoxicological Biomonitoring. CR813138. Final Project Report. U.S. Environmental Protection Agency, Gulf Breeze, FL.
Scott GI, Fulton MH, Crosby MC, Key PB, Daughomah J, Waldren JT, Strozier ED, Louden CJ, Chandler GT, Bidleman RF, Jackson KL, Hampton TW, Hoffman T, Shulz A, Bradford M (1992) Agricultural Insecticide Runoff Effects on Estuarine Organisms: Correlating Laboratory and Field Toxicity Tests, Ecophysiology Bioassays, and Ecotoxicological Biomonitoring. CR813138–2. Final Project Report. U.S. Environmental Protection Agency, Gulf Breeze, FL.
Sheahan D, Bates H, Hurst M, Matthiessen P, Smith A, White C, Williams R. (1999) The Impact of Pesticides on River Ecology: A Study of Headwater Streams. Environment Agency R&D Technical Report, Bristol, UK.
Sheffield SR, Lochmiller RL (2001) Effects of field exposure to diazinon on small mammals inhabiting a semi enclosed prairie grassland ecosystem. I. ecological and reproductive effects. Environ Toxicol Chem 20:284–296.
Shelford VE (1911) Physiological animal geography. J Morphol 22:551–618.
Shelford VE (1913) Animal Communities in Temperate America. University of Chicago, Chicago.
Sherman PW, Runge MC (2002) Demography of a population collapse: the Northern Idaho ground squirrel (*Spermophilus brunneus brunneus*). Ecology 83:2816–2831.
Sherratt TN, Roberts G, Williams P, Whitfield M, Biggs J, Shillabeer N, Maund SJ (1999) A life-history approach to predicting the recovery of aquatic invertebrate populations after exposure to xenobiotic chemicals. Environ Toxicol Chem 18:2512–2518.
Shilova SA (1990) Ecological consequences of disturbances in the population structure of small mammals. In: Krivolutsky DA (ed) Bioindications of Chemical and Radioactive Pollution. CRC Press, Boca Raton, pp 107–129.
Shires SW, Bennett D (1985) Contamination and effects in freshwater ditches resulting from an aerial application of cypermethrin. Ecotoxicol Environ Saf 9:145–158.

Sibly RM, Calow P (1989) A life-cycle theory of responses to stress. Biol J Linn Soc 37:101–116.

Slobodkin LB (1994) The connection between single species and ecosystems. In: Sutcliffe DW (ed) Water Quality and Stress Indicators: Linking Levels of Organization. Freshwater Biological Association, Ambleside, UK.

Smith MN (2000) The hornet robberfly *Asilus crabroniformis*: land use and livestock grazing regimes at sites in England. English nature rep no. 387.

Smith MN (2001) The current distribution of the hornet robberfly *Asilus crabroniformis* Linnaeus (Diptera, Asilidae) in England and Wales. Dipterists Dig 8:79–84.

Snell TW, Serra M (2000) Using probability of extinction to evaluate the ecological significance of toxicant effects. Environ Toxicol Chem 19:2357–2363.

Solomon KR, Baker DB, Richards RP, Dixon KR, Klaine SJ, La Point TW, Kendall RJ, Weisskopf CP, Giddings JM, Giesy JP, Hall LW Jr, Williams WM (1996) Ecological risk assessment of atrazine in North American surface waters. Environ Toxicol Chem 15:31–76.

Sotherton N, Holland J (2003) Indirect effects of pesticides on farmland wildlife. In: Hoffman DJ, Rattner BA, Burton Jr GA, Cairns Jr J (eds) Handbook of Ecotoxicology. CRC/Lewis, Boca Raton, pp 1173–1195.

Soulé ME (1987) Viable Populations for Conservation. Cambridge University Press, Cambridge.

Southwood RE, DJ Cross (2002) Food requirements of grey partridge *Perdix perdix* chicks. Wildl Biol 8:175–183.

Sparling DW, Linder G, Bishop CA (eds) (2000) Ecotoxicology of Amphibians and Reptiles. SETAC Press, Pensacola, FL.

Sparling DW, Fellers GM, McConnell LL (2001) Pesticides and amphibian population declines in California, USA. Environ Toxicol Chem 20:1591–1595.

Stansley W, Roscoe DE (1999) Chlordane poisoning of birds in New Jersey, USA. Environ Toxicol Chem 18:2095–2099.

Stearns SC (1992) The Evolution of Life Histories. Oxford University Press, Oxford.

Stearns SC, Crandall RE (1984) Plasticity of age and size at sexual maturity: a life history response to unavoidable stress. In: Potts G, Wootton R (eds) Fish Reproduction. Academic Press, London, pp 13–34.

Stewart PM, Scribailo RW, Simon TP (1999) The use of aquatic macrophytes in monitoring and in assessment of biological integrity. Environ Sci Forum 96:275–302.

Stowe TJ, Newton AV, Green RE, Mayes E (1993) The decline of the corncrake *Crex crex* in Britain and Ireland in relation to habitat. J Appl Ecol 30:53–62.

Sullivan DS, Sullivan TP, Bisalputra T (1981) Effects of Roundup® herbicide on diatom populations in the aquatic environment of a coastal forest. Bull Environ Contamin Toxicol 26:91–96.

Suter GW II (1993) Ecological Risk Assessment. Lewis, Boca Raton.

Tagatz ME (1976) Effect of mirex on predator-prey interaction in an experimental estuarine ecosystem. Trans Am Fish Soc 105:546–549.

Talent LG, Dumont JN, Bantle JA, Janz DM, Talent SG (2002) Evaluation of western fence lizards (*Sceloporus occidentalis*) and eastern fence lizard (*Sceloporus undulates*) as laboratory reptile models for toxicological investigations. Environ Toxicol Chem 21:899–905.

Tanaka Y (2000) Extinction of populations by inbreeding depression under stochastic environments. Pop Ecol 42:55–62.

Tanaka Y (2003) Ecological risk assessment of pollutant chemicals: extinction risk based on population-level effects. Chemosphere 53:421–425.
Taub FB (1997) Are ecological studies relevant to pesticide decisions? Ecol Appl 7:1083–1085.
ter Braak CJF, Smilauer P (1998) CANOCO Reference Manual and Users' Guide to CANOCO for Windows: Software for Canonical Community Ordination (Version 4). Microcomputer Power, Ithaca, NY.
Tew TE, MacDonald DW, Rands MRW (1992) Herbicide application affects microhabitat use by arable wood mice *Apodemus sylvaticus*. J Appl Ecol 29:532–539.
Thompson HM (2002) Sub-lethal effects in honeybees: their significance and use in pesticide risk assessment. DEFRA Funded R and D Project PN0944 final report, pp 13.
Thompson HM, Hunt LV (1999) Extrapolating from honeybees to bumblebees in pesticide risk assessment. Ecotoxicology 8:147–166.
Tones SJ, Ellis SA, Breeze VG, Fowbert J, Miller PCH, Oakley JN, Parkin CS, Arnold DJ (2001) Review and evaluation of test species and methods for assessing exposure of non-target plants and invertebrates to crop pesticide sprays and spray drift (desk study). DEFRA project number PN 0937. p. 76.
Touart LW, Maciorowski AF (1997) Information needs for pesticide registration in the United States. Ecol Appl 7:1086–1093.
US EPA (2002) Methods for evaluating wetland condition. #12 Using amphibians in bioassessments of wetlands. EPA-822-R-02-022. U.S. Environmental Protection Agency, Washington, DC.
van Apeldoorm RC, Celada C, Nieuwenhuizen W (1994) Distribution and dynamics of the red squirrel (*Sciurus vulgaris*) in a landscape with fragmented habitat. Land Ecol 9:227–235.
Van den Brink PJ, Van Wijngaarden RPA, Lucassen WGH, Brock TCM, Leeuwangh P (1996) Effects of the insecticide Dursban® 4E (active ingredient chlorpyrifos) in outdoor experimental ditches: II. Invertebrate community responses and recovery. Environ Toxicol Chem 15:1143–1153.
Van der Hoeven N, Gerritsen AA (1997) Effects of chlorpyrifos on individuals and populations of *Daphnia pulex* in the laboratory and field. Environ Toxicol Chem 16:2438–2447.
Van Geest GJ, Zwaardemaker NG, Van Wijngaarden, RPA, Cuppen JGM (1999) Effects of a pulsed treatment with the herbicide Afalon (active ingredient linuron) on macrophyte-dominated mesocosms. II. Structural responses. Environ Toxicol Chem 18:2866–2874.
van Langevelde F (2000) Scale of habitat connectivity and colonization in fragmented nuthatch populations. Ecography 23:614–622.
Van Vlaardingen P, Traas T, Aldenburg T (2003) ETX-2000: Normal Distribution Based Hazardous Concentration and Potentially Affected Fraction. RIVM, Bilthoven, The Netherlands.
Wagner C, Lokke H (1991) Estimation of ecotoxicological protection levels from NOEC toxicity data. Water Res 25:1237–1242.
Wake DB (1991) Declining amphibian populations. Science 253:860.
Wang G, Edge WD, Wolff JO (2001) Demographic uncertainty in ecological risk assessments. Ecol Model 136:95–102.
Wardhaugh KG, Holter P, Longstaff B (2001) The development and survival of three species of coprophagous insect after feeding on the faeces of sheep treated with

controlled-release formulations of ivermectin or albendazole. Aust Vet J 79: 125–132.

Watson J, Watson A, Paull D, Freudenberger D (2003) Woodland fragmentation is causing the decline of species and functional groups of birds in southeastern Australia. Pac Conserv Biol 8:261–270.

Wickham JD, Wu J, Bradford DF (1997) A conceptual framework for selecting and analyzing stressor data to study species richness at large spatial scales. Environ Manag 21:247–257.

Williams RJ, Brooke DN, Clare RW, Matthiessen P, Mitchell RDJ (1996) Rosemaund Pesticide Transport Study 1987–1993. Report No. 129. Institute of Hydrology, Wallingford, UK.

Wilson AM, Vickery JA, Browne SJ (2001) Numbers and distribution of northern lapwings *Vanellus vanellus* breeding in England and Wales in 1998. Bird Study 48:2–17.

Wilson JD, Morris AJ, Arroyo BE, Clark SC, Bradbury RB (1999) A review of the abundance and diversity of invertebrate and plant foods of granivorous birds in northern Europe in relation to agricultural change. Agric Ecosyst Environ 75:13–30.

Wingfield Gibbons D, Reid JB, Chapmand RA (1993) The New Atlas of Breeding Birds in Britain and Ireland: 1988–1991. T & AD Poyser, London.

Withgott J (2002) Amphibian decline. Ubiquitous herbicide emasculates frogs. Science 296:447–448.

Woin P (1998) Short- and long-term effects of the pyrethroid insecticide fenvalerate on an invertebrate pond community. Ecotoxicol Environ Saf 41:137–156.

Zhao Y, Newman MC (2004) Shortcomings of the laboratory derived LC50 for predicting mortality in field populations: exposure duration and latent mortality. Environ Toxicol Chem 23:2147–2153.

Manuscript received June 23, accepted July 2, 2005.

Rev Environ Contam Toxicol 187:67–101

Sulfonamides in the Environment as Veterinary Drugs

Premasis Sukul and Michael Spiteller

Contents

I. Introduction

It is common practice to use liquid manure from livestock and sewage sludge from wastewater treatment plants in agriculture with the objective of producing sustainable nutrient recycling. Unfortunately, this practice provokes entry of various hazardous components into different compartments of the environment and is, thus, responsible for contamination of food and groundwater resources. In recent years, the extensive use of veterinary drugs in intensive animal husbandry has caused considerable environmental contamination (Boxall et al. 2004a). Since 1950, probably more than 1 million tonnes of antibiotics have been used worldwide as veterinary

Communicated by George W. Ware.

P. Sukul (✉) · M. Spiteller
Institute of Environmental Research, University of Dortmund, Otto Hahn Strasse 6, 44221 Dortmund, Germany

drugs in animal husbandry (Mazel and Davies 1999). However, unlike pesticides and other priority pollutants, the behavior and effects of veterinary medicines in the environment have not been extensively studied until recently. Currently, increased attention has been drawn toward pharmaceuticals, particularly veterinary drugs, and their residual presence in the environment all over the world (Stan and Heberer 1997; Buser et al. 1998; Halling-Sørensen et al. 1998; Ternes 1998; Daughton and Ternes 1999; Stumpf et al. 1999; Zuccato et al. 2000; Jones et al. 2001; Heberer 2002; Kolpin et al. 2002; Boyd et al. 2003; Calamari et al. 2003).

To assess the environmental impact of veterinary medicines released into the environment, it is necessary to (1) identify those factors and processes controlling the degradability of compounds in manure, slurry, sediment, soil, and water; (2) identify those factors and processes controlling the leaching of compounds into the environment; (3) determine the effect of compounds on soil fauna and flora; and (4) monitor the environmental distribution of a range of compounds at the semifield and field scale. Therefore, it is essential to carry out basic research directed toward risk analysis. It is somewhat difficult to make fate assessment feasible for all the existing huge number of compounds present in the environment. It is rather more advantageous to assess structurally related compounds instead of individual ones. Sulfonamides (SAs) represent one such example of a structurally related group of active compounds. SAs may be generated through the application of other active compounds in the environment. Chlorsulfuron (Strek 1998a,b) and sulfosulfuron (Saha and Kulshrestha 2002), which are good examples of sulfonylurea herbicides, are metabolized to SAs. In the 1930s, SAs were discovered to act as antibacterial agents. Since then, more than 5000 different SAs have been developed and remain widely used (Rang et al. 1995). SAs have also been reported for other uses; e.g., phenylsulfonamides are used as corrosion inhibitors or in polymer production (Krepper et al. 1999).

SAs, used in aquaculture (Ternes 1998), in agriculture as herbicides (Hirsch et al. 1999; Battaglin et al. 2000), in animal husbandry as a preventative measure (Hirsch et al. 1999; Holm et al. 1995a,b), and in the treatment of respiratory and urinary tract infections in humans (Jones et al. 2001), are found in the environment at concentrations ranging from 0.13 to 1.9 μg/L (Holm et al. 1995a,b; Halling-Sørensen et al. 1998; Hirsch et al. 1999; Kolpin et al. 2002). Minimum inhibition concentrations of SAs for *Escherichia coli* are typically below 1 mg/kg (Neuman 1981). SAs are used throughout the world to treat domestic and farm animals. Once administered to an animal, the drugs are absorbed and partially metabolized before being excreted in urine and feces. The resulting manure and slurry are released directly to the environment or collected and stored before being applied to the land. Once either is released to the land, the medicines, with their metabolites present in manure and slurry as residues, may interact with different soil components and/or may be washed off into surface waters or leached to groundwaters, thus impacting on environmental and human health. A monitoring program in the United States showed that water samples were contaminated with sulfamethoxazole (Kolpin et al. 2000). In Germany, the presence of sulfamethazine, at a maximum concentration of 11 µg/kg, was highlighted by a soil-monitoring program (Kleefisch and Kues 1997). In this review, we examine those research publications on SAs, placing emphasis on their fate, sorption behavior, photodegradation, and residual behavior in or on soil and water.

II. Estimated Usage

In 1985, the estimated use of antibiotics in livestock in the U.S. was 8300 t (Vicari et al. 1999). In the European Union (EU), of the total active ingredients used in animal husbandry, 90% and 8% represent antibiotics and antiparasitics, respectively, with the rest representing other active ingredients (EMEA 1999). Of the total antibiotics used in animal husbandry, SAs account for a high proportion (Table 1). In Europe, two-thirds of all pharmaceutical antibiotics are used in human medicine, with one-third being used for veterinary purposes (FEDESA 2001).

III. Occurrence

Veterinary drugs enter the environment mainly through the following routes (Hamscher et al. 2004): (a) drugs can settle in aquaculture sediments after direct application of antibiotics via fish feed and fish excrement; (b) antibiotics reach the soil via liquid manure, applied as fertilizers with the aim of sustainable nutrient recycling, and from the use of manure and sludge already contaminated with antibiotics, as fertilizer to crop plants is considered to be the primary route of entry of antibiotics in soil; and (c) another possible route, of soil contamination by antibiotics, may be by excrement of animals grazing outdoors. Further, the entry of antibiotics in dust-bound form into the environment via exhaust air from a manufactur-

Table 1. Annual usage of sulfonamides in animal husbandry in some countries.

Country/ usage/year	Usage (t)	Usage % of total therapeutic antibiotics used	Reference
EU/1999	78.0	2.0	FEDESA 2001
France/1980	139.0	22.0	Espinasse 1993
Sweden/1996	2.2	11.0	Mudd et al. 1998
Denmark/1997	13.0	23.0	Halling-Sørensen et al. 2002
UK/2000	94.0	22.0	NOAH 2002

Source: Thiele-Bruhn (2003).

ing site (Hamscher et al. 2003b) should not be overlooked, as they may have the potential to contaminate the soil after deposition. However, they are primarily responsible for health hazards through inhalation of these dust particles. Another possible route of entry of antibiotics into the environment and, thus, the contamination of surface water, groundwater, and soil, is through disposed wastes of hospitals and manufacturing sites. Once applied to soil with manure and slurry, veterinary medicines, as well as being adsorbed to the soil, are transported to surface water via overload flow and to groundwater by leaching (Fig. 1). In surface waters, these substances may partition to sediments and/or undergo abiotic (photodegradation and/or hydrolysis) and biotic (by aerobic and anaerobic organisms) degradation.

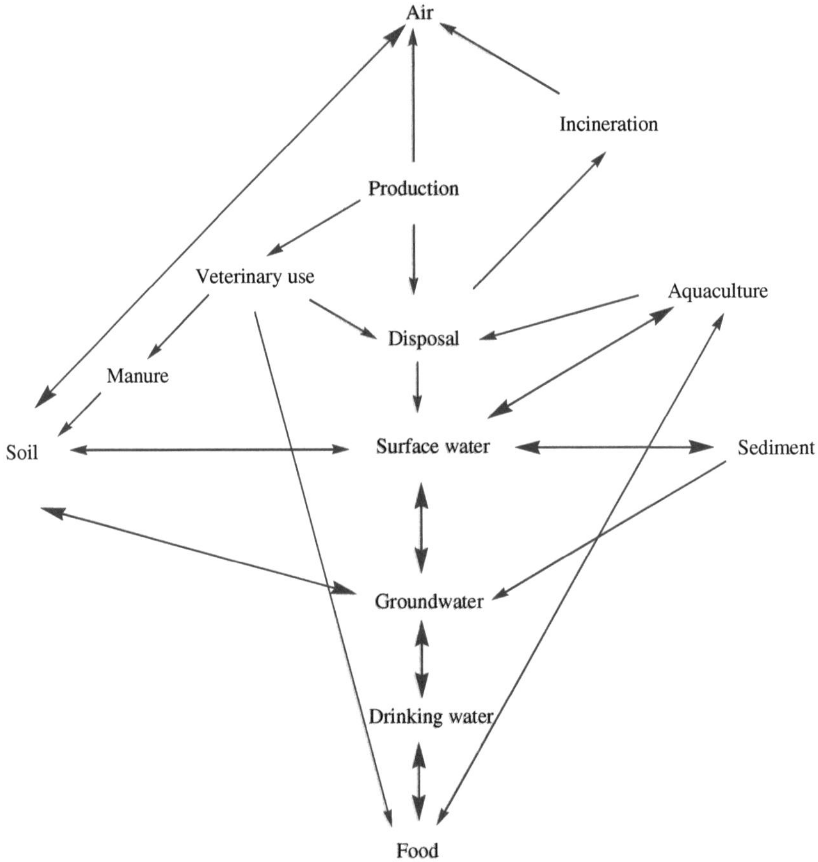

Fig. 1. Occurrence and distribution of veterinary drugs in different environmental compartments.

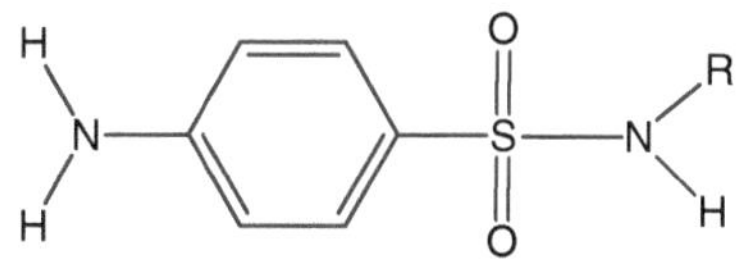

Fig. 2. General Structure of sulfonamides.

IV. Chemistry, Physicochemical Properties, and Mode of Action

SAs belong to a large group of structurally related antibiotics and are N-substituted derivatives of the substance sulfanilamide. SAs contain a 4-aminobenzene sulfonamide core and differ between each other in the N-substituent of the sulfonamide linkage (Fig. 2). They exhibit different physicochemical properties due to the differing side moieties. The properties of a few very important sulfonamide veterinary drugs are presented in Table 2, along with their chemical structure. In Table 3, the physicochemical properties of a few SAs are shown, namely sulfanilamide, sulfadiazine, sulfamidine, sulfadimethoxine, sulfapyridine, sulfamethoxazole. SAs are fairly water soluble, and polar compounds that ionize depending on pH of the medium (Thiele-Bruhn et al. 2004). In general, SAs contain polar functional groups on a nonpolar core and therefore show sensitivity toward bases and acids. SAs generally possess two pK_a values arising from (a) protonation of the amino group at pH 2–3 and (b) deprotonation at pH 5–11

Table 2. Properties of a few important sulfonamide compounds.

Serial no.	Properties		References
Sulfadimidine (CAS No. 57-68-1)			
1.	Chemical structure	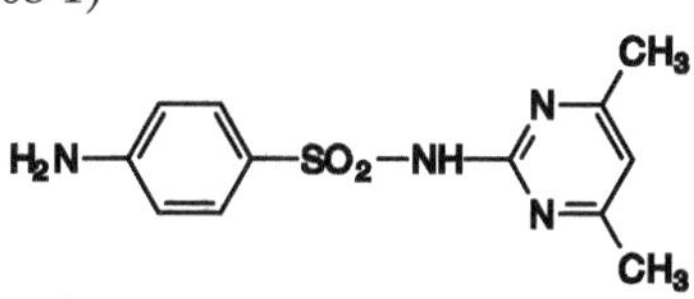	
2.	Nomenclature	2-(4-Aminobenzenesulfonamido)-4,6-dimethylpyrimidine	
3.	Other names	Sulfadine, sulfamethazine, sulfamethiazine, sulfadimethylpyrimidine, sulfodimesin, etc.	
4.	Formula	$C_{12}H_{14}N_4O_2S$	
5.	Molecular weight	278.33	ACD
6.	Water solubility	1500 mg/L	Thiele-Bruhn et al. 2004
7.	Melting point	178°–179°C	Frisk 1943
8.	Boiling point	526.2° ± 52.0°C at 760 Torr	ACD
9.	Flash point	272.1° ± 55.2°C	ACD

Table 2. *Continued*

Serial no.	Properties		References
10.	Bioconc. factor	2.26 at pH 4, 1.78 at pH 7, 1.00 at pH 10	ACD
11.	Vapor pressure	3.64 E – 11 Torr at 25°C	ACD
12.	Density	1.4655 g/cm^3	Deo et al. 1980
13.	Enthalpy of vapor	80.05 ± 3.0 kJ/mol	ACD
14.	Molar solubility	Soluble at pH 1, sparingly soluble at pH 4, slightly soluble at pH 7, very soluble at pH 10	ACD
15.	K_{OC} value	1 at pH 1, 61.2 at pH 4, 48.2 at pH 7, 1 at pH 10	ACD
16.	pK_a	7.45 ± 0.50 (most acidic) 2.79 ± 0.24 (most basic)	ACD

Sulfapyridine (CAS No. 144-83-2)

1.	Chemical structure	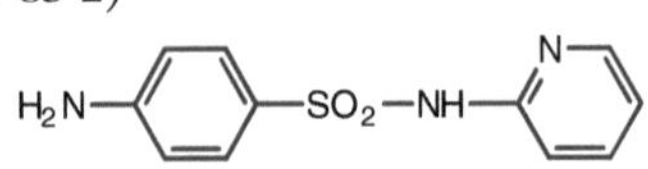	
2.	Nomenclature	4-Amino-*N*-2-pyridinylbenzenesulfonamide	
3.	Other names	Sulfidine, pyriamid, pyridazol, ronin, dagenan, etc.	
4.	Formula	$C_{11}H_{11}N_3O_2S$	
5.	Molecular weight	249.29	ACD
6.	Water solubility	270 mg/L	Thiele-Bruhn et al. 2004
7.	Melting point	188°–191°C 191°–193°C	Saxena and Seydel 1980; Frisk 1943
8.	Boiling point	473.5° ± 50.0°C at 760 Torr	ACD
9.	Flash point	240.2° ± 54.2°C	ACD
10.	Bioconc. factor	1.00 at pH 1, 4, 7, 8, 10	ACD
11.	Vapor pressure	3.90 E – 9 Torr at 25°C	ACD
12.	Enthalpy of vapor	73.67 ± 3.0 kJ/mol	ACD
13.	Molar solubility	Very soluble at pH 1, slightly soluble at pH 4, 7, 8, very soluble at pH 10	ACD
14.	K_{OC} value	1 at pH 1, 22.8 at pH 4, 24.0 at pH 7, 1 at pH 10	ACD
15.	pK_a	8.54 ± 0.30 (most acidic) 2.90 ± 0.19 (most basic)	ACD

Sulfamethoxazole (CAS No. 723-46-6)

1.	Chemical structure	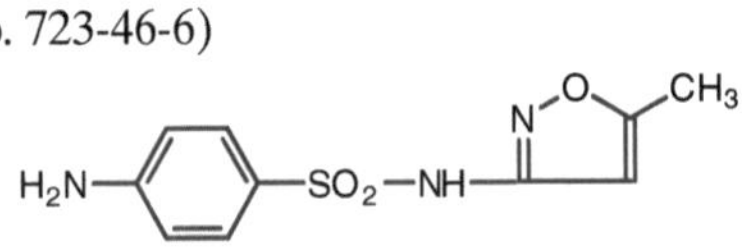	

Table 2. *Continued*

Serial no.	Properties		References
2.	Nomenclature	4-Amino-*N*-(5-methyl-3-isoxazolyl) benzenesulfonamide	
3.	Other names	Radonil, sulfamethalazole, sulfisomezole, sinomin, etc.	
4.	Formula	$C_{10}H_{11}N_3O_3S$	
5.	Molecular weight	253.28	ACD
6.	Melting point	169°C	Sekhon et al. 2001
7.	Boiling point	482.1° ± 55.0°C at 760 Torr	ACD
8.	Flash point	245.4° ± 56.7°C	ACD
9.	Bioconc. factor	1.00 at pH 1, 7, 8, 10; 2.73 at pH 4	ACD
10.	Vapor pressure	1.87 E − 9 Torr at 25°C	ACD
11.	Enthalpy of vapor	74.70 ± 3.0 kJ/mol	ACD
12.	Molar solubility	Slightly soluble at pH 1, 4; soluble at pH 7; very soluble at pH 8, 10	ACD
13.	K_{OC} value	21.1 at pH 1, 71.0 at pH 4, 4.44 at pH 7, 1.00 at pH 8, 10	ACD
14.	pK_a	8.81 ± 0.50 (most acidic) 1.39 ± 0.10 (most basic)	ACD

Sulfanilamide (CAS No. 63-74-1)

1.	Chemical structure	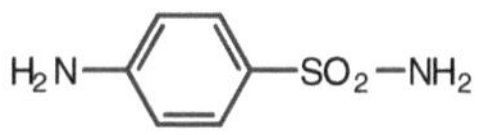	
2.	Nomenclature	4-Aminobenzenesulfonamide	
3.	Other names	Sulfonylamide, sulfamine, sulfonamide-P, sulfamoylaniline, etc.	
4.	Formula	$C_6H_8N_2O_2S$	
5.	Molecular weight	172.21	ACD
6.	Water solubility	7500 mg/L	Thiele-Bruhn et al. 2004
7.	Melting point	165°–167°C	Frisk 1943
8.	Boiling point	400.5° ± 45.0°C at 760 Torr	ACD
9.	Flash point	196.0° ± 51.7°C	ACD
10.	Bioconc. factor	1.00 at pH 1, 4, 7, 8, 10	ACD
11.	Vapor pressure	1.27 E − 6 Torr at 25°C	ACD
12.	Enthalpy of vapor	65.13 ± 3.0 kJ/mol	ACD
13.	Molar solubility	Very soluble at pH 1, 4, 7, 8, 10	ACD
14.	K_{OC} Value	1.23 at pH 1, 9.60 at pH 4, 9.66 at pH 7, 9.59 at pH 8, 5.39 at pH 10	ACD
15.	pK_a	10.10 ± 0.10 (most acidic) 1.85 ± 0.10 (most basic)	ACD

Table 2. *Continued*

Serial no.	Properties		References
Sulfadiazine (CAS No. 68-35-9)			
1.	Chemical structure	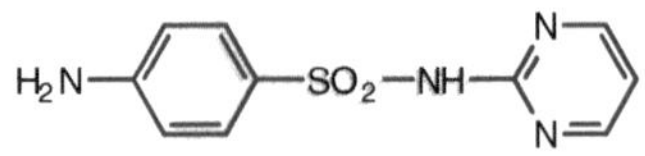	
2.	Nomenclature	4-Amino-*N*-(2-pyrimidinyl) benzenesulfonamide	
3.	Other names	Sanodiazine, sulfadiazin, sulfapyrimidine, sulfazin, etc.	
4.	Formula	$C_{10}H_{10}N_4O_2S$	
5.	Molecular weight	250.28	ACD
6.	Water solubility	77 mg/L	Thiele-Bruhn et al. 2004
7.	Melting point	255°–256°C	Roblin et al. 1942
8.	Boiling point	512.6° ± 52.0°C at 760 Torr	ACD
9.	Flash point	263.8° ± 55.2°C	ACD
10.	Bioconc. factor	1.00 at pH 1, 4, 7, 8, 10	ACD
11.	Vapor pressure	1.28 E – 10 Torr at 25°C	ACD
12.	Enthalpy of vapor	78.38 ± 3.0 kJ/mol	ACD
13.	Molar solubility	Soluble at pH 1, slightly soluble at pH 4, soluble at pH 7, very soluble at pH 8, 10	ACD
14.	K_{OC} value	2.48 at pH 1, 20.0 at pH 4, 4.95 at pH 7, 1.00 at pH 8, 10	ACD
15.	pK_a	6.50 ± 0.30 (most acidic) 1.57 ± 0.10 (most basic)	ACD
Sulfadimethoxine (CAS No. 122-11-2)			
1.	Chemical structure	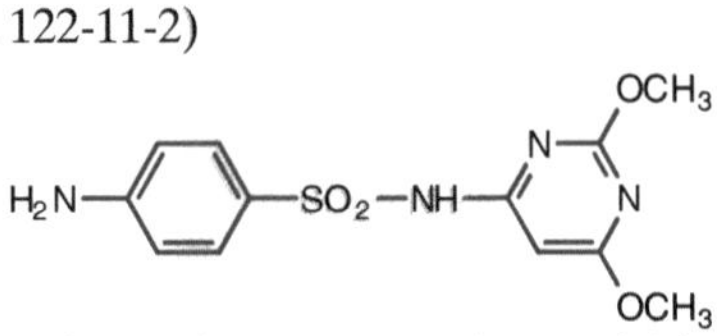	
2.	Nomenclature	4-(*p*-Aminobenzenesulfonamido)-2,6-dimethoxypyrimidine	
3.	Other names	Ultrasulfon, sulfadimetoxin, sulfadimoxine, sulfadimethoxydiazine, etc.	
4.	Formula	$C_{12}H_{14}N_4O_4S$	
5.	Molecular weight	310.33	ACD
6.	Water solubility	340 mg/L	Thiele-Bruhn et al. 2004
7.	Melting point	202°–204°C	
8.	Boiling point	548.5° ± 60.0°C at 760 Torr	ACD
9.	Flash point	285.5° ± 59.2°C	ACD
10.	Bioconc. factor	1.00 at pH 1, 8, 10; 7.21 at pH 4; 1.06 at pH 7	ACD

Table 2. *Continued*

Serial no.	Properties		References
11.	Vapor pressure	4.41 E – 12 Torr at 25°C	ACD
12.	Enthalpy of vapor	82.81 ± 3.0 kJ/mol	ACD
13.	Molar solubility	Slightly soluble at pH 1, 8, sparingly soluble at pH 4, 7, very soluble at pH 10	ACD
14.	K_{OC} value	3.23 at pH 1, 141 at pH 4, 20.8 at pH 7, 2.48 at pH 8, 1.00 at pH 10	ACD
15.	pK_a	6.21 ± 0.50 (most acidic) 2.44 ± 0.48 (most basic)	ACD

Sulfamethoxypyridazine (CAS No. 80-35-3)

1.	Chemical structure	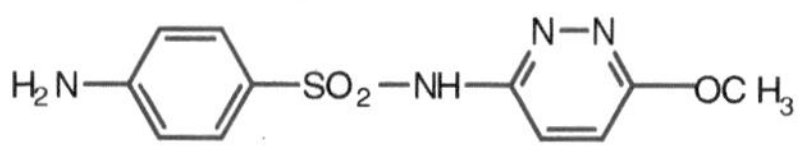	
2.	Nomenclature	3-Sulfanilamido-6-methoxypyridazine	
3.	Other names	Sulfalex, sulfapyridazine, sulfdurazin, sulfozona, etc.	
4.	Formula	$C_{11}H_{12}N_4O_3S$	
5.	Molecular weight	280.30	ACD
6.	Melting point	182.5°–183.5°C	Clark et al. 1958
7.	Boiling point	564.9° ± 60.0°C at 760 Torr	ACD
8.	Flash point	295.4° ± 59.2°C	ACD
9.	Bioconc. factor	1.00 at pH 1, 7, 8, 10; 1.52 at pH 4	ACD
10.	Vapor pressure	8.81 E – 13 Torr at 25°C	ACD
11.	Enthalpy of vapor	84.86 ± 3.0 kJ/mol	ACD
12.	Molar solubility	Soluble at pH 1; slightly soluble at pH 4, 7, 8; very soluble at pH 10	ACD
13.	K_{OC} value	1.02 at pH 1, 46.4 at pH 4, 29.2 at pH 7, 6.46 at pH 8, 1.00 at pH 10	ACD
14.	pK_a	7.19 ± 0.30 (most acidic) 2.18 ± 0.50 (most basic)	ACD
15.	Density	1.48318 g/cm^3	Haridas and Singh 1986

Sulfabenzamide (CAS No. 127-71-9)

1.	Chemical structure	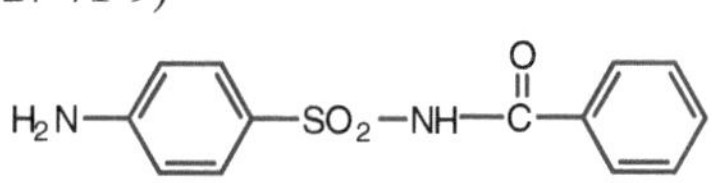	
2.	Nomenclature	*N*-(*p*-Aminobenzenesulfonyl) benzamide	
3.	Other names	Sulfabenzide, sulfabenzamid, sulfabenzid, sulfabenzoylamide, etc.	
4.	Formula	$C_{13}H_{12}N_2O_3S$	
5.	Molecular weight	276.31	ACD

Table 2. *Continued*

Serial no.	Properties		References
9.	Bioconc. factor	2.04 at pH 1, 2.80 at pH 4, 1.00 at pH 7, 8, 10	ACD
12.	Molar solubility	Sparingly soluble at pH 1, 4; very soluble at pH 7, 8, 10	ACD
13.	K_{OC} value	45.7 at pH 1, 62.7 at pH 4, 1.00 at pH 7, 8, 10	ACD
14.	pK_a	4.18 ± 0.70 (most acidic) 1.11 ± 0.10 (most basic)	ACD
Sulfamerazine (CAS No. 127-79-7)			
1.	Chemical structure	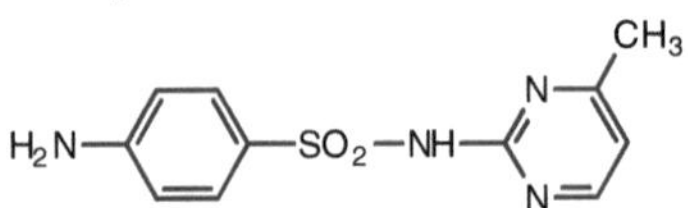	
2.	Nomenclature	2-(4-Aminobenzenesulfonamido)-4-methylpyrimidine	
3.	Other names	Sulfameradine, sumedine, sulfamerazin, sulfamethyldiazine, etc.	
4.	Formula	$C_{11}H_{12}N_4O_2S$	
5.	Molecular weight	264.30	ACD
6.	Melting point	234°–235°C	
7.	Boiling point	519.1° ± 52.0°C at 760 Torr	ACD
8.	Flash point	267.8° ± 55.2°C	ACD
9.	Bioconc. factor	1.00 at pH 1, 7, 8, 10; 1.05 at pH 4	ACD
10.	Vapor pressure	7.03 E − 11 Torr at 25°C	ACD
11.	Density	1.3425 g/cm^3	Deo et al. 1980
11.	Enthalpy of vapor	79.18 ± 3.0 kJ/mol	ACD
12.	Molar solubility	Soluble at pH 1, 8, slightly soluble at pH 4, 7, very soluble at pH 10	ACD
13.	K_{OC} value	1.73 at pH 1, 35.9 at pH 4, 17.8 at pH 7, 3.20 at pH 8, 1.00 at pH 10	ACD
14.	pK_a	6.98 ± 0.30 (most acidic) 1.58 ± 0.10 (most basic)	ACD
Sulfacetamide (CAS No. 144-80-9)			
1.	Chemical structure	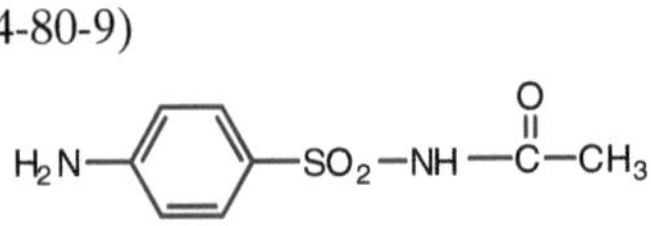	
2.	Nomenclature	*N*-(*p*-Aminophenylsulfonyl) acetamide	
3.	Other names	Sulfacet, sulfacyl, sulfacetimide, urosulfon, etc.	
4.	Formula	$C_8H_{10}N_2O_3S$	
5.	Molecular weight	214.24	ACD
6.	Melting point	181°–183°C	Kravchenya 1989

Table 2. *Continued*

Serial no.	Properties		References
7.	Bioconc. factor	1.00 at pH 1, 4, 7, 8, 10	ACD
8.	Density	1.416 g/cm^3	Alberola et al. 1975
9.	Molar solubility	Soluble at pH 4; very soluble at pH 1, 7, 8, 10	ACD
10.	K_{OC} value	4.16 at pH 1; 7.55 at pH 4; 1.00 at pH 7, 8, 10	ACD
11.	pK_a	5.78 ± 0.70 (most acidic) 0.93 ± 0.10 (most basic)	ACD
Carbutamide (CAS No. 339-43-5)			
1.	Chemical structure	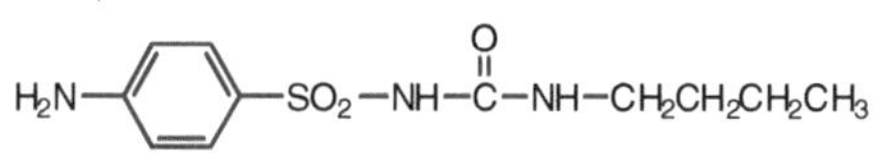	
2.	Nomenclature	1-Butyl-3-sulfanilylurea	
3.	Other names used	Bucarban, bucrol, carbutamid, diaboral, etc.	
4.	Formula	$C_{11}H_{17}N_3O_3S$	
5.	Molecular weight	271.34	ACD
6.	Bioconc. factor	1.00 at pH 1, 7, 8, 10; 2.80 at pH 4	ACD
7.	Molar solubility	Slightly soluble at pH 1, 7, sparingly soluble at pH 4, soluble at pH 8, very soluble at pH 10	ACD
8.	K_{OC} value	25.7 at pH 1, 72.4 at pH 4, 6.21 at pH 7, 1.00 at pH 8, 10	ACD
9.	pK_a	1.27 ± 0.10 (most basic)	ACD
Sulfameter (CAS No. 651-06-9)			
1.	Chemical structure	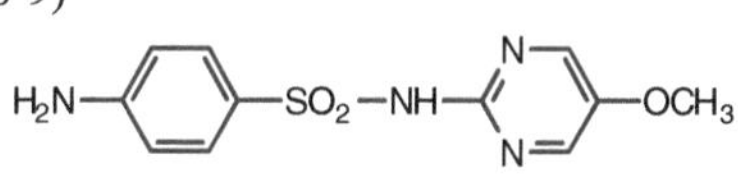	
2.	Nomenclature	2-(4-Aminobenzenesulfonamido)-5-methoxypyrimidine	
3.	Other names	Sulfamethoxypyrimidine, sulfamethoxyine, sulfamethoxydiazine, sulfamethoxydin, sulfametin, etc.	
4.	Formula	$C_{11}H_{12}N_4O_3S$	
5.	Molecular weight	280.30	ACD
6.	Melting point	210°C	Yang et al. 2003
7.	Boiling point	539.4° ± 56.0°C at 760 Torr	ACD
8.	Flash point	280.0° ± 57.2°C	ACD
9.	Bioconc. factor	1.00 at pH 1, 7, 8, 10; 1.22 at pH 4	ACD
10.	Vapor pressure	1.05 E − 11 Torr at 25°C	ACD
11.	Enthalpy of vapor	81.68 ± 3.0 kJ/mol	ACD
12.	Molar solubility	Soluble at pH 1, 8, slightly soluble at pH 4, 7, very soluble at pH 10	ACD

Table 2. *Continued*

Serial no.	Properties		References
13.	K_{OC} value	4.11 at pH 1, 40.0 at pH 4, 13.2 at pH 7, 1.90 at pH 8, 1.0 at pH 10	ACD
14.	pK_a	6.69 ± 0.30 (most acidic) 1.48 ± 0.10 (most basic)	ACD
Sulfadoxine (CAS No. 2447-57-6)			
1.	Chemical structure	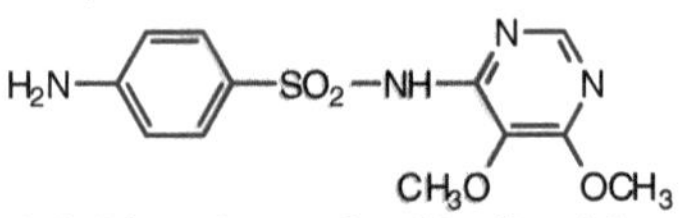	
2.	Nomenclature	4,5-Dimethoxy-6-sulfanilamidopyrimidine	
3.	Other names	Sulfadoxin, sulformethoxine, sulformetoxin, sulforthomidine, sanasil, etc.	
4.	Formula	$C_{12}H_{14}N_4O_4S$	
5.	Molecular weight	310.33	ACD
6.	Boiling point	522.8° ± 60.0°C at 760 Torr	ACD
7.	Flash point	270.0° ± 59.2°C	ACD
8.	Bioconc. factor	1.00 at pH 1, 4, 7, 8, 10	ACD
9.	Vapor pressure	5.01 E – 11 Torr at 25°C	ACD
10.	Enthalpy of vapor	79.63 ± 3.0 kJ/mol	ACD
11.	Molar solubility	Slightly soluble at pH 4, 7, soluble at pH 8, very soluble at pH 1, 10	ACD
12.	K_{OC} value	31.30 at pH 4; 4.60 at pH 7; 1.0 at pH 1, 8, 10	ACD
13.	pK_a	6.16 ± 0.50 (most acidic) 3.15 ± 0.49 (most basic)	ACD

Table 3. Average and representative physicochemical properties of sulfonamide antibiotics.

Properties	Sulfonamides (representative examples, as stated in the text)
Molar mass (g/mol)	177.2–300.3
Water solubility (mg/L)	7.5–1500
Log K_{ow}	–0.1–1.7
pK_a	2–3/4.5–10.6
Henry's constant (PaL/mol)	1.3×10^{-12}–1.8×10^{-8}

Reproduced from Thiele-Bruhn (2003).

(Ingerslev and Halling-Sørensen 2000). The amphoteric SAs have been observed to behave as weak acids and to form salts in strongly acidic or basic solutions. The antibacterial activity of SAs is greatly reduced when the amino nitrogen is substituted (Thiele-Bruhn 2003). SAs compete with *p*–aminobenzolic acid in the enzymatic synthesis of dihydrofolic acid and thus inhibit the growth and reproduction of bacteria (Yang et al. 2004). Values of log K_{ow} for SAs range between –0.1 and 1.7 (see Table 3), suggesting that they are not hydrophobic.

V. Biotransformation in Mammals

In mammals, the metabolism of antibiotics is biphasic. With the help of monooxygenases, reductases, and hydrolases, most antibiotics undergo transformation in the first step, and in the second step the molecules become more hydrophylic after a covalent conjugation, lose their biological activity, and become easily excretable through urine and feces (Halling-Sørensen et al. 1998; Daughton and Ternes 1999; Thiele-Bruhn 2003).

The metabolic pathway for sulfadiazine adminstered to rats, alone or with trimethoprim, has been elucidated (Woolley and Sigel 1979). The metabolic products obtained from sulfadiazine, along with the possible pathways, are shown in Fig. 3. Following a single oral dose (30 mg/kg), 87% and 15% radioactivity were obtained in urine and feces, respectively. Sulfadiazine and N^4-acetylsulfadiazine (Fig. 3, no. 3) accounted for 56% and 19% of the

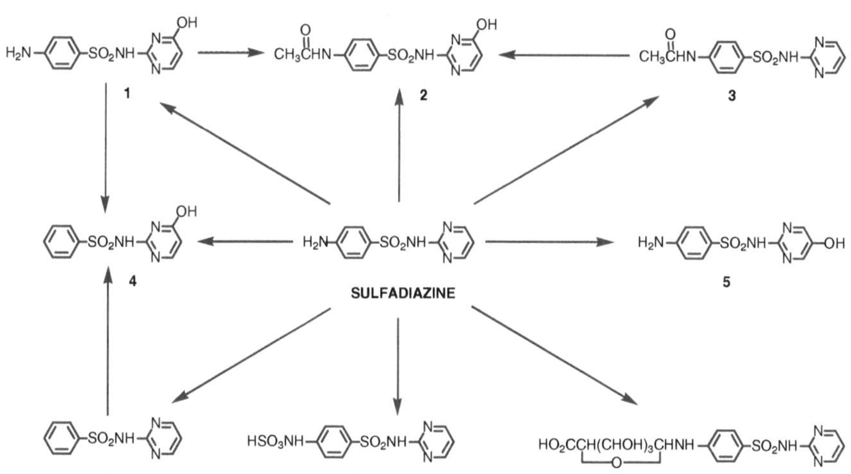

Fig. 3. Biotransformation of sulfadiazine. **1**, N^1-2-(4-Hydroxypyrimidinyl)sulfanilamide; **2**, N^4-acetyl-N^1-2-(4-hydroxypyrimidinyl)sulfanilamide; **3**, N^4-acetyl-N^1-2-pyrimidinylsulfanilamide; **4**, N^1-2-(4-hydroxypyrimidinyl) benzenesulfonamide; **5**, N^1-2-(5-hydroxypyrimidinyl)sulfanilamide; **6**, N^1-2-pyrimidinylbenzenesulfonamide; **7**, N^4-sulfonate-N^1-2-pyrimidinylsulfanilamide; **8**, N^4-glucuronide-N^1-2-primidinylsulfanilamide. (Reproduced with permission from Springer, from Sigel 1983.)

radioactivity in urine, respectively, in the first 24 hr after application. N^4-Glucuronide (Fig. 3, no. 8) was the other major metabolite in urine, accounting for ≥5% of the radioactivity. The excretion pattern of radioactivity was not influenced significantly by trimethoprim. In another experiment, using neonatal calf tissue, plasma, and urine following oral administration of ^{14}C-sulfadiazine, the presence of two deaminated metabolites (Fig. 3, nos. 4, 6) was established as biotransformation products (Woolley et al. 1980). Deaminosulfadiazine represented most of the solvent-extractable radioactivity when total ^{14}C-sulfadiazine residues approached tolerance levels. The metabolic rate of SAs in animals is considered to be high (>80% metabolism; Boxall et al. 2002b). Of the amount of SAs applied, ~40%–90% is excreted after consumption (Langhammer 1989). However, the rate of excretion depends on the compound, the dose, the species treated, and the mode of application (Kümmerer et al. 2000; Haller et al. 2002). Sulfadiazine is excreted to about 50% as the parent compound and 50% as acetyl or glucuronyl conjugates in the urine and feces (Löscher et al. 2002). The N^4–acetyl derivatives of SAs, deacetylated to the parent compound in vivo and in vitro, are considered to be among the more important metabolites in food-producing animals (Krishida et al. 2005).

VI. Abiotic Degradation

A. Photodegradation

When the active ingredients are spread on soil surface they are exposed to many biotic and abiotic factors. Sunlight, one of the important abiotic factors, leads to the degradation of many antibiotics. Photochemical degradation, considered to be a factor in determining the environmental fate of pharmaceuticals and personal care products (Mill and Maybey 1985; Moore 1987; Zepp and Cline 1977; Boreen et al. 2003, 2004), may occur through direct photolysis or through indirect pathways with the help of sensitized photoprocesses, such as reactions with transient excited species, e.g., singlet oxygen ($^1O_2$, O_2), hydroxide radical (·OH), and other reactive species formed in sunlit natural waters (Blough and Zepp 1995; Zhou and More 1997; Mill 1999; Huber et al. 2003; Latch et al. 2003; Packer et al. 2003). To be photoreactive, a compound must absorb light energy directly or indirectly. Compounds with no ultraviolet (UV) absorption above 290 nm do not break down by direct photochemical mechanisms because the ozone layer in the upper atmosphere absorbs all radiation emitted by the sun below 290 nm (Watkins 1979). Only light with a wavelength greater than 290 nm reaches the surface of the earth. Thus, the UV absorption spectrum of a compound gives basic information about the possibility of direct photodegradation, as only substances with an UV absorption in the range of sunlight can undergo direct interaction. Humic substances can absorb strongly in the UV region of sunlight and can, therefore, act as sensitizers and initiators of degradation processes of those chemicals that do not undergo direct photolysis.

Photodecomposition of any compound declines with water depth (Lunestad et al. 1995) and, in soil, only a very small portion of the soil surface is exposed to light. Photodegradation may have little effect on antibiotics in soils because they may enter into soil pores, especially in presence of slurry, and may even be fixed to soil particles and receive protection from sunlight (Thiele-Bruhn 2003). The process of fixation of compounds to the surface or in pores of the soil matrix has also been observed to protect compounds from biodegradation (Samuelsen et al. 1992; Gavalchin and Katz 1994). Although photodegradation is restricted to only a few millimeters of soil depth, it might help to minimize the concentration of antibiotic compounds on the soil surface before their entry into the greater depth of soil and, thus, also minimize the ultimate groundwater contamination.

Because of the microbial degradation of SAs, photolysis on soil is unlikely to be a major route of degradation for SAs under normal circumstances. However, abiotic degradation by photolysis on soil surfaces is also of interest to determine their impact on the environment where soil microbial activity is low. SAs have aquatic uses (Boxall et al. 2004a), and they could potentially enter into surface water by runoff from soil treated with manure and slurry containing SAs. Therefore, determination of the rate and route of photolytic degradation in water is crucial in defining the environmental impact of sulfonamide application. It is also important to keep in mind that the rate and route of photolysis of SAs can be significantly different in natural waters containing humic substances and other photosensitizers, as evidenced with florasulum, a sulfonamide herbicide, and other pesticides (Frimmel and Hessler 1994; Tsao and Eto 1994; Hapeman et al. 1998; Mansour et al. 1997; Krieger et al. 2000). The possibility of forming photoproducts that are more tightly bound to soil surface or to particulate materials present in water than the metabolic degradates cannot be ruled out (Krieger et al. 2000). However, most SAs are not readily photodegradable (Fjelde et al. 1993; Lunestad et al. 1995).

The rate of photolysis of SA drugs, such as especially with reference to those with varying five-membered heterocyclic substituents such as sulfamethoxazole, sulfisoxazole, sulfamethizole, and sulfathiazole, depends on the chemical nature of the heterocyclic group and pH of the solution (Boreen et al. 2004). Their study revealed that the photodegradation of all four compounds in natural water samples was attributed solely to direct photolysis, with sulfanilic acid as a common photoproduct, but with a wide range of reaction rates. However, direct photolysis may be hindered by the presence of higher concentrations of dissolved organic carbon (DOC) or nitrate enrichment in water due to the formation of large concentrations of reactive species. In general, the direct photolysis rate constants were found to be highly pH dependent, but the trend was not consistent for all SAs.

There was a correlation between a higher direct photolysis rate constant and an increased absorption of sunlight in sulfamethoxazole. The direct photolysis of sulfamethoxazole as a function of pH was also observed in another study (Moore and Zhou 1994). The influence of pH on rate of direct photolysis may be attributed to the alteration of the protonation state of the compounds, as well as to their absorption spectrum. The position of maximum absorbance and the shape of the spectrum changed with changing pH. However, these spectral changes occurred largely below 300 nm, which is, of course, not relevant to absorbance of sunlight irradiation. A few SAs in the experiments of Boreen et al. (2004), showed distinct changes in the magnitude of absorption above 300 nm that affected their rate of photoabsorption. The SAs containing an identical backbone structure but differing in the heterocyclic R substituent do not exhibit very similar direct photolysis behavior. For example, sulfamethoxazole and sulfisoxazole photodegraded readily under acidic conditions, while sulfamethizole and sulfathiazole photodegraded in basic medium and, surprisingly, sulfamoxole was found to degrade in aqueous solutions in the absence of light at all pH values ($t_{1/2} = 4$ hr at pH 5; Boreen et al. 2004.). This finding suggests that the differences in the photochemical behavior must be governed by the heterocyclic substituent.

The photoproducts expected from irradiation of SAs in aqueous solution are shown in Fig. 4. The main photoproduct of SAs in aqueous solution is sulfanilic acid, which is generated through δ-cleavage (Chignell et al. 1980, 1981; Weiss et al. 1980; Motten and Chignell 1983; Boreen et al. 2004; Zhou and Moore 1994). Aniline may be generated as a photoproduct through γ-cleavage and has been identified from the photolysis of sulfamethoxazole and sulfathiazole (Weiss et al. 1980; Zhou and Moore 1994). However, it was surprisingly not detected in the experiment of Boreen et al. (2004).

Continuing uncertainties about the pathways and consequences of the photostability of aryl SAs were partly resolved by the results of comprehensive product analysis in the photolysis of aqueous *N*–tosylglysine (Hill et al. 1999). The results confirmed the functioning of intramolecular electron or hydrogen transfer to promote the widely reported S–N cleavage. It was observed that the major route involved the known behavior of excited aryl SAs as an electron acceptor (Jones et al. 1999) and led to the generation of an intermediate from which the observed products were derived by loss of CO_2 alone or with concomitant S–N cleavage and, more speculatively, via cyclizations and hydrolysis. It would appear from this study that electron or hydrogen transfer to the sulfonyl group is prominent in the photochemistry of aryl SAs.

While studying the photodegradation in aqueous solution, it is better to take the sorption parameters of the compounds to particles within water into consideration, as there is every possibility that photodecomposition may be hindered due to sorption, although it is also true that SAs have little

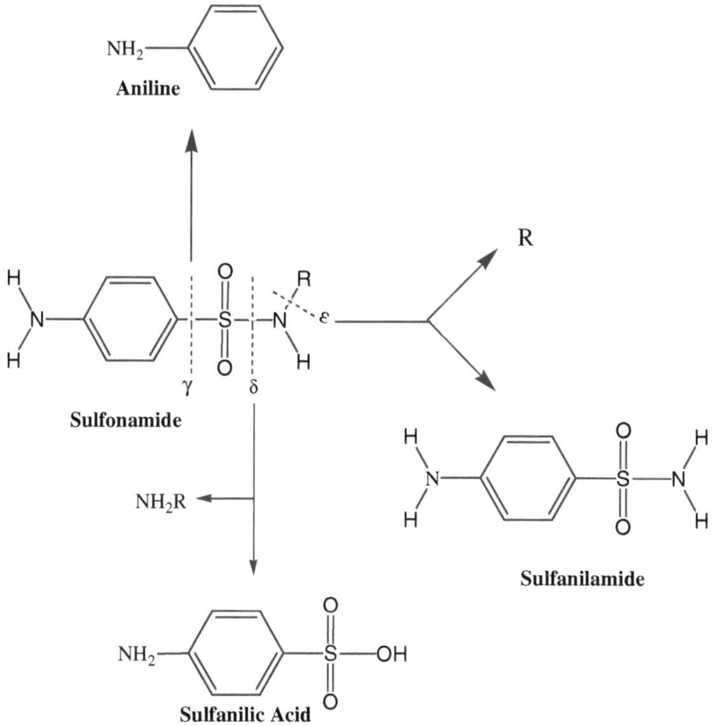

Fig. 4. Possible photodegradation mode of sulfonamides. *Source*: Weiss et al. (1980); Motten and Chignell (1983); Zhou and Moore (1994); Boreen et al. (2004).

affinity for particles because they have relatively low sorption coefficients (sulfathiazole, log $k_{ow} = 0.05$; $k_{d\,solid} = 4.9$; Tolls 2001). It has been suggested that the sulfonamide, which is more photolabile than others of similar structure, should be preferentially used, obviously so long as the photoproducts are benign and the compound maintains similar efficacy to the alternatives.

B. Hydrolysis

Hydrolysis is another significant process of abiotic elimination of antibiotics (Halling-Sørensen 2000). On acid hydrolysis, the sulfonamide bond breaks to produce sulfanilic acid and the appropriate amino derivatives as the common degradation products. At the same time, it is also possible to get sulfanilamide and the appropriate hydroxyl derivative, which can decompose further, depending on the nature of the nitrogen heterocycle (Kilmes and Mokry 1996). In another experiment, sulfamoxole was found to break down in aqueous solution with $t_{1/2}$ of 4 hr at pH 5 (Boreen et al. 2004). The acid hydrolysis of sulfisoxazole is depicted in Fig. 5.

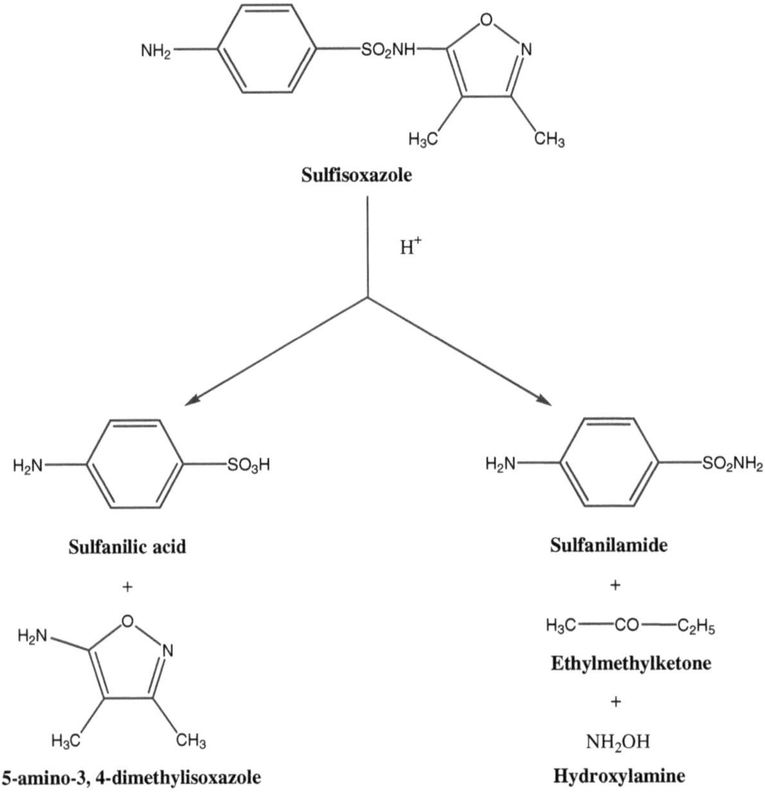

Fig. 5. Products generated after acid hydrolysis of sulfisoxazole. *Source*: Klimes and Mokry (1996).

VII. Residues

A. Manure

A low degradation rate for SAs in slurry has been observed, and it is interesting that the excreted acetyl conjugates of SAs are again converted to the parent compounds (Berger et al. 1986; Langhammer 1989). Sulfamethazine (40 mg/kg; Langhammer et al. 1988), ^{4}N-acetylsulfamethiazine (2.6 mg/kg; Haller et al. 2002), sulfadiazine (1.1 mg/kg; Höper et al. 2002), sulfathiazole (12.4 mg/kg; Haller et al. 2002) and sulfamerazine, sulfamethoxypyridazine, sulfamethoxazole, and sulfadimethoxine (each at <0.02 mg/kg; Höper et al. 2002) were detected in liquid manure samples from pigs and calves. Analysis of six grab-samples, taken in Switzerland from manure pits on farms where medicinal feed had been applied, revealed total sulfonamide (sulfamethiazine, *N*-acetyl sulfamethiazine, and sulfathiazole) concentrations of up to 20 mg/kg liquid manure. However, sulfaguanidine, sulfadiazine, sulfamethoxazole, and sulfamethoxine were not detected in any of the samples

(Haller et al. 2002). The authors proposed that with manure slurry being applied in the field as fertilizer with a maximum dose rate of $50\,m^3/ha$ (EMEA 1997), sulfonamide residues in soil could reach 1 kg/ha, which is the same order of magnitude as the application rate of modern pesticides. This result clearly shows the importance of investigating the fate of SAs in the environment, as this could play an important role in the rise of resistant gene pools. Of the six samples, five contained the metabolite N^4-acetylsulfamethazine, which is itself not antimicrobial. However, this metabolite may be converted into its parent compound in manure (Berger et al. 1986). Thus, it is not only the parent compound that should be the subject of risk assessment but also the metabolites.

SAs cannot be characterized as readily biodegradable. Results from aerated and activated sludge reactors showed that SAs were degraded only after an adaptation period, and the biodegradation rate was found to be identical for several SAs. Primary degradation of different ones occurred after a lag period of 6–12 d in activated sludge at 20° ± 2°C and 34–47 d at 6° ± 1°C (Ingerslev and Halling-Sørensen 2000).

It has been observed that degradation of the same compound may differ among the types of manure or slurry. For sulfachloropyridazine, DT_{50} were found to be <8 d, <3 mon, and >8 d in broiler feces, laying hen feces, and pig slurry, respectively (van Dijk and Keukens 2000; Boxall et al. 2003, 2004b). After collection from livestock farms, manures and slurry are usually stored for a considerable period of time before application to soils. In the U.K., the storage time of slurry varies from 0 to 50 mon, with an average of 9 mon (WRc–NSF 2000). Thus, storage time should also play a significant role in the dissipation of veterinary drugs.

B. Soil

As SAs have been found to be persistent in liquid manures (see Section VII. A), soils treated with liquid manures are likely to carry these compounds. Several SAs were detected in farmed soils under conventional fertilization with liquid manure (maximum concentrations detected: sulfamethazine, 11 µg/kg; sulfadiazine, sulfathiazole, sulfamerazine, sulfamethoxypyridazine, sulfamethoxazole, and sulfadimethoxine, each at 1 µg/kg Höper et al. 2002). In a separate study, 15 µg/kg sulfadimidine was found in soil (Christian et al. 2003). It has been established that SAs, especially sulfachloropyridazine, dissipate quickly, attaining their limit of detection within only 3 mon after slurry application to soils (Boxall et al. 2003). SAs as a class are less sorptive, impersistent, and leachable (Boxall et al. 2004a). Cleavage of the amine–sulfur bond, followed by formation of several intermediates during mineralization of SAs in pure cultures of three bacterial species, has been reported (Walker 1978; Balba et al. 1979); this may also occur in soils.

It is essential to put enormous efforts into deducing the metabolic pathways of SAs, not only in native soils, but also in soils with simultaneous application of slurry, which might influence the degradation of SAs in light of the kinetics and the nature of the metabolites. Further, the large input of additional carbon sources may cause a priming effect (Kuzyakov et al. 2000) on the existing microbial activity. Because of the additional input of carbon sources in the form of manure, one can expect differences in the behavior of xenobiotics, viz. (a) fast mineralization of active ingredients due to stimulation of microbial growth; (b) bound residue formation through sorption; and (c), because of sorption, microbe–xenobiotic interaction may be encountered to a lesser extent, leading to reduced biodegradability as sorption reduces bioavailability. A prestudy on aldicarb, used as an agricultural insecticide, revealed a lower biodegradation rate of aldicarb in slurry-amended soil than that in soil without slurry application (Rouchaud et al. 1993). Acclimation of degrading organisms for SAs in soils has also been reported (Ingerslev and Halling-Sørensen 2000). After respiking the soils with the same or similar active ingredient, the SAs degraded more quickly.

The fate of sulfachloropyridazine was investigated in a treated pig manure slurry and a macroporous tile drained clay soil. Maximum concentrations of 365 and 613 μg/kg were detected in the soil and drainflow, respectively (Kay et al. 2004, 2005). The authors proposed to opt for a tillage operation before slurry application to reduce the concentration in the drainflow, as preferential flow via the desiccation cracks and worm channels to the tile drains might be the most important route for translocation of the SAs. Monitoring data showed that it dissipated from soil over a period of several months. For sulfachloropyridazin, DT_{50} and DT_{90} were found to be 29 and 89 d, respectively, in soil.

C. Groundwater, Surface Water, and Sediments

From agricultural land, water reaches groundwater by percolation and surface water by runoff. While doing so, it may carry the contaminants and, thus, contaminate the surface water and groundwater. Therefore, soil can act as a source of SAs for the aqueous environment via surface runoff and leaching (Lindsey et al. 2001; Sacher et al. 2001; Thiele-Bruhn 2003; Müller et al. 2003).

The presence of sulfamethazine at a maximum concentration of 0.24 μg/L in groundwater at 1.4 m below a soil surface fertilized with animal slurry established the leaching behavior of the drug from soil surface to groundwater (Hamscher et al. 2003a, 2005; Höper et al. 2003). It might be the case that, because of their hydrophilicity, SAs could be transferred into the aquatic environment. Sulfamethazine concentrations in a lake with intensive animal husbandry surroundings were found to be higher than that in the effluents of a wastewater treatment plant in the same area (Alder et al. 2001). This observation also suggests that, in this case, sulfamethazine came not from human medicines but originated from animal manure. Usually, the origin of antibiotic contamination in surface water and groundwater is considered to be point and nonpoint-source discharges of municipal wastewater and agricultural wastewater from animal feeding operations (Halling-Sørensen et al. 1998). Yang et al. (2004) also established the great influence of agriculture on the occurrence of SAs in river waters. Sulfadimethoxane, sulfamethazine, and sulfamethoxazole were detected in the Poudre River (northern Colorado) at those sites that experienced more influence from agricultural than from human use. SAs were not found at any sites on the river that had the greatest influence of wastewater treatment plants. However, sulfamethoxazole was first detected at a site influenced by a wastewater treatment plant, and its concentration exhibited a general increase from 0.12 to 0.17 μg/L, keeping a parity with the increase of the agricultural influence. It is believed that, in surface water, biodegradation of antibiotics is slower than in sewage systems as a result of a lower density and diversity of microbes (Alexy et al. 2004). Reported half-lives of primary degradation for sulfadiazine and sulfadimethoxine are >21 d in water (Boxall et al. 2004a) and >50 d in sediments (Boxall et al. 2002b). Minor degradation of sulfadimethoxine was noticed over a period of 180 d in aquatic sediments, with a characteristic demethylation process under laboratory condition (Samuelsen et al. 1994). It is observed that rate of degradation of SAs is affected by the depth of the sediments. Sulfadiazine showed half-lives of 50 and 100 d in marine sediments of 0–1 and 5–7 cm depths, respectively (Hektoen et al. 1995).

VIII. Sorption

Literature related to detailed studies on the sorption behaviors of SAs such as sulfachloropyridazine (Boxall et al. 2002a; Tolls et al. 2002), sulfadimidine (Langhammer and Büning-Pfaue 1989; Thiele et al. 2002; Tolls et al. 2002; Thiele-Bruhn et al. 2004), sulfamethazine (Langhammer 1989), sulfapyridine (Thiele 2000; Thiele et al. 2002; Thiele-Bruhn et al. 2004), sulfadiazine (Thiele et al. 2002; Tolls et al. 2002; Thiele-Bruhn et al. 2004), sulfanilamide (Thiele et al. 2002; Thiele-Bruhn et al. 2004), sulfadimethoxine (Thiele et al. 2002; Thiele-Bruhn et al. 2004), sulfaisoxazole, and sulfathiazole (Tolls et al. 2002) in soils of different texture and organic carbon content are available. Sorption influences the mobile and bioavailable portion of antibiotics by governing their distribution and transfer between the phases. Thus, sorption also influences the concentration in groundwater and surface water.

Sorption of most of the antibiotics is not solely characterized by their hydrophobic properties, but also by other mechanisms such as ion exchange, cation binding at clay surfaces, surface complexation, and hydrogen bonding, which also play a significant role (Tolls 2001; Thiele-Bruhn 2003; Hamscher et al. 2004). Thus, sorption of SAs is influenced by the quantity, composition, and structure of soil colloids (Thiele 2000).

Sorption behavior is normally represented as the sorption coefficient (K_d), which is defined as the ratio of the concentration of a compound in the sorbent phase and in the aqueous phase at equilibrium. The concentration in the aqueous phase can only attain the concentration of the freely dissolved compound. The total concentration in the soil solution cannot be considered, as it includes fractions that are associated with DOM and adsorbed to suspended particles.

The association of solutes with DOM is usually described by considering DOM as a third phase in the system of soil solids, water, and DOM. For neutral hydrophobic organic chemicals, K_d is influenced by organic carbon content. Use of a organic carbon-normalized sorption coefficient (K_{oc}, $K_{oc} = K_d/\%$ organic carbon) is a normal practice, assuming that the chemical will almost entirely partition into organic fraction of the soil solid. However, this is applicable only for neutral hydrophobic compounds. Although the neutral form of weak acids, such as SAs, can adsorb via hydrophobic interactions, the use of a K_{oc} value is inappropriate because of the significant involvement of other interaction mechanisms (Tolls 2001; Thiele-Bruhn et al. 2004). However, the adsorption of SAs to soil organic matter has been established (Thiele et al. 2002).

Distribution coefficients for the adsorption of SAs to soil components and aquatic sediments vary from 0.6 to 4.9 (Tolls 2001). The phenomenon of adsorption–desorption is influenced markedly by soil pH, as the ionization of compounds depends on the pH of the medium. The adsorption coefficients of SAs increased from <1 to 30 when the soil pH was decreased in the range of 8 to 4 (Boxall et al. 2002a; Tolls et al. 2002), which might be related to the ionization of the amphoteric SAs. It has also been proposed that the aminophenyl group in SAs plays a significant role in the sorption processes (Thiele 2000). It has been established that the sorption of SAs is influenced by their molecular structures and physicochemical properties, available functional groups at organic–mineral interacting sites, and the significant role of the voids and cavities in the three-dimensional structure of soluble organic matter (SOM) and its combination with the mineral matrix (Thiele-Bruhn et al. 2004). The influence of SOM on adsorption depends not only on the quantity of organic carbon, but also on its composition. In general, the chemical composition of the SOM compounds correlated with SAs sorption indicates that the polar regions of complex SOM structures contribute to site-specific binding of SAs. The release of unaltered SAs is also expected as a follow-up desorption process because SA–SOM complexes are primarily formed by weak van der Waals forces and hydrogen bonding. Besides adsorption, fixation of SAs due to their diffusion into the inter- and/or intralayer clay matrix, as seen for other classes of antibiotics (Pinck et al. 1962; Nowara et al. 1997), cannot be ruled out.

Considering the case of sulfanilamide, sulfadimidine, sulfadiazine, sulfadimethoxine, and sulfapyridine, the adsorption isotherm of the SAs was best fitted by the Freundlich equation, with correlation coefficients (r^2) ranging from 0.87 to 0.99 and showing a strong sorption nonlinearity ($1/n \leq 0.76$; Thiele-Bruhn et al. 2004). Among the SAs, the adsorption increased with the introduction of functional groups to the sulfonyl phenylamine structure in the order amino < diazine < diazine plus methoxy = diazine plus methyl < pyridine, and with the increase of aromaticity and electronegativity of functional groups attached to the sulfonyl phenylamine core. The adsorption coefficients of SAs are small compared to those of tetracyclines and fluoroquinolones. Because of their low K_d values, SAs are considered to be very mobile and highly bioavailable in soil with no bioaccumulation (Lawrence et al. 2000). Accordingly, significant portions of sulfadimidine leached through a soil column (Langhammer 1989), while in field experiments, sulfachloropyridazine was found in the drainflow of a clay loam soil after application of spiked manure (Boxall et al. 2002a). It was proposed that anionic SAs adsorb to positively charged surfaces of pedogenic oxides, which are the most abundant in the clay fraction (Thiele-Bruhn et al. 2004). Sulfamethazine and sulfathiazole showed a strong sorption behavior to soil (Langhammer et al. 1990); 55% and 1% sulfamethazine, repectively, was detectable in soil within 2 hr and after 32 d of its incorporation into the soil.

Addition of manure to soil led to a pronounced effect on the soil sorption of selected SAs (sulfanilamide and sulfadiazine), resulting in an increased sulfonamide solution concentration and mobility at a lower and environmentally relevant concentration range (Thiele-Bruhn and Aust 2004). Sorption of SAs decreased in a mixture of soil and slurry (50:1, w/w); this was due to competitive adsorption of DOM from manure. An increase in total organic matter and ionic strength and a decrease in pH occurred with an increase in pig slurry amendment; 47%–87% of SAs remained nonadsorbed in the differently manured soils. Dissolved fractions of SAs reached a maximum at a soil:slurry ratio of 9:1. Sulfadimidine, sulfadimethoxine, and sulfapyridine did not show any change in their adsorption–desorption behavior in the presence of increased manure input, perhaps because of their lower polarity. However, a higher adsorption of SAs in soil with higher organic matter content has also been reported (Langhammer and Büning-Pfaue 1989), possibly resulting from particulate organic matter, viz., soil organomineral particles (Schulten and Leinweber 2000).

IX. Predicted Environmental Concentration

The widespread release of active molecules suggests a potential risk for the environment and the equilibrium of the ecosystem. The predicted environmental concentration (PEC) in soil is an important decision parameter in the risk assessment procedure for antibiotics. The amount of antibiotics actually released in the environment principally depends on the dosage used, the animal type treated, the methods of application, the degree of metabolism in animals, and also the codes of good agricultural practices of fertilizing land. PEC calculation gives an idea on the presence and absence of any negative impact on the environment from residues of concern. In case PEC exceeds the trigger value set by Commission of European Communities (1992), experimental testing in Phase II becomes necessary. Here, an attempt has been made to estimate and demonstrate the PEC value for sulfadiazine taking beef-bullock and adult pig as reference animals (Table 4). The PEC calculation is based on a worst case scenario assuming 100% excretion of the parent compound. The calculation is made with the help of a balancing model (Spaepen et al. 1997). The result shows that the PEC value differs with target species, although the dose, the degree of metabolism, and the code of good agricultural practices remain the same, which may be explained by examining the data on their body weight and yearly production of phosphorus, maximum application of liquid manure per hectare per year based on maximum allowed phosphate concentration in soil following good agricultural practice. Maximum application of liquid manure from beef bullock becomes 2.49 times more than that of adult pig, which makes the PEC in soil for beef bullock (0.501 mg/kg) 2.9 times more than that of adult pig (0.173 mg/kg). The PEC value should be compared with predicted no-effect concentration (PNEC). Concern for the environment is recognized in cases where PEC/PNEC >1. In such cases, further assessment on the effects of the compound on the fauna and flora within the environmental compartments that are likely to be affected is required. In the present study, however, the PEC/PNEC value could not be enumerated because data on PNEC for sulfadiazine were not available.

X. Conclusions

In the last few years, scientists have become very concerned about the use of SAs in the veterinary field and their expected exposure to different compartments of the environment. However, data generated on the behavior and fate of SAs under field conditions are not sufficient to perform an environmental risk assessment. A realistic approach is needed for the generation of data on the fate of SAs in soils and manures under environmentally relevant conditions. Unlike pesticides used in agriculture, veterinary drugs enter the soil and adjacent environmental compartments not directly but by way of contaminated manure and sludge used as fertilizer. Manure is rich in organic matter, which influences the sorption through its quality and quantity, and may also influence degradation, as it harbors a large amount of microbial biomass. Therefore, it is pertinent to include the effect of manure in test methods for determining the fate of SAs in soil. As SAs are detected in soils and manure, there is a special need to intensify studies on their sorption and leaching behaviors.

Groundwater contamination with veterinary drugs through local emission cannot be ruled out. SAs (5 mg/L) were found in the groundwater downgradient from a landfill originating from the disposed waste of a pharmaceutical production site (Holm et al. 1995a,b). Research publications related to the presence of veterinary drugs in landfill leachates are scarce. Thus, this area should be receive further consideration and investigation.

Table 4. Predicted environmental concentration (PEC) calculation in soil for sulfadiazine when applied to beef bullock and adult pig.

Parameters	Calculated or suggested values		References
	Beef bullock	Adult pig	
Weight of the animal (BW, kg)	500	130	Maton et al. 1983
ID (individual dose rate, mg/kg body wt.)[a]	30	30	
T (no. of individual treatments per animal)[a]	10	10	
Total amount of active ingredient (Q, mg/yr/place, $Q = ID \times BW \times T \times N$)	150,000	39,000	Spaepen et al. 1997
Yearly output of excreta (P_E, kg/place/yr)	9,185	2,829	Vetter and Steffens 1986; Het Bestuur van de Stichting Ontwikkelings- en Saneringsfonds voor de Landbouw 1987
Concentration of active ingredient in liquid manure, mg/kg liquid manure ($\mathbf{C_E}$, $C_E = Q/P_E$)	16.33	13.79	Spaepen et al. 1997
Maximum allowed phosphate concentration (A_P, kg/ha/yr)	210	210	Ministers für Umwelt, Raumordnung und Landwirtshaft des Landes Nordrhein-Westfalen 1985
Annual phosphate production per animal (P_P, kg P_2O_5/place/yr)	16.24	12.45	Vetter and Steffens 1986
Maximum application of liquid manure/ha/yr (M, $M = A_P \times P_E/P_P$)	118,771.55	47,718.07	Spaepen et al. 1997
Applied active ingredient, mg/ha/yr ($\mathbf{C_{SA}}$, $C_{SA} = M \times C_E$)	1,939,655.17	657,831.32	Spaepen et al. 1997
W, kg/ha (soil weight/ha/depth of input)[b]	3,750,000	3,750,000	
PEC in soil, mg/kg ($PEC = C_{SA}/W + M$)	0.501	0.173	Spaepen et al. 1997

[a]ID and T are considered as per recommendation of Tribrisen®, a commercially available formulation.
[b]Calculation is based on soil depth of 25 cm and density of 1.5 gm/cc.

The development and spread of antibiotic-resistant microorganisms remains to be proven. An in-depth investigation is required to determine whether microorganisms exposed to antibacterial agents in the environment develop antibiotic resistance. Efforts should also be made to develop a better understanding of the transfer of SAs into biota and into food chains.

Information relating to the degradation of transformation products of SAs is limited and poorly understood, which might be important for a true environmental risk assessment. It is well known that pesticide application to underdrained clay soil poses a threat to the environment due to rapid movement to surface water and groundwater through soil macropores and field drains. Limited information is available on this topic with SAs. It is important to obtain precise information on their physicochemical properties and the geochemical processes that affect their mobility and degradation.

Unlike pesticides and other organic pollutants, SAs, as veterinary drugs, enter the soil along with the manure slurry, considered to be their main route of entry in soil. Therefore, the additional carbon source, which must have a direct influence on the soil microbial biomass and sorptive properties of the soil, may promote or inhibit the degradation and dissipation of the active ingredients in soil. It is imperative to conclude that the complex nature of SAs and their association with manure makes it difficult to fit the potential risk assessment models.

There is a normal practice of using OECD guideline No. 106 (OECD 2000) to determine the sorption coefficients of xenobiotics in soil by batch-equilibrium experiments. However, this practice does not reflect the influence of manure or sludge on the sorption of antibiotics that enter soil by way of contaminated manure or sludge. Therefore, it seems more realistic to study the sorption of antibiotics to soil in the presence of manure or sludge.

SAs can enter the food chain via uptake by plants, translocation into groundwater, or soil surface runoff into surface water (Migliore et al. 1995, 1996; Boxall et al. 2002b; Thiele-Bruhn 2003). Additionally, the development of resistant bacteria caused by the application of antibiotics to farm animals and their presence in the feces, milk, meat, and eggs has already been observed (Huysman et al. 1993; Gavalchin and Katz 1994; Teuber 1999, 2001). Therefore, entry of resistant genes to human pathogens through this pathway may put a question mark to the expected therapeutic value of the active ingredients. Further, as SAs are not readily biodegradable (Ingerslev and Halling-Sørensen 2000) and have medium to high mobility (Tolls 2001), there is a possible exposure pathway for aquatic organisms, and they may also be encountered in drinking water. However, tillage before slurry application could be adopted as a management practice to reduce the transport of SAs to water bodies (Kay et al. 2004). SAs might exert impact on organic matter turnover and nutrient recycling in soil, resulting in an indirect influence on crop plants because they lead to an alteration of functional and structural diversity in microbial populations in soil. Thus, reductions in the use of antibiotics should be encouraged.

Summary

SAs, a structurally related group of antibiotics containing a similar 4-aminobenzene sulfonamide backbone, are used in agriculture, aquaculture, animal husbandry, and also as human medicines. Competing with *p*-aminobenzoic acid in the enzymatic synthesis of dihydrofolic acid, SAs inhibit the growth and reproduction of bacteria. Once released to the environment, SAs distribute themselves among different environmental compartments, along with their degradation products, and are transported to surface water and groundwater. The physicochemical properties, the dosage applied and the nature of the environmental components with which they interact, govern the whole process. SAs, as a class, are less sorptive, impersistent, and leachable. They cannot be characterized as readily biodegradable. Their adsorption to soil increases with the aromaticity and electronegativity of functional groups attached to the sulfonyl phenyl amine core. Preferential flow in clay soils has been identified as a mechanism responsible for surface water contamination by SAs.

Acknowledgments

This study has been performed within the frame of the research project investigations on "Veterinary medicines in soils: basic research for risk analysis" under Deutsche Forschungsgemeinschaft (DFG). Therefore, the authors gratefully acknowledge the financial support of DFG, Germany. The authors also deeply appreciate the help rendered by Dr. Andrew J. Hall during the preparation of the manuscript.

References

ACD: Calculated using Advanced Chemistry Development (ACD/Labs) Software Solaris V 4.67 (© 1994–2005 ACD/Labs).

Alberola S, Rambaud J, Jeanjean B, Sabon F, Pauvert B (1975) Radiocrystallographic study of sulfacetamide. Trav Soc Montpellier 35(4):335–338.

Alder CA, McArdell CS, Golet EM, Ibric S, Molnar E, Nipales NS, Giger W (2001) Occurrence and fate of fluoroquinolone, macrolide, and sulfonamide antibiotics during waste water treatment and in ambient water in Switzerland. In: Doughton CG, Jones-Lepp I (eds) Pharmaceuticals and Personal Care Products in the Environment: Scientific and Regulatory Issues. Symposium Series 791. American Chemical Society, Washington, DC, pp 56–69.

Alexy R, Schöll A, Kümpel T, Kümmerer K (2004) What do you know about antibiotics in the environment? In: Kümmerer K (ed) Pharmaceuticals in the Envi-

ronment: Sources, Fate, Effects and Risks, 2nd Ed. Springer-Verlag, Berlin, pp 209–221.
Balba MT, Khan MR, Evans WC (1979) The microbial degradation of a sulfanilamide-based herbicide (Asulam). Biochem Soc Trans 7:405–407.
Battaglin WA, Furlong ET Burkhardt MR, Peter CJ (2000) Occurrence of sulfonylurea, sulfonamide, imidazolinone, and other herbicides in rivers, reservoirs and ground water in the Midwestern United States, 1998. Sci Total Environ 248(2–3):123–133.
Berger K, Petersen B, Büning-Pfaue H (1986) Persistenz von Arzneimitteln aus Gülle in der Nahrungskette. Archiv Lebensmittelhyg 37:99–102.
Blough NV, Zepp RG (1995) Reactive oxygen species in chemistry. In: Foote, et al. (eds) Structure Energetics and Reactivity in Chemistry Series, vol 2. Blackie, London, pp 280–333.
Boreen AL, Arnold WA, McNeill K (2003) Photodegradation of pharmaceuticals in the aquatic environment: a review. Aquat Sci 65(4):320–341.
Boreen AL, Arnold WA, McNeill K (2004) Photochemical fate of sulfa drugs in the aquatic environment: sulfa drugs containing five membered heterocyclic groups. Environ Sci Technol 38(14):3933–3940.
Boxall ABA, Blackwell PA, Cavallo R, Kay P, Tolls J (2002a) The sorption and transport of a sulfonamide antibiotic in soil system. Toxicol Lett 131:19–28.
Boxall ABA, Fogg LA, Kay P, Blackwell PA, Pemberton EJ (2002b) Review of veterinary medicines in the environment. Environment Agency R + d Tech Rep P 6-$_{012}$/8. Environment Agency, Bristol, UK.
Boxall ABA, Blackwell PA, Boleas S, Halling-Sørensen B, Ingerslev F, Jacobsen AM, Kay P, Montforts MHM, Soeborg T, Ter Laak T, Tarzona JV, Tolls T (2003) Environmental risk assessment of veterinary medicines in slurry. Final report of the EU ERAVMIS Project EVK$_1$–CT, 1999–2003. Cran field University, UK.
Boxall ABA, Fogg LA, Blackwell PA, Kay P, Pemberton EJ, Croxford A (2004a) Veterinary medicines in the environment. Rev Environ Contam Toxicol 180:1–91.
Boxall ABA, Kay P, Blackwell PA, Fogg LA (2004b) Fate of veterinary medicines applied to soils. In: Kümmerer K (ed) Pharmaceuticals in the Environment: Sources, Fate, Effects and Risks, 2nd Ed. Springer-Verlag, Berlin, pp 165–180.
Boyd GR, Reemtsma H, Grimm DA, Mitra S (2003) Pharmaceuticals and personal care products (PPCPs) in surface and treated waters of Louisiana, USA and Ontario, Canada. Sci Total Environ 311:135–149.
Buser HR, Müller MD, Theobald N (1998) Occurrence of the pharmaceutical drug clofibric acid and the herbicide Mecoprop in various Swiss lakes and in the North Sea. Environ Sci Technol 32:188–192.
Calamari D, Zuccato E, Castiglioni S, Bagnati R, Fanelli R (2003) Strategic survey of therapeutic drugs in the rivers Po and Lambro in Northern Italy. Environ Sci Technol 37:1241–1248.
Chignell CF, Kalyanaraman B, Mason RP, Sik RH (1980) Spectroscopic studies of cutaneous photosensitizing agents. I. Spin trapping of photolysis products from sulfanilamide, 4-aminobenzoic acid, and related compounds. Photochem Photobiol 32(5):563–571.
Chignell CF, Kalyanaraman B, Sik RH, Mason RP (1981) Spectroscopic studies of cutaneous photosensitizing agents. II. Spin trapping of photolysis products from

sulfanilamide and 4-aminobenzoic acid using 5,5-dimethyl-1-pyrroline-1-oxide. Photochem Photobiol 34(2):147–156.
Christian T, Schneider RJ, Färber HA, Skutlarek D, Meyer MT, Goldbach HE (2003) Determination of antibiotic residues in manure, soil and surface waters. Acta Hydrochim Hydrobiol 31:36–44.
Clark JH, English JP, Jansen GR, Marson HW, Rogers MM, Taft WE (1958) 3-Sulfanilamido-6-alkoxypyridazines and related compounds. J Am Chem Soc 80:980–983.
Commission of European Communities (1992) Commission Directive 92/18/EEC of 20 March 1992 modifying the Annex to Council Directive 81/852/EEC on the approximation of the laws of Member States relating to analytical, pharmaco-toxicological and clinical standards and protocols in respect of the testing of veterinary medical products. OFF J Eur Communities L97:1.
Daughton CG, Ternes TA (1999) Pharmaceuticals and personal product care products in the environment: agents of subtle change? Special report. Environ Health Perspect (Suppl) 107(6):907–938.
Deo N, Tiwari RK, Singh TP (1980) Crystallization and x-ray crystal data of sulfonamides. J Sci Res 2(2):137–139.
EMEA (1997) European Aency for the Evaluation of Medical Products. EMEA/CVMP/055/96 – Final. EMEA, London.
EMEA (1999) Antibiotic resistance in the European Union associated with therapeutic use of veterinary medicines. European Agency for the Evaluation of Medicinal Products, London, UK, p 9.
Espinasse J (1993) Responsible use of antimicrobials in veterinari medicine: perspective in France. Vet Microbiol 35:289–301.
FEDESA (2001) Antibiotic use in farm animals does not threaten human health. FEDESA/FEFANA press release. European Federation of Animal Health, Brüssels, July 13.
Fjelde S, Lunestad BT, Samuelsen OB, Ervik A (1993) Photostability of oxytetracycline, nitrofurazolidone, oxolinic acid, flumequine, sulfadiazine, sulfadimethoxine, trimethoprim, and ormethoprim in seawater. Residues Vet Drugs Food Proc Residue Euro Conf 2nd 1:285–288.
Frimmel FH, Hessler DP (1994) Photochemical degradation of triazine and amide pesticides in natural waters. In: Helz GR, Zepp RG, Crosby DG (eds) Aquatic and Surface Photochemistry. Lewis, Boca Raton, pp 137–147.
Frisk AR (1943) Sulfanilamides: chemotherapeutic evaluation of N1-substituted sulfanilamides. Acta Med Scand, Suppl 142:1–199.
Gavalchin J, Katz SE (1994) The persistence of fecal-borne antibiotics in soil. J Assoc Offic Anal Chem Int 77:481–485.
Haller MY, Müller SR, McArdell CS, Alder AC, Suter MJ-F (2002) Quantification of veterinary antibiotics (sulfonamides and trimethoprim) in animal manure by liquid chromatography–mass spectrometry. J Chromatogr A 952:111–120.
Halling-Sørensen B (2000) Algal toxicity of antibacterial agents used in intensive farming. Chemosphere 40:731–739.
Halling-Sørensen B, Nielsen SN, Lanzky PF, Ingerslev F, Holten Lützhoft HC, Jorgensen SE (1998) Occurrence, fate and effect of pharmaceutical substances in the environment: a review. Chemosphere 36:357–393.

Halling-Sørensen B, Nielsen SN, Jensen J (2002) Environmental Assessment of Veterinary Medical Products in Denmark. Danish Environmental Protection Agency, Kobenhavn, Denmark.
Hamscher G, Powelzick HT, Höper H, Nau H (2003a) Different behavior of tetracyclines and sulfonamides in sandy soils fertilized with animal slurry. In: Abstracts, 13th Annual Meeting of SETAC Europe, May 27–June 1, 2003, Hamburg, p 72.
Hamscher G, Powelzick HT, Sczesny S, Nau H, Hartung J (2003b) Antibiotics in dust originating from a pig fattening farm: a new source of health hazard for farmers? Environ Health Perspect 111:1590–1594.
Hamscher G, Powelzick HT, Höper H, Nau H (2004) Antibiotics in soil: routes of entry, environmental concentrations, fate and possible effects. In: Kümmerer K (ed) Pharmaceuticals in the Environment: Sources, Fate, Effects and Risks, 2nd Ed. Springer-Verlag, Berlin, pp 139–147.
Hamscher G, Powelzick HT, Höper H, Nau H (2005) Different behavior of tetracyclines and sulfonamides in sandy soils after repeated fertilization with liquid manure. Environ Toxicol Chem 24(4):861–868.
Hapeman CJ, Biboulian S, Anderson BG, Torrents A (1998) Structural influences of low-molecular-weight dissolved organic carbon mimics on the photolytic fate of atrazine. Environ Toxicol Chem 17:975–981.
Haridas M, Singh TP (1986) Crystal and molecular structure of sulfamethoxypyridazine. Ind J Chem Sect A Inorg Phys Theor Anal 25A(8):707–713.
Heberer T (2002) Occurrence, fate, and removal of pharmaceutical residues in the aquatic environment: a review of recent research data. Toxicol Lett 131:5–17.
Hektoen H, Berge JA, Hormazabal V, Yndestad M (1995) Persistence of antibacterial agents in marine sediments. Aquaculture 133:175–184.
Het Bestuur van de Stichting Ontwikkelings- en Saneringsfonds voor de Landbouw (1987) Bestuursbesluit van de Stichting Ontwikkelings-en Saneringsfonds voor de Landbouw. Staatscourant 245, nr. 391. Koninklijke Vermande, Leliestad, The Netherlands.
Hill RR, Jeffs EG, Roberts DR, Wood SA (1999) Photodegradation of aryl sulfonamides: N–tosylglysine. Chem Commun 17:1735–1736.
Hirsch R, Ternes T, Heberer K, Kratz K-L (1999) Occurrence of the antibiotics in the aquatic environment. Sci Total Environ 225:109–118.
Holm JV, Ruegge K, Bjerg PL, Christensen TH (1995a) Occurrence and distribution of pharmaceutical organic compounds in the groundwater downgradient of a landfill (Grinsted, Denmark). Environ Sci Technol 29:1415–1420.
Holm JV, Bjerg PL, Ruegge K, Christensen TH (1995b) Response to comment on "Occurrence and distribution of pharmaceutical organic compounds in the groundwater downgradient of a landfill (Grinsted, Denmark)." Environ Sci Technol 29(12):3074.
Höper H, Kues J, Nau H, Hamscher G (2002) Eintrag und Verbleib von Tieraruneimittelwirkstoffen in Böden. Bodenschutz 7:141–148.
Höper H, Hamscher G, Powelzick HT, Schäfer W, Nau H (2003) Occurrence and fate of tetracyclines and sulfonamides in soils fertilized with animal manure. In: Proceedings of the Workshop: Diffuse input of chemicals into soil and ground water. Dresden, Feb. 26–28, 2003. Mitteilung Institut Grundwasserwirtschaft TU Dresden 3:117–124.

Huber MM, Cannonica S, Park G-Y, von Gunten U (2003) Oxidation of pharmaceuticals during ozonation and advanced oxidation processes. Environ Sci Technol 37:1016–1024.
Huysman E, van Renterghem B, Verstraete W (1993) Antibiotic resistant sulphite-reducing *Clostridia* in soil and ground water as indicator of manuring practices. Water Air Soil Pollut 69:243–255.
Ingerslev F, Halling-Sørensen B (2000) Biodegradability properties of sulfonamides in activated sludge. Environ Toxicol Chem 19(10):2467–2473.
Jones G, Lu LN, Fu H, Farahat CW, Oh C, Greenfield SR, Gosztola DJ, Wasielewski MR (1999) Intramolecular electron transfer across amino acid spacers in the picosecond time regime. charge-transfer interaction through peptide bonds. J Phys Chem B 103(3):572–581.
Jones OAH, Voulvoulis N, Lester JN (2001) Human pharmaceuticals in the aquatic environment: a review. Environ Technol 22:1383–1394.
Kay P, Blackwell PA, Boxall ABA (2004) Fate of veterinary antibiotics in a macroporous tile drained clay soil. Environ Toxicol Chem 23(5):1136–1144.
Kay P, Blackwell PA, Boxall ABA (2005) A lysimeter experiment to investigate the leaching of veterinary antibiotics through a clay soil and comparison with field data. Environ Pollut 134:333–341.
Kilmes J, Mokry M (1996) TLC detection of photolytic and hydrolytic degradation products of selected antibacterial sulfonamides. J Planar Chromatogr 9:61–64.
Kleefisch B, Kues J (1997) Das Boden-Dauerbeobachtungsprogramm von Niedersachsen. Methodik und Ergebnisse. Arbeitshefte Boden 2:1–122.
Kolpin DW, Meyer MT, Barber LB, Zaugg SD, Furlong ET, Buxton HT (2000) A national reconnaissance for antibiotics and hormones in streams of the United States. Presented at SETAC 21st Annual Meeting in North America, Nashville, TN, November 12–16, 2000.
Kolpin DW, Furlong ET, Meyer MT, Thurman EM, Zaugg SD, Barber LB, Buxton HT (2002) Pharmaceuticals, hormones, and other organic wastewater contaminants in U.S. streams. Environ Sci Technol 36:1202–1211.
Kravchenya NA (1989) Selective N-acylation of 4-aminobenzenesulfonamide. Khimiko-Farmtsevticheskii Zh 23(4):454–456.
Krepper TP, Kirchhöfer F, Leichter I, Maes A, Wilken RD (1999) Metabolism studies of phenylsulfonamides relevant for water works. Environ Sci Technol 33:945–950.
Krieger MS, Yoder RN, Gibson R (2000) Photolytic degradation of florasulam on soil and in water. J Agric Food Chem 48:3710–3717.
Krishida K, Nishinari K, Furusawa N (2005) Liquid chromatographic determination of sulfamonomethoxine, sulfadimethoxine, and their N^4–acetyl metabolites in chicken plasma. Chromatographia 61(1/2):81–84.
Kümmerer K, Al Ahmed A, Mersch-Sundermann V (2000) Biodegradability of some antibiotics, elimination of the genotoxicity and affection of wastewater bacteria in a simple test. Chemosphere 40:701–710.
Kuzyakov Y, Friedel JK, Stahr K (2000) Review of mechanisms and quantification of priming effects. Soil Biol Biochem 32:1485–1498.
Langhammer JP (1989) Untersuchungen zum Verbleib antimikrobiell wirksamer Arzeistoffe als Rückstände in Gülle und im landwirtschaftlichen Umfeld. Ph.D. thesis, Rheinische Friedrich-Wilhelms-Universität Bonn, Germany.
Langhammer JP, Büning-Pfaue H (1989) Bewertung von Arzneistoff-Rückständen aus der Gülle im Boden. Lebensmittelchem Gerichtl Chem 43:103–113.

Langhammer JP, Büning-Pfaue H, Winkelmann J, Körner E (1988) Chemotherapeutika: Rückstände und Resistenzverhalten bei der Bestandsbehandlung von Sauen post partum. Tierärztl Umschau 43:375–382.
Langhammer JP, Führ F, Büning-Pfaue H (1990) Verbleib von Sulfonamid-Rückständen aus der Gülle in Boden und Nutzpflanze. Lebensmittelchem Gerichtl Chem 44:93.
Latch DE, Stender BL, Packer JL, Arnold WA, McNeil K (2003) Photochemical fate of pharmaceuticals in the environment: climetidine and ranitidine. Environ Sci Technol 37(15):3342–3350.
Lawrence MAM, Davies NA, Edwards PA, Taylor MG, Simkiss K (2000) Can adsorption isotherms predict sediment bioavailability? Chemosphere 41:1091–1100.
Lindsey ME, Meyer M, Thurman EM (2001) Analysis of trace levels of sulfonamide and tetracycline antimicrobials in groundwater and surface water using solid-phase extraction and liquid chromatography/mass spectrometry. Anal Chem 73:4640–4646.
Löscher W, Ungemach FR, Kroker R (2002) Pharmakotherapie bei Haus- und Nutztieren. 5. Auflage. Parey, Buchverlag, p 550.
Lunestad BT, Samuelsen OB, Fjelde S, Ervik A (1995) Photostability of eight antibacterial agents in seawater. Aquaculture 134:217–225.
Mansour M, Feicht EA, Behechti A, Scheunert I (1997) Experimental approaches to studying the photostability of selected pesticides in water and soil. Chemosphere 35:39–50.
Maton A, Daelemans J, Lambrecht J (1983) De Huisvesting van Dieren. Rijksstation Voor Landbouwtechniek, Merelbeke, Belgium.
Mazel D, Davies J (1999) Antibiotic resistance in microbes. Cell Mol Life Sci 56:742–754.
Migliore LG, Brambilla S, Cozzolino S, Gaudio L (1995) Effect on plants of sulfadimethoxine used in intensive farming (*Panicum miliaceum, Pisum sativumand* and *Zea mays*). Agric Ecosyst Environ 52:103–110.
Migliore LG, Brambilla S, Casoria P, Civitareale SC, Gaudio L (1996) Effect of sulphadimethoxine contamination on barley (*Hordeum disticum* L., Poaceae, Liliopsida). Agric Ecosyst Environ 60:121–128.
Mill T (1999) Predicting photoreaction rates in surface waters. Chemosphere 38(6):1379–1390.
Mill T, Maybey W (1985) Photochemical transformations. Environ Exposure Chem 1:175–216.
Ministers für Umwelt, Raumordnung und Landwirtshaft des Landes Nordrhein-Westfalen (1985) Umweltschutz und Landwirtschaft Teil 2: Die Gülleverordnung. Schriftenreihe des Ministers für Umwelt, Raumordnung und Landwirtshaft des Landes Nordrhein-Westfalen, Düsseldorf, Germany.
Moore DE (1987) Principle and practice of drug photodegradation studies. J Pharmacol Biomed Anal 5(5):441–453.
Moore DE, Zhou W (1994) Photodegradation of sulfamethoxazole: a chemical system capable of monitoring seasonal changes in UVB intensity. Photochem Photobiol 59 (5):497–502.
Motten AG, Chignell CF (1983) Spectroscopic studies of cutaneous photosensitizing agents. III. Spin trapping of photolysis products from sulfanilamide analogs. Photochem Photobiol 37(1):17–26.

Mudd AJ, Lawrence K, Walton J (1998) Study of Sweden's model on antimicrobial use shows usage has increased since 1986 ban. Feedstuffs 10.
Müller SR, Singer H, Stoob K, Burkhardt M, Hartmann N, Götz C, Stamm C, Waul C (2003) Occurrence and fate of antibiotics in manure, soil and water. Mitt Lebensm Hyg 94(6):574–578.
Neuman M (1981) Antibiotika Kopendium. Verlag Hans Huber, Bern, Stuttgart, Vienna.
NOAH (2002) Sales of antimicrobiological products used as veterinary medicines, growth promoters and coccidiostats in the UK in 2000. National Office of Animal Health, UK.
Nowara A, Burhenne J, Spiteller M (1997) Binding of fluoroquinolone carboxylic acid derivatives to clay minerals. J Agric Food Chem 45:1459–1463.
OECD (2000) Organization for Co-operation and Development (OECD) guidelines for testing of chemicals, test guideline 106. Adsorption/desorption using a batch equilibrium method. Revised draft document. OECD, Paris, pp 1–26.
Packer JL, Werner JJ, Latch DE, McNeil K, Arnold WA (2003) Photochemical fate of pharmaceuticals in the environment: naproxen, diclofenac, clofibric acid and ibuprofen. Aquat Sci 65(4):342–351.
Pinck LA, Soulides DA, Allison FE (1962) Antibiotics in soils: 4. Polypeptides and macrolides. Soil Sci 94:129–131.
Rang HP, Dule MM, Ritter JM (1995) Pharmacology, 3rd Ed. Churchill Livingstone, London.
Roblin RO Jr, Winnek PS, English JP (1942) Chemotherapy IV. Sulfanilamidopyrimidines. J Am Chem Soc 64:567–570.
Rouchaud J, Gustin F, Roisin C, Grevy L, Raimond Y (1993) Effects of organic fertilizers on aldicarb soil biodegradation in sugar beet crops. Arch Environ Contam Toxicol 24:67–74.
Sacher F, Lange FT, Brauch HJ, Blankenhorn I (2001) Pharmaceuticals in groundwaters: analytical methods and results of a monitoring program in Baden-Württemberg, Germany. J Chromatogr A 938:199–210.
Saha S, Kulshrestha G (2002) Degradation of sulfuron, a sulfnylurea herbicide, as influenced by abiotic factors. J Agric Food Chem 50(16):4572–4575.
Samuelsen OB, Torsvik V, Ervik A (1992) Long-range changes in oxytetracycline concentration and bacterial resistance toward oxytetracycline in a fish farm sediment after medication. Sci Total Environ 114:25–36.
Samuelsen OB, Lunestad BT, Ervik A, Fjelde S (1994) Stability of antibacterial agents in an artificial marine aquaculture sediment studied under laboratory conditions. Aquaculture 126:283–290.
Saxena AK, Seydel JK (1980) QSAR studies of potential antimalarial sulfonamide partitioning into erythrocytes. Eur J Med Chem 15(3):241–246.
Schulten H-R, Leinweber P (2000) New insights into organic-mineral particles: composition, properties and models of molecular structure. Biol Fertil Soils 30: 399–432.
Sekhon BS, Sahai HK, Randhawa HS (2001) On the coordinating behaviour of sulphamethoxazole towards some metal ions. Proc Natl Acad Sci India Sect A Phys Sci 71(3):195–203.
Sigel CW (1983) Disposition and metabolism of trimethoprim, tetroxoprim, sulfamethoxazole and sulfadiazine. Handb Exp Pharmacol 64:163–184.

Spaepen KRI, Van Leemput LJJ, Wislocki PG, Verschueren C (1997) A uniform procedure to estimate the predicted environmental concentration of the residues of veterinary medicines in soil. Environ Toxicol Chem 16(9):1977–1982.
Stan H-J, Heberer T (1997) Pharmaceuticals in the environment. Analysis 25:20–23.
Strek HJ (1998a) Fate of chlorsulfuron in the environment. 1. Laboratory evaluations. Pestic Sci 53:29–51.
Strek HJ (1998b) Fate of chlorsulfuron in the environment. 2. Field evaluations. Pestic Sci 53:52–70.
Stumpf M, Ternes TA, Wilken RD, Rodgrigues SV, Bauman W (1999) Polar drug residues in sewage and natural waters in the state of Rio de Janeiro, Brazil. Sci Total Environ 225(1–2):135–141.
Ternes TA (1998) Occurrence of drugs in German sewage treatment plants and rivers. Water Res 32:3245–3260.
Teuber M (1999) Spread of antibiotic resistance with food-borne pathogens. Cell Mol Life Sci 56:755–763.
Teuber M (2001) Veterinary use and antibiotic resistance. Curr Opinion Microbiol 4:493–499.
Thiele S (2000) Adsorption of the antibiotic pharmaceutical compound sulfapyridine by a long-term differently fertilized loess Chernozem. J Plant Nutr Soil Sci 163:589–594.
Thiele S, Seibicke T, Leinweber P (2002) Sorption of sulfonamide antibiotic pharmaceuticals in soil particle size fractions. SETAC 12th Meeting, May 12–16, 2002, Vienna.
Thiele-Bruhn S (2003) Pharmaceutical antibiotic compounds in soils: a review. J Plant Nutr Soil Sci 166:145–167.
Thiele-Bruhn S, Aust M-O (2004) Effects of pig slurry on the sorption of sulfonamide antibiotics in soil. Arch Environ Contam Toxicol 47:31–39.
Thiele-Bruhn S, Seibicke T, Schulten H-R, Leinweber P (2004) Sorption of sulfonamide pharmaceutical antibiotics on whole soils and particle-size fractions. J Environ Qual 33:1331–1342.
Tolls J (2001) Sorption of veterinary pharmaceuticals in soils: a review. Environ Sci Technol 35 (17):3397–3406.
Tolls J, Gebbink W, Cavallo R (2002) pH-dependence of sulfonamide antibiotic sorption: data and model evaluation. SETAC 12th Annual Meeting, May 12–16, 2002, Vienna.
Tsao R, Eto M (1994) Effect of some natural photosensitizers on photolysis of some pesticides. In: Helz GR, Zepp RG, Crosby DG (eds) Aquatic and Surface Photochemistry. Lewis, Boca Raton, pp 137–147.
van Dijk J, Keukens HJ (2000) The stability of some veterinary drugs and coccidiostats during composting and storage of laying hen and broiler faeces. In: van Ginkel LA, Ruiter A (eds) Residues of Veterinary Drugs in Food. Proceedings of the Eur Residue IV Conference, Veldhoven, The Netherlands, May 8–10, 2000.
Vetter H, Steffens G (1986) Wirtschaftseigene Düngung. Verlagsunion AGRAR: DLG-Verlag, Frankfurt am Main.
Vicary A, Landy R, Genthner F, Morales R (1999) Antimicrobials used in animal feedlots: targetting research on microbial resistance. Presented at the 20th SETAC Meeting, Philadelphia, PA, Nov. 15–19, 1999.
Walker N (1978) A soil *Flavobacterium* sp. that degrades sulfonamides and asulam. J Appl Bacteriol 45:125–129.

Watkins DAM (1979) Photochemical studies on pesticides. Pestic Sci 10:181–182.
Weiss B, Durr H, Hass HJ (1980) Photochemistry of sulfonamides and sulfonylureas: a contribution to the problem of light-induced dermatoses. Angew Chem 19(8):648–650.
Woolley JL, Sigel CW (1979) Metabolism and disposition by the rat of ^{35}S-sulfadiazine alone and in the presence of trimethoprim. Drug Metab Dispos 7(2):94–99.
Woolley JL, Sigel CW, Wels CM (1980) Novel deaminated sulfadiazine metabolites in neonatal calf tissue, plasma, and urine following oral treatment with ^{14}C-sulfadiazine. Life Sci 27:1819–1826.
WRc–NSF (2000) The development of a model for estimating the environmental concentrations (PECs) of veterinary medicines in soil following manure spreading (project code VM_{0295}). Final Project Report to MAFF.
Yang L, Yang X, Liu J, Li Y, Lou Q, Liu Q (2003) Synthesis, characterization and susceptibility of bacteria against sulfamethoxydiazine complexes of copper(II), zinc(II), nickel(II), cadmium(II), chromium(III) and iron(III). J Coord Chem 56(13):1131–1139.
Yang S, Cha J, Carlson K (2004) Quantitative determination of trace concentrations of tetracycline and sulfonamide antibiotics in surface water using solid-phase extraction and liquid chromatography/ion trap tandem mass spectrometry. Rapid Commun Mass Spectrom 18:2131–2145.
Zepp RG, Cline DM (1977) Rates of direct photolysis in aquatic environment. Environ Sci Technol 11(4):359–366.
Zhou W, Moore DE (1994) Photochemical decomposition of sulfamethoxazole. Int J Pharmacol 110(1):55–63.
Zhou W, Moore DE (1997) Photosensitizing activity of the anti-bacterial drugs sulfamethoxazole and trimethoprim. J Photochem Photobiol B 39(1):63–72.
Zuccato E, Calamari D, Natangelo M, Fanelli R (2000) Presence of therapeutic drugs in the environment. Lancet 355:1789–1790.

Manuscript received June 29; accepted July 13, 2005.

Rev Environ Contam Toxicol 187:103–131

Terminology of Gonadal Anomalies in Fish and Amphibians Resulting from Chemical Exposures

Markus Hecker, Margaret B. Murphy, Katherine K. Coady, Daniel L. Villeneuve, Paul D. Jones, James A. Carr, Keith R. Solomon, Ernest E. Smith, Glen Van Der Kraak, Timothy Gross, Louis Du Preez, Ronald J. Kendall, and John P. Giesy

Communicated by George W. Ware.

M. Hecker (✉), M.B. Murphy, K.K. Coady, P.D. Jones, J.P. Giesy.
Department of Zoology, National Food Safety and Toxicology Center, Institute for Environmental Toxicology, Michigan State University, East Lansing, MI 48824-1222, U.S.A.

D.L. Villeneuve
U.S. EPA Mid-Continent Ecology Division, 6201 Congdon Blvd., Duluth, MN, 55804-2595 U.S.A.

J.A. Carr
Department of Biological Sciences, Texas Tech University, Lubbock, TX 79409, U.S.A.

K.R. Solomon
Centre for Toxicology and Department of Environmental Biology, University of Guelph, Guelph, Ontario, N1G 2W1, Canada

E.E. Smith
The Institute of Environmental and Human Health, and Department of Environmental Toxicology, Texas Tech University, Lubbock, TX 79409, U.S.A.

G. Van Der Kraak
Department of Integrative Biology, University of Guelph, Ontario, NIG 2W1, Canada

T. Gross
Ecotoxicology Laboratory, University of Florida, Department of Physiological Sciences, College of Veterinary Medicine, and USGS Florida Caribbean Science Center, 7920 NW 71st St., Gainesville, FL 32653, U.S.A.

L. Du Preez
School of Environmental Sciences and Development, North-West University, Private Bag X6001, Potchefstroom 2520, South Africa

R.J. Kendall
The Institute of Environmental and Human Health, and Department of Environmental Toxicology, Texas Tech University, Lubbock, TX 79409, U.S.A.

J.P. Giesy
Biology and Chemistry Department, City University of Hong Kong, Kowloon, Hong Kong, SAR, China

Contents

I. Introduction

During the past decade, the scientific community and the public have become increasingly aware that some chemicals have the potential to interfere with endocrine systems in both vertebrate and invertebrate wildlife species (WHO 2002; Ankley et al. 1998). One aspect of these effects has been the observation of gonadal abnormalities in fish, amphibians, reptiles, birds, and mammals, including humans (Kavlock et al. 1996). To date, most research in this field has focused on demasculinization or feminization effects on male animals (Sumpter et al. 1996; Gimeno et al. 1998a,b; Crain et al. 1999; Jobling et al. 1998; Kloas et al. 1999; Hayes et al. 2002). Evidence for this estrogenic or antiandrogenic type of "endocrine disruption" has come largely from studies of teleost fish, either in controlled laboratory experiments where they have been exposed to specific chemicals or in the wild where organisms have been exposed to mixtures of compounds (Jobling et al. 1998; Harries et al. 1999; Minier et al. 2000; Hecker et al. 2002; Matthiessen et al. 2002). More recently, attention has shifted toward other groups of animals living in or closely associated with aquatic environments, such as alligators (Crain et al. 1999) and amphibians (Kloas et al. 1999; Hayes et al. 2002; Hecker et al. 2004).

The ontogeny of most fish and amphibians is characterized by at least some degree of sexual plasticity. Larvae, and sometimes adults, retain vestigial tissues, such as ovarian ducts in adult male *Rana pipiens* (Lee 1969) and Bidder's organ on the top of testes in adult male *Bufo bufo*. Both fish and amphibians have tissues at some point in their development that have the potential of developing into an intermediate gonadal state that contains both ovarian and testicular tissue. This situation has been termed ambisexuality (Gallien 1974). Although ambisexuality is not evident for most vertebrate species, it is the general rule for many groups of amphibians and reflects their basic bisexual constitution; this explains the sexual plasticity that is often observed in fish and amphibians. One result of this plasticity is that exposure of many of these species to estrogen or androgen agonists or antagonists or compounds that block specific steroidogenic enzymes can result in the stimulation of germ cells such that individuals develop complete or incomplete gonads of the gender opposite to that of their genotype, sometimes even causing complete phenotypic sex reversal in the adult life stage. Individuals are often responsive to exposure to exogenous hormonally active substances during specific periods of development, a fact that often complicates and confounds the interpretation of responses among species. Determining what is abnormal and assigning causality of observed "abnormal" effects is complicated by an incomplete understanding of ontogeny of many aquatic species and what is "normal" (Sumpter and Johnson 2005).

Although numerous studies have evaluated the effects of exogenous chemicals on the endocrine systems and gonadal development of nonmammalian vertebrates such as fish, amphibians, and reptiles, the terminology used to describe the observed effects has been adopted mainly from terms used in clinical diagnosis. Given differences in sexual differentiation, development, and plasticity between mammals and nonmammalian vertebrates, broadly defined clinical terms often are not sufficiently specific to describe effects in these groups of animals. As a consequence, researchers have freely applied these clinical terms to describe both normal and abnormal gonadal development in vertebrates, which has resulted in a confusing variety of terms used to describe the same effects. Here we provide a glossary of the terminology that has been used to describe gonadal abnormalities in a series of studies (Table 1). The use of these terms is further complicated by differences in the terms applied to gross and histological evaluations. When making gross observations of gonads rather than histological observations, it is difficult to determine whether abnormalities are present at the cellular level. For this reason, different terminologies have been applied to these two situations. Some of these divergent terms are illustrated in Figs. 1–3.

To be able to compare and discuss the results of different studies and to reduce ambiguous classification in ecoepidemiological observation and reporting, it is necessary to reconcile the different terminologies used. The purpose of this review is to provide a brief synopsis of the types of gonadal abnormalities seen in fish and amphibians as a consequence of exposure to endocrine-active compounds in the laboratory and field. The failure of researchers to use a common terminology in describing these changes has led to confusion in the literature and has often complicated efforts to describe cause-and-effect relationships in relation to either the causative agents or the etiology of the gonadal changes. We provide a description of some common terminologies that should be used in describing alterations in normal gonadal development in fish and amphibians.

Table 1. Terminology used in the literature to describe gonadal deformities in fish and frogs.

Description	Term	Species	Reference
Gonadal anomalies based on gross morphology in fish and frogs			
Ambiguous sex: ovarian and testicular tissue mixed in same gonad	Hermaphrodite	*Xenopus laevis* (Amphibia)	Hayes et al. 2003
Ambiguous sex: ovary and testis in the same animal but segregated laterally or rostrally/caudally	Hermaphrodite	*X. laevis* (Amphibia)	Hayes et al. 2002
Rostral/caudal or left/right separation of testicular and ovarian tissue	Intersex	*X. laevis* and *Rana clamitans*, (Amphibia)	Carr et al., 2003; Coady et al. 2004
Masses of ovarian and testicular tissue separated left/right or rostral/caudal	Intersex	*X. laevis* (Amphibia)	Carr et al. 2003
Cooccurrence of both ovarian and testicular tissue in a single gonad	Mixed sex	*R. clamitans* (Amphibia)	Coady et al. 2004
Male gonads almost completely filled with oocytes, limited number of labules	Sex reversal	*R. pipiens* (Amphibia)	Hayes et al. 2003
Abnormal segmentation in gonad and/ or segments of gonad separated by undifferentiated tissue	Multiple gonads	*X. laevis* and *R. clamitans* (Amphibia)	Hayes et al. 2002; Coady et al. 2004
Gonads segmented and with sections connected by thin strands of connective tissue	Discontinuous gonad	*X. laevis* and *R. clamitans* (Amphibia)	Carr et al. 2003; Coady et al. 2004; Smith et al. 2003; Jooste et al. 2005
Large size discrepancies between gonad pairs or unusually large or small gonads	Size irregularities	*X. laevis* and *R. clamitans* (Amphibia)	Hecker et al. 2003; Smith et al. 2005; Jooste et al. 2005
Retarded gonadal development, underdeveloped testes	Testicular dysgenesis	*R. pipiens* (Amphibia)	Hayes et al. 2003

Table 1. *Continued*

Description	Term	Species	Reference
Gonadal anomalies based on histology in fish and amphibians			
Few to absent germ cells, testicular tubules poorly developed	Testicular dysgenesis	*R. pipiens* (Amphibia)	Hayes et al. 2003
Oocytes present in the testes	Testicular oocytes	*X. laevis* (Amphibia)	Smith et al. 2005; Jooste et al. 2005
Oocytes present in the testes (stage of oocytes not stated)	Testicular oogenesis	*R. pipiens* (Amphibia)	Hayes et al. 2003
Cooccurrence of both ovarian and testicular tissue in a single gonad	Mixed sex	*X. laevis, R. clamitans*, (Amphibia)	Coady et al. 2004, 2005; Carr et al. 2003
Rostral/caudal or left/right separation of testicular and ovarian tissue	Intersex	*X. laevis, R. clamitans*, (Amphibia)	Coady et al. 2004, 2005; Carr et al. 2003
Occurrence of various degrees of oocytes in testicular tissue	Intersex	*Rutilus rutilus, Gobio gobio, Abramis brama, Platichthys flesus* (Pisces); *R. sylvatica, R. pipiens* (Amphibia)	Jobling et al. 1998; Allen et al. 1999; Van Aerle et al. 2001; Hecker et al. 2002; Mackenzie et al. 2003
Occurrence of oocytes in testicular tissue regardless of their number	Ovotestis	*P. flesus, A. brama* (Pisces); *R. sylvatica, R. pipiens* (Amphibia)	Allen et al. 1999; Vethaak et al. 2002; Mackenzie et al. 2003
Over 45% of testicular tissue is female	Ovotestis	*A. brama* (Pisces)	Hecker et al. 2002
Mature female tissue containing scattered testicular tissue	Ovotestis	*Oryzias latipes* (Pisces)	Getsfrid et al. 2004
Mature testicular tissue containing scattered ovarian follicles	Testis-ova	*O. latipes* (Pisces)	Getsfrid et al. 2004

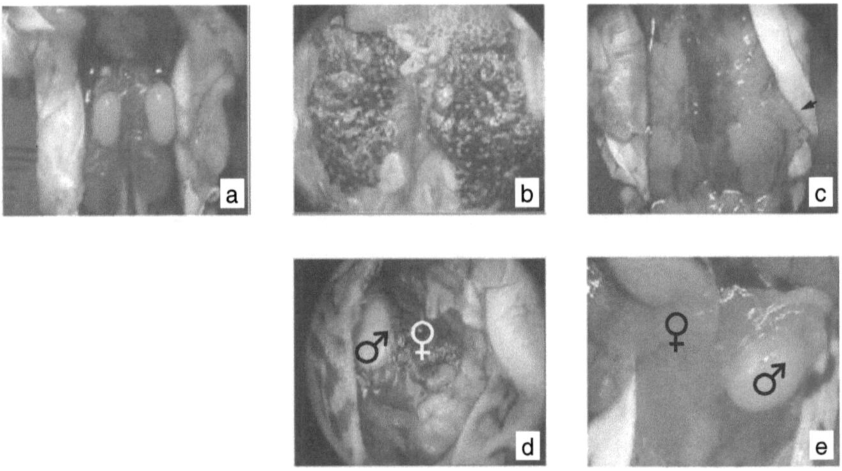

Fig. 1. Photomicrographs show normal and abnormal gonadal morphology of male and female green frogs (*Rana clamitans*): (*a*) normal testes; (*b*) normal ovaries; (*c*) immature ovaries (juvenile); (*d*) mixed gonadal tissue (*left*, testis; *right*, ovary) with oviduct below ovarian tissue; (*e*) mixed gonadal tissue (juvenile: *left*, immature ovarian tissue; *right*, testis).

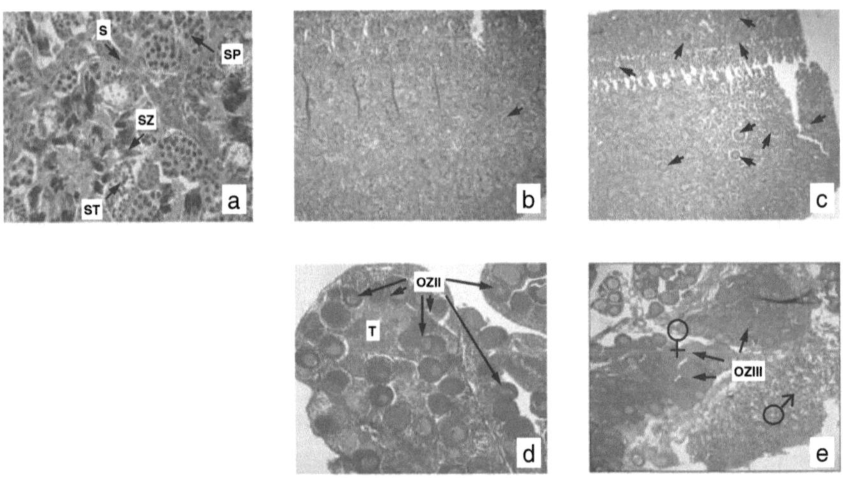

Fig. 2. Photomicrographs of cross sections taken of normal and rudimentary hermaphroditic gonads of green frogs (*Rana clamitans*): (*a*) normal testis (***SP***, spermatocytes; ***ST***, spermatids; ***S***, Sertoli cells; ***SZ***, spermatozoa); (*b*) individual testicular oocytes (*arrows*: stage II, previtellogenic stage); (*c*) multiple testicular oocytes (*arrows*: stage II oocytes, previtellogenic stage); (*d*) ovotestis (***OZII***, stage II oocytes, previtellogenic stage; ***T***, testicular tissue); (*e*) true hermaphrodite (testicular and ovarian tissue left/right separated by connective tissue; ***OZIII***, stage III oocytes, early vitellogenic stage).

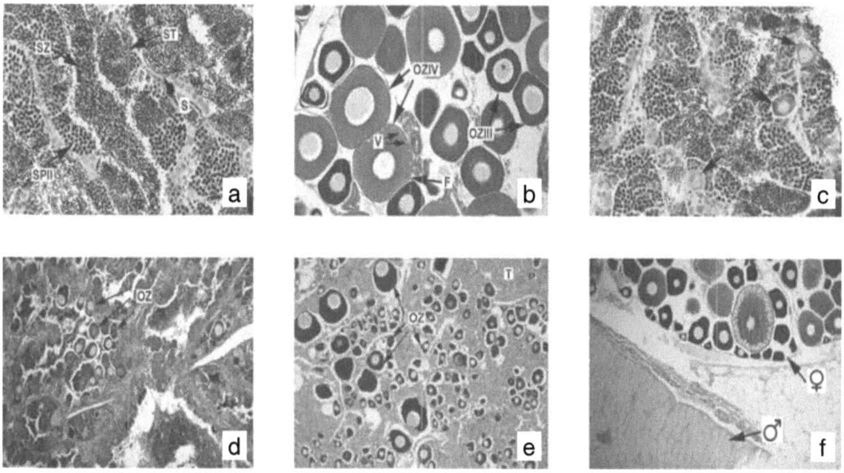

Fig. 3. Photomicrographs of cross sections taken of normal and rudimentary hermaphroditic gonads of adult fish (bream, *Abramis brama*, Cyprinidae): (*a*) normal testis (***SPII***, secondary spermatocytes; ***ST***, spermatids; ***SZ***, spermatozoa; ***S***, Sertoli cells); (*b*) normal female gonad (***OZIII***, stage III oocytes, perinucleolus stage; ***OZIV***, stage IV oocytes, cortical alveoli stage; ***V***, vacuoles; ***F***, follicle layer); (*c*) individual testicular oocytes (stage I, perinucleolus stage); (*d*) multiple testicular oocytes (***OZ***, stage I oocytes, perinucleolus stage); (*e*) ovotestis (***OZ***, stage I oocytes; ***T***, testicular tissue); (*f*) true hermaphrodite (testicular and ovarian tissue left/right separated by connective tissue), only primary spermatocytes present, oocytes are between stage I and IV.

II. Endocrine-Active Chemical Effects on Gonadal Development

A number of natural and synthetic environmental pollutants, referred to as endocrine-disrupting chemicals (EDCs), have been reported to interact with endocrine systems in fish and amphibians (reviewed in WHO 2002). Both laboratory and field studies have identified a suite of changes in gonadal morphology that have been attributed to chemical exposure. Often these changes relate to the development of hermaphroditic conditions, including the presence of oocytes within otherwise normal testicular tissue or development of an oviduct within the testis. These changes have been variously described using terms such as testicular oocytes (TOs), ovotestis, or intersex gonads, when describing similar conditions. A second major gonadal effect involves degenerative changes including reductions in the size or number of testicular cells and regression or atresia of ovarian follicle cells. A summary of different histological effects on the gonadal tissue in fish can be found in Blazer (2002). To be able to compare and discuss the results of different studies it is necessary to reconcile the different terminologies used. Therefore, in this review, where appropriate, the terminology used by Blazer (2002) has been redefined. To our knowledge, there has been no comparable review on histological effects on the gonads of fish or amphibians nor any attempts to standardize the terminology used to describe the observed phenomena.

Numerous studies have been conducted in the wild to describe effects of natural and man-made chemicals on gonad morphology and histology in fish and amphibians (Jobling et al. 1998; Hecker et al. 2002; Matthiessen et al. 2002; Vethaak et al. 2002; Hayes et al. 2003; Murphy et al. 2005). Most of the evidence for estrogenic effects in fish including the occurrence of TOs and ovotestis has been linked with discharges of estrogenic compounds such as natural and synthetic estrogens, alkylphenols, or bisphenol A via sewage treatment plants (STP) (Purdom et al. 1994; Jobling et al. 1998; Minier et al. 2000; Hecker et al. 2002; Mikaelian et al. 2002). On rare occasions there have also been reports of masculinization of females exposed to pulp and paper mill effluents (Cody and Bortone 1997; Mikaelian et al. 2002). It has been hypothesized that pulp and paper mill effluents contain plant sterols that can be degraded to androgens by microorganisms (Marshek et al. 1972). In frogs, only two studies have reported the occurrence of TOs or mixed male and female gonadal tissue in animals from the wild (Reeder et al. 1998; Hayes et al. 2003). Although it has been hypothesized by both these authors that the use of pesticides such as the triazine herbicide atrazine may be responsible for the observed effects, to date the exposure to this herbicide has not be conclusively linked to these effects. The natural occurrence of rudimentary hermaphroditism (for definition, see Section III.A) such as the occurrence of TOs or ovotestis is largely unknown for most fish and amphibian species, and therefore conclusions concerning the causes of these phenomena must be drawn with care. This concern is especially true regarding the ambisexual nature of fish and amphibians that results in a sexual plasticity that is not yet fully understood.

By comparison, fewer studies are available on such effects under controlled laboratory exposure conditions. Most laboratory studies have investigated endocrine responses at lower levels of biological organization, including molecular and biochemical responses such as induction of vitellogenin, plasma sex steroid concentrations, steroidogenic enzymes, and hormone receptor binding. The laboratory studies that describe EDC effects on gonad morphology and histology mostly focused on the effects of estrogenic or antiandrogenic compounds that bind to steroid receptors in an agonistic or antagonistic manner. The chemicals studied include the natural and synthetic hormones estradiol (E2) and ethinylestradiol (EE2), alkylphenols and their ethoxylates [nonylphenol (NP), octylphenol (OP), nonylphenol-ethoxylate (NPEO), octylphenol-ethoxylate (OPEO)], methoxychlor, *o, p'*-DDT, *o, p'*-DDE, PCBs, and PAHs, and in a few cases also androgens such as testosterone (Table 2).

The compounds that have been found to be the most potent at feminizing male gonads (e.g., development of TOs) are the natural or synthetic sex steroids E2 and EE2 (Piferrer and Donaldson 1992; Gimeno et al. 1998a,b; Seki et al. 2002; Mackenzie et al. 2003; Leino et al. 2004). In contrast, weak

Table 2. Effects of steroids and endocrine-active compounds on the gonads of fish and frogs.

Compound	Effect	Species	Class	Developmental stage	Reference
17β-Estradiol	Development of testicular oocytes in males	*Cyprinus carpio*	Pisces	Juvenile	Gimeno et al. 1998a
		C. carpio	Pisces	Adult	Gimeno et al. 1998b
		Oryzias latipes	Pisces	Juvenile	Metcalfe et al. 2001
		Xenopus laevis	Amphibia	Juvenile	Carr et al. 2003
		Rana pipiens	Amphibia	Juvenile	Mackenzie et al. 2003
		Rana sylvatica	Amphibia	Juvenile	Mackenzie et al. 2003
	Ovotestis	*O. latipes*	Pisces	Juvenile	Metcalfe et al. 2001
		R. pipiens	Amphibia	Juvenile	Mackenzie et al. 2003
	Occurrence of separate ovary and testis in the same individual	*X. laevis*	Amphibia	Juvenile	Carr et al. 2003
	Development of oviduct in testis	*R. pipiens*	Amphibia	Juvenile	Mackenzie et al. 2003
		R. sylvatica	Amphibia	Juvenile	Mackenzie et al. 2003
	Degeneration of testicular cells	*C. carpio*	Pisces	Adult	Gimeno et al. 1998b
		Pimephales promelas	Pisces	Adult	Miles-Richardson et al. 1999a
	Hypertrophy of sertoli cells	*Poecilia reticulata*	Pisces	Adult	Kinnberg and Toft 2003
	Reduction of number of mature oocytes	*P. promelas*	Pisces	Adult	Miles-Richardson et al. 1999a
	Increase in the number of atretic oocytes	*P. promelas*	Pisces	Adult	Miles-Richardson et al. 1999a

Table 2. *Continued*

Compound	Effect	Species	Class	Developmental stage	Reference
17α-Ethinyl estradiol	Development of testicular oocytes in males	*O. latipes*	Pisces	Adult	Seki et al. 2002
		O. latipes	Pisces	Juvenile	Metcalfe et al. 2001
		Onkorhynchus tshawytscha	Pisces	Juvenile	Piferrer and Donaldson 1992
		R. pipiens	Amphibia	Juvenile	Mackenzie et al. 2003
		R. sylvatica	Amphibia	Juvenile	Mackenzie et al. 2003
	Ovotestis	*Oryzias latipes*	Pisces	Juvenile	Metcalfe et al. 2001
		R. pipiens	Amphibia	Juvenile	Mackenzie et al. 2003
		R. sylvatica	Amphibia	Juvenile	Mackenzie et al. 2003
	Occurrence of separate ovary and testis in the same individual	*R. pipiens*	Amphibia	Juvenile	Mackenzie et al. 2003
	Development of oviduct in testis	*R. pipiens*	Amphibia	Juvenile	Mackenzie et al. 2003
	Degeneration of testicular tissue	*O. latipes*	Pisces	Adult	Seki et al. 2002
	Enlargement of seminiferous tubuli	*R. sylvatica*	Amphibia	Juvenile	Mackenzie et al. 2003
	Increase of the number of atretic oocytes and regression of ovarian tissue in females	*O. latipes*	Pisces	Adult	Seki et al. 2002
		P. promelas	Pisces	Adult	Laenge et al. 2001
Testosterone	Shrinking of ovary and degeneration of oocytes	*R. clamitans*	Amphibia	Juvenile	Foote and Witschi 1939
	Development of a rete testis with vas deference and formation of sex cords in females	*R. clamitans*	Amphibia	Juvenile	Foote and Witschi 1939

Table 2. *Continued*

Compound	Effect	Species	Class	Developmental stage	Reference
4-*tert*-Nonylphenol	Development of testicular oocytes in males	*O. latipes*	Pisces	Adult	Kang et al. 2003
		R. pipiens	Amphibia	Juvenile	Mackenzie et al. 2003
		R. sylvatica	Amphibia	Juvenile	Mackenzie et al. 2003
	Ovotestis (one gonad, other gonad normal-looking testis or ovary)	*R. pipiens*	Amphibia	Juvenile	Mackenzie et al. 2003
	Ovotestis (both gonads)	*R. pipiens*	Amphibia	Juvenile	Mackenzie et al. 2003
		R. sylvatica	Amphibia	Juvenile	Mackenzie et al. 2003
	Formation of an ovarian type cavity in testis	*R. pipiens*	Amphibia	Juvenile	Mackenzie et al. 2003
	Increase in necrosis of germ cells and spermatozoa and hypotrophy of Sertoli cells in males	*P. promelas*	Pisces	Adult	Miles-Richardson et al. 1999b
	Increase in apoptosis in spermatocytes, Sertoli cells, and Leydig-homologue cells	*O. latipes*	Pisces	Adult	Weber et al. 2002
4-*tert*-Octylphenol	Shorter time to gonadal differentiation in males and females	*Rana catesbeiana*	Amphibia	Juvenile	Mayer et al. 2003
	Rupture of spermatozeugmata resulting in increased numbers of free sperm	*P. reticulata*	Pisces	Adult	Kinnberg and Toft 2003
	Enlargement of sperm ducts and increase in number of spermatozeugmata	*P. reticulata*	Pisces	Adult	Kinnberg and Toft 2003

Table 2. *Continued*

Compound	Effect	Species	Class	Developmental stage	Reference
4-*tert*-Pentylphenol	Development of testicular oocytes in males	*C. carpio*	Pisces	Juvenile	Gimeno et al. 1998a
	Development of oviduct in testis	*C. carpio*	Pisces	Juvenile	Gimeno et al. 1998a
	Reduction of seminiferous tubuli diameter and atrophy of germinal epithelium	*C. carpio*	Pisces	Adult	Gimeno et al. 1998b
Bisphenol A	Development of testicular oocytes in males	*O. latipes*	Pisces	Juvenile	Metcalfe et al. 2000 Yokota et al. 2000
	Decreased number of spermatozoa in males	*O. latipes*	Pisces	Juvenile	Metcalfe et al. 2000 Yokota et al. 2000
	Increase in testicular fibrosis	*O. latipes*	Pisces	Juvenile	Metcalfe et al. 2000 Yokota et al. 2000
	Rupture of spermatozeugmata resulting in increased numbers of free sperm	*P. reticulata*	Pisces	Adult	Kinnberg and Toft 2003
o,p-'DDT	Development of testicular oocytes in males	*O. latipes*	Pisces	Juvenile	Metcalfe et al. 2000
o,p-'DDE	Increase in atresia of oocytes and decrease in oocyte maturation	*O. latipes*	Pisces	Juvenile	Papoulias et al. 2003
PCBs (diverse)	Development of testicular oocytes in males	*X. laevis*	Amphibia	Juvenile	Qin et al. 2003
	Reduction of number of seminiferous tubuli	*X. laevis*	Amphibia	Juvenile	Qin et al. 2003
Atrazine	Development of testicular oocytes in males	*X. laevis*	Amphibia	Juvenile	Hayes et al. 2002
			Amphibia	Juvenile	Carr et al. 2003
		R. pipiens	Amphibia	Juvenile	Hayes et al. 2003
	Ovotestis	*X. laevis*	Amphibia	Juvenile	Hayes et al. 2002
		R. pipiens	Amphibia	Juvenile	Hayes et al. 2003
	Decreased testicular development	*R. pipiens*	Amphibia	Juvenile	Hayes et al. 2003

estrogen receptor agonists such as alkylphenols, bisphenol A, *o,p'*-DDE, and some PCBs could not be consistently linked to the occurrence of morphological feminization of the testis such as increases in the number of TOs or formation of an oviduct in the testis (Miles-Richardson et al. 1999b; Metcalfe et al. 2001; Papoulias et al. 2003; Pickford et al. 2003; Levy et al. 2004). Similar inconsistent findings have been observed for different studies using atrazine (Hayes et al. 2002, 2003; Carr et al. 2003; Coady et al. 2004).

However, the estrogenic or weak feminizing effects attributed to either steroidal estrogens or the less potent xenoestrogens were usually less prominent than the effects caused by degeneration of the gonads of both males and females. These effects include atresia of oocytes (Miles-Richardson et al. 1999a; Laenge et al. 2001; Seki et al. 2002) and general degeneration of testicular or ovarian tissue in fish (Gimeno et al. 1998b; Miles-Richardson et al. 1999a; Seki et al. 2002; Laenge et al. 2001) and reduction of maturation processes in fish (Gimeno et al. 1998b; Miles-Richardson et al. 1999a,b; Laenge et al. 2001; Seki et al. 2002; Papoulias et al. 2003) and frogs (Hayes et al. 2003; Pickford and Morris 2003). Overall, with the exception of the effects of natural or synthetic estrogens such as E2 or EE2, compounds with weaker estrogenic properties seldom cause feminization, but rather at sufficiently great concentrations cause degeneration of the testes or ovaries of fish and amphibians. Although in some studies exposure to weak estrogens including alkylphenols and bisphenol A have been reported to cause an increase in the occurrence of TOs, these studies were generally conducted during early previtellogenic stages, during which TOs were separated from the surrounding testicular tissue and were not associated with functional impairments of the gonads (see following section on TOs). As described in subsequent sections, the occurrence of a small number of TOs appears to be a normal phenomenon in many fish or amphibian species. Thus, it is possible that the occurrence of individual organisms with TOs as observed in some of the field and laboratory studies with relatively small sample sizes (e.g., 1 fish with TOs in a group of 7 results in a 14% incidence; Kang et al. 2003; 1 frog in a sample size of 4 or 5 individuals results in an incidence of 20% and 25%, respectively; Reeder et al. 1998) is a function of the natural variability of this phenomenon rather than a direct chemical effect.

III. Terminology

A. Hermaphroditism

From a clinical perspective, hermaphroditism is a form of intersexuality in which both male and female gonadal tissues are present in the same individual. There are several forms of hermaphroditism. In some cases, complete and functioning male and female reproductive tracts are found in the same individual. This situation is termed "functional" or "synchronous" hermaphroditism. In these situations, organisms may be able to fertilize themselves or require fertilization by a separate individual. Although functional hermaphroditism has been well documented in fish (summarized in Van Thienhoven 1983; Devlin and Nagahama 2002), the occurrence of this condition in amphibians is not well known. "Sequential" hermaphroditism, where an individual functions first as a male and subsequently as a female (protandrous) or first as a female and subsequently as a male (protogynous), is a typical condition in many teleost fishes but is rare in amphibians (Van Tienhoven 1983). Certain toads (e.g., *B. bufo*) have rudimentary and inactive ovarian tissue dorsal to the testis that is known as the Bidder's organ. This organ can develop into a functional ovary after castration of the toad and thus represents a form of protandry (Pancak-Roessler and Norris 1991).

A condition that is far more typical in amphibians is "developmental hermaphroditism," which is a common pattern in undifferentiated races of frogs that can also be observed in certain undifferentiated gonochoristic species of fish (secondary gonochorists). Examples of undifferentiated amphibian species are *Rana temporaria, R. esculenta, B. bufo*, and to some extent *X. laevis* (Gallien 1974). Typical examples of teleost species that possess a bipotential gonad that then develops either into an ovary or a testis are the European and Japanese eels (*Anguilla anguilla* and *A. japonica*) (reviewed in Devlin and Nagahama 2002). In amphibians, typically only males of most undifferentiated species go through a nonfunctional hermaphroditic gonadal stage as a natural part of their early sexual development and then develop a functional testis whereas females usually directly develop an ovary (Witschi 1921; Hsu and Liang 1971; Gamapurohit et al. 2000). In undifferentiated fish species, both males and females can go through a hermaphroditic stage before the gonad develops into either an ovary or a testis (Devlin and Nagahama 2002). This condition, termed developmental hermaphroditism (Table 3), is the case for the zebrafish (*Danio rerio*), which is widely used for studies of developmental biology. The African clawed frog (*X. laevis*), which is commonly used for toxicity testing in laboratory studies and that has previously been thought to directly develop either into a male or female, also appears to exhibit some developmental plasticity. A microcosm study conducted in South Africa found that recently metamorphosed *X. laevis* had great incidences of female germ cells dispersed throughout developing testicular tissue (Jooste et al. 2005). When these frogs were grown for several months past metamorphosis, the incidence of TOs gradually decreased with maturation. As a result of this hermaphroditic developmental pattern, it appears that the occasional occurrence of TOs in the testis, even into adulthood, is a natural phenomenon under certain conditions, which accords with earlier reports on the temporary occurrence of TOs in gonads of young male *X. laevis* (Gallien 1974).

However, the extent to which TOs were observed in developing *X. laevis* was less than that observed in other rudimentary hermaphroditic species

Table 3. Suggested terminology to describe gonadal abnormalities based on histology in fish and amphibians.

Term	Diagnostic description
Ambisexuality	Natural structural state that occurs in larvae and in some species also in adults, and which has the potential of developing into an intermediate gonadal state containing both ovarian and testicular tissue
Testicular oocytes	Oocytes present in the testes regardless of maturation stage
Rudimentary testicular oocytes	Oocytes present in the testes during early development; oocytes disappear during further development into adulthood
Testicular oogenesis	Genesis of oocytes by an intact testes, e.g., via stimulation through exogenous or endogenous estrogens.
Testicular dysgenesis	Abnormal development of testicular tissue during sexual differentiation or development (e.g., abnormal hormone synthesis or action during reproductive tract development)
Segmented gonads	Gonads are segmented into discrete subunits with obvious gonadal tissue separated by thin pieces of connective or nongonadal tissue
Intersex	Phenotypic sex different from genotypic sex
Mixed gonadal tissue	Testicular and ovarian tissues occur in the same individual; phenotypic sex is unclear
Unilateral	Ovarian and testicular tissue is not separated; tissue is mixed in a single gonadal structure (ovotestis)
Bilateral	Separation of ovarian and testicular tissue left/right; one gonad = male; other gonad = female
Ovotestis	Occurrence of male and female germ cells in the same gonad at an incidence >30%; this condition describes mixed sex condition but does not indicate if individual is functional or nonfunctional hermaphrodite
Hermaphroditism	
Functional/synchronous	Individual functions simultaneously as male and female; both male and female reproductive organs are present and fully functional
Developmental	Ovotestis present during early sexual development; during further development gonad develops into either functional male or functional female tissue; rudimentary oocytes may be present during later developmental stages but without functional relevance
Rudimentary	Nonfunctional from of hermaphroditism in which few to many germ cells of the other sex are present in the gonad
Castration	Removal of the Gonads or their destruction as by external influence resulting in a nonfertile organism
Sex reversal	Difference between chromosomal sex and phenotypic sex

such as *R. temporaria* (Witschi 1921), *R. catesbeiana* (Hsu and Liang 1971), or *R. curtipes* (Gamapurohit et al. 2000), and no all-female developmental stages were observed for *X. laevis*. This finding indicates that in *X. laevis* this phenomenon is unlikely to be caused by "true developmental hermaphroditism" but rather may be the result of some primordial germ cells remaining as remnants of rudimentary tissue during early development. These cells may continue to divide for a period and then atrophy when the testis becomes active and starts producing testosterone and Müllerian-inhibiting substance. Also, a series of other studies did not report testicular oocytes in male developing *X. laevis* (Chang and Witschi 1956; Villalpando and Merchant-Lerois 1990; Miyata et al. 1999; Miyata and Kubo 2000). It is unclear if these differences are due to differences in methodology [e.g., differences in numbers of observations per gonad (see Section III.C, Testicular Oocytes), diet, or housing conditions] or are a function of a greater plasticity in *X. laevis* gonadal differentiation than previously thought. We would term this condition rudimentary testicular oocytes (Table 3).

In cases where both male and female gonadal tissues are present in the same organism, but the condition cannot be assigned to a specific developmental history such as an undifferentiated pattern, we are in favor of the term rudimentary hermaphroditism (van Tienhoven 1983; Coady et al. 2004, 2005). For some species, including the green frog (*R. clamitans*), the details of the developmental pattern are not fully understood (Coady et al. 2004). Therefore, it currently remains unclear whether the condition of rudimentary hermaphroditism described in some reports is caused by the natural developmental pattern of the animal or "natural" exogenous factors such as temperature or nutrition or is due to chemicals in the environment. It should be remembered that, for a variety of ranid species, rudimentary hermaphroditism was reported as early as the late 19th century, which is long before the widespread occurrence of many of the synthetic chemicals in the environment that have been suggested as potential causes of gonadal abnormalities. Examples of species for which sexual abnormalities such as TOs, bilaterally or unilaterally mixed ovarian and testicular tissues, ovotestis, or the development of male or female ducts in individuals from the opposite sex have been observed are summarized in Table 4. Because most of the modern pesticides including DDT, lindane, atrazine, etc. were adopted for widespread use only after 1945, in this analysis, we considered only reports of surveys conducted before World War II. In conclusion, care must be taken when correlating phenomena such as rudimentary hermaphroditism to pesticide use. In this context, there is a need for additional studies in pristine environments to broaden our knowledge on the natural occurrence of rudimentary hermaphroditism in amphibian species that then can serve as reference scenarios for the assessment of the effects of pesticide exposure.

It was the opinion of Coady et al. (2004, 2005) that it was important to differentiate between the diverse morphologies of rudimentary hermaph-

Table 4. Examples for records of gonadal abnormalities in amphibians before 1945, with findings adapted to terminology described in Table 3.

Species	Description of abnormality	Age	Year	Reference
Rana pipiens	Male frog with oviducts having a female-like granular layer	Adult	1928	Mac Lean 1929
	Testicular oocytes in both testes	Juvenile	1928	Christensen 1929
	Unilateral mixed gonadal tissue (left gonad = ovotestis; right gona = testis)	Adult	1928	Christensen 1929
	Bilateral mixed gonadal tissue (right gonad = testis; left gonad = ovary)	Adult	<1924	Neal 1924
	Bilateral mixed gonadal tissue (both gonads ovotestis)	Adult	1928	Christensen 1929
		Adult	1930	Evans 1931
R. temporaria	Male frog with oviducts	Juvenile and adult	1929, 1969	Lloyd 1929; Lee 1969
	Unilateral mixed gonadal tissue (one gonad = ovotestis; other gonad = ovary)	Adult	<1921	Cited in Crew 1921
	Unilateral mixed gonadal tissue (one gonad = ovotestis; other gonad = testis)	Juvenile	<1921	Cited in Crew 1921
	Unilateral mixed gonadal tissue (one gonad = ovotestis; other gonad = testis)	Adult	<1921 and 1925	Cited in Crew 1921; Eggert 1929
	Bilateral mixed gonadal tissue (both gonads ovotestes)	Juvenile	<1921	Cited in Crew 1921
	Bilateral mixed gonadal tissue (both gonads ovotestes)	Adult	<1921	Cited in Crew 1921
R. esculenta	Male frog with oviducts	Adult	<1921	Cited in Crew 1921
	Normal testis with ovarian cavity	Juvenile	<1921	Cited in Crew 1921
	Bilateral mixed gonadal tissue (both gonads ovotestis)	Adult	<1921	Cited in Crew 1921

Table 4. *Continued*

Species	Description of abnormality	Age	Year	Reference
R. catesbeiana	Bilateral mixed gonadal tissue (left gonad = testis flanked by ovarian tissue along both sides of the body cavity)	Sex unknown	1919	Clemens 1921
R. virescens	Male frog with oviducts	Adult	<1894	Sumner 1894
R. cantabrigensis (now *R. sylvatica*)	Bilateral mixed gonadal tissue (both gonads ovotestis)	Adult	1928	Cheng 1929
R. fusca	Male frog with oviducts	Adult	<1921	Cited in Crew 1921
	Bilateral mixed gonadal tissue (both gonads ovotestis)	Adult	<1921	Cited in Crew 1921
Bufo vulgaris	Unilateral mixed gonadal tissue (one gonad ovotestis; other gonad ovary)	Adult	<1921	Cited in Crew 1921
	Bilateral mixed gonadal tissue (both gonads ovotestis)	Adult	<1921	Cited in Crew 1921
B. cinereus	Bilateral mixed gonadal tissue (both gonads ovotestis)	Adult	<1921	Cited in Crew 1921
B. lentiginosus	Bilateral mixed gonadal tissue (both gonads ovotestis)	Adult	<1921	Cited in Crew 1921
Pelobates fuscus	Unilateral mixed gonadal tissue (left gonad ovotestis; right gonad testis)	Adult	<1921	Cited in Crew 1921
Hyla aurea	Male frog with oviducts	Adult	<1921	Cited in Crew 1921
	Testicular oocytes in both testes	Adult	<1921	Cited in Crew 1921

roditism by indicating whether the hermaphrodism was a situation in which both ovarian and testicular tissue occurred in a single gonad or if they occurred in separate distinct regions of the gonad or between gonad pairs. Clinicians have defined the condition where mixed ovarian and testicular tissues occur (ovotestis, see following) as "true hermaphroditism." These clinical definitions must be consistent because they are used for clinical diagnoses. However, different terms are used in the ecotoxicological literature, and some terms are used interchangeably. For instance, in the literature describing the effects of xenobiotics on fish, the terms intersex and ovotestes seem to be used interchangeably. The more recent literature (Sumpter and Johnson 2005) on the effects of weak estrogen agonists on gonadal development in fish seems to use the term intersex to refer to more than one type of hermaphrodite.

B. Ovotestis

The term ovotestis was coined specifically to describe the condition in which oocytes are present in the testes. However, the term ovotestis has been used differently by different authors. Although some authors assigned the term ovotestis to male gonads regardless of the incidence of oocytes (Allen et al. 1999; Vethaak et al. 2002), others have defined ovotestis as a condition where a large portion (>45%) of the gonad was composed of oocytes (Hecker et al. 2002). Alternatively, Getsfried et al. (2004) distinguished between ovotestis and testis-ova by defining the former condition as mature ovarian tissue with scattered testicular tissue and the latter as mature testicular tissue with scattered ovarian follicles. The term testis-ova has also been used to describe the condition where TOs were observed in Japanese medaka (*Oryzias latipes*) (Metcalfe et al. 2001; Yokota et al. 2000). However, in clinical terminology, there is no differentiation between ovotestis and testis-ova, and therefore, in the subsequent sections, we refer to both conditions as ovotestis.

According to its medical definition, ovotestis refers to the histology of gonadal tissue that contains both ovarian follicles and testicular tubular elements. Ovotestes are usually compartmentalized, with connective tissue separating the ovarian from the testicular components. However, on rare occasions, an intermixture of these elements may occur (Whitman-Elia and Queenan 2002). Although the definition of the term ovotestis indicates the occurrence of oocytes regardless of their number in gonadal tissue, it is used in clinical terminology in the context of true hermaphroditism. Given the plasticity in sexual differentiation in many lower vertebrates such as fish or frogs, which often includes development of the testis from earlier all-female stages or via developmental hermaphroditic stages, residual occurrences of single or low numbers of oocytes appear to be common for some species, and thus the definition of "ovotestis" as described in clinical terminology seems inappropriate here. Therefore, for nonmammalian vertebrates, we are in favor of a more distinct definition of this term, defining ovotestis as gonadal tissue that is more than 30% female (see Table 3). In cases where only a single oocyte or few oocytes occur in testicular tissue, we prefer to use the more neutral term "testicular oocytes," expressed as an incidence (Coady et al. 2005; Murphy et al. 2005). Although the term ovotestis is generally used in context of a histopathological condition, it can be also used to describe the occurrence of male and female tissue at the gross morphological level.

C. Testicular Oocytes

The most common gonadal anomaly observed at the histological level among fish and amphibians is "rudimentary hermaphroditism" (Jobling et al. 1998; Minier et al. 2000; Hecker et al. 2002; Coady et al. 2004, 2005; Murphy et al. 2005). Rudimentary hermaphroditism can be characterized either by rostral–caudal or left–right separation of testicular and ovarian tissue, which we would term mixed gonadal tissue. However, when oocytes are present, not as a tissue, but as a few oocytes dispersed in testicular tissue, the condition has been termed TOs.

Determination of TOs is only possible at the histological level. In most cases, TOs were characterized by having an intact nucleus, nucleoli within the nucleus, and a surrounding squamous epithelial layer (Coady et al. 2004, 2005). Most of the oocytes observed in newly metamorphosed frogs in studies by Coady et al. (2004, 2005) were stage I previtellogenic oocytes, but the TOs observed in frogs collected from the field were mostly stage II (see Fig. 2; Murphy et al. 2005). It is not clear from the descriptions in a paper by Hayes et al. (2003) what criteria were used to identify TOs in that study. However, the oocyte stages presented for male adult leopard frogs (*R. pipiens*) by Hayes et al. (2003) appeared to be either stage I or stage II oocytes. Similar observations have been made in studies with male fish where early previtellogenic oocyte stages were relatively common while later vitellogenic oocytes only occurred under extreme exposure situations or were very rare (Jobling et al. 1998; Minier et al. 2000; Hoffmann 2005) (see Fig. 3). Features such as the presence of a follicle cell layer (see Dumont 1972, plate 2 and fig. 3) surrounding the oocyte can often be observed in TOs (Jooste et al. 2005; Hoffmann 2005; Smith et al. 2005). This characteristic is important because it can explain the potential source of estradiol that could be produced in the testes. The squamous and apparently inactive epithelium surrounding the stage I TOs reported in studies by Coady et al. (2004) and Jooste et al. (2005) suggests negligible 17β-estradiol (E2) production from TOs. Because of the presence of vitellogenic oocytes in the testes, some authors have implied that these male frogs produce substantial amounts of E2 (Hayes et al. 2003). However, these authors did not report whether a follicle cell layer was observed, even though this is a criterion for identifying oocytes (Dumont 1972). This inconsistency was not reconciled by Hayes et al. (2003), but the criteria for identifying oocytes (large nucleus present, multiple nucleoli, basophilic cytoplasm, squamous epithelial cell layer surrounding oocyte, size less than 300 μm) were followed in the papers by Coady et al. (2004, 2005), Jooste et al. (2005), Smith et al. (2005), and Murphy et al. (2005).

There are two possible ways in which the occurrence of TOs can be reported, depending on the type of observation made. In some studies only a subset of sections (e.g. 3–6 longitudinal sections per gonad pair; Bögi et al. 2003) was analyzed. In such cases, it is impossible to present the data as an exact incidence, and the occurrence of testicular oocytes per animal should be reported as equal to or greater than the observed number of TOs. Given the uncertainties resulting from this estimation, especially in cases where very large numbers of TOs are observed (e.g., ovotestis), it is preferable to serially section the entire gonad and make observations on every section (Coady et al. 2004, 2005; Du Preez et al. 2005; Murphy et al. 2005).

D. Intersex, Mixed Sex, and Mixed Gonadal Tissue

In the more recent literature on endocrine disruption, especially as it pertains to wildlife, the terms intersex and hermaphrodite have been used interchangeably to describe a number of conditions including the presence of complete testes and/or ovaries, or the presence of testicular tissue in the ovaries of females or ovarian tissue in the gonads of males, termed ovotestis, or the occurrence of TOs (Jobling et al. 1998; Vethaak et al. 2002; Hecker et al. 2002; Hayes et al. 2002; Carr et al. 2003; Coady et al. 2004, 2005). The term "intersex," used to describe masses of ovarian and testicular tissue in the same gonad by some authors (Carr et al. 2003; Coady et al. 2004, 2005), is the equivalent of the term "hermaphrodite" used by others (Hayes et al. 2003). In still other cases, the presence of ovarian tissue in the testes has been termed mixed sex (Coady et al. 2004). Descriptions of conditions observed during gonadal development of frogs and summaries of the terms applied to them are provided in Table 2.

"Intersexuality" is a well-described clinical condition that, for descriptive purposes, is typically divided into several different subclassifications based on the degree of gonadal development (Van Tienhoven 1983). The term intersex does not seem appropriate because it implies that an individual is of some intermediate sex. Based on the phenotype of the gonads, this may true, but without karyotyping the sex of the individual cannot be determined at the genotypic level, and the individual most likely would be genetically either a male or female that is expressing some male and female primary sexual characteristics. Given the fact that both genotype and phenotype are rarely determined, we suggest that it is more appropriate to describe specifically the anomaly that is observed, i.e., the phenotype of the individual. We therefore suggest that the more appropriate term for the situation where both testicular and ovarian tissue is present in the same individual would be "mixed gonadal tissue" (see Table 3). Although the term "mixed sex" could also be applied, we do not favor it because it is the mixture of tissues, not of the sex of the individual, that is being reported. For this reason, we suggest that the term "mixed gonadal tissue" be used together with a description of the specific type of abnormality (e.g., left–right separation of male and female gonadal tissue would be bilateral mixed gonadal tissue) (Table 3). Mixed gonadal tissue can be used to describe conditions at both the histological and gross-morphological level. Examples for mixed gonadal tissue in fish and frogs are given next (see Figs. 1–3).

E. Gonadal Dysgenesis

Malformed or incompletely formed gonads, particularly the testes, have been reported to occur in frogs (Hayes et al. 2003; Coady et al. 2004, 2005). The presence of underdeveloped testes with poorly structured, closed lobules, or no lobules at all, and only a few or the complete absence of germ cells has been termed dysgenesis by some authors (Hayes et al. 2003), whereas others use the term dysgenesis to refer to a "size irregularity" in which there was a demonstrable difference in the size of the left and right testes (Coady et al. 2004, 2005). Dysgenesis is a clinical term that refers to abnormal development of testicular or ovarian tissue during sexual differentiation or fetal development that can be caused by genomic effects as well as by environmental factors including pollution (Skakkebaek et al. 2001). Without quantifying the size of the testis or number (or fractional volume) of testicular cells it would be difficult to accurately measure the degree of dysgenesis in fish or frogs; however, large discrepancies in size could be subjectively identified. Other authors have used specific terms to describe testes that were misshapen or of different size (Smith et al. 2005). We propose that more specific descriptive terms be used to describe these types of atypical gonads if there is no histological evidence during development that actual dysgenesis has occurred. Terms such as "asymmetrical gonads" or "irregularly shaped gonads" are suggested instead of the less descriptive and inherently more confusing term "dysgenesis," which seems to relate more to a process than a description of the actual anomaly.

F. Segmented Gonads

In some individuals, there appear to be multiple gonads or the gonads are segmented into discrete subunits with obvious gonadal tissue separated by thin pieces of connective or nongonadal tissue. This condition has been termed "discontinuous gonads" by some authors (Carr et al. 2003; Coady et al. 2004, 2005) but referred to as "multiple gonads" by others (Hayes et al. 2003). We propose that the term segmented gonads is superior to the term multiple gonads because there is no evidence that multiple gonads arise from separate primordial cells.

Summary

Given the recent increase in the number of studies describing the ability of chemicals to exert endocrine-disrupting effects, not only in fish but in a variety of other oviparous groups such as amphibians and reptiles, there is an urgent need to harmonize the terminology currently used in describing pathological changes of the gonads. In addition to difficulties in comparing results from different studies, there is also the risk of miscommunication by using terms that imply a certain clinical relevance which may not be true for the species examined. Especially in the case of the recent and controversial issue about potential effects of the triazine herbicide atrazine on amphibians, clinical terminology has been utilized beyond its true meaning by using terms such as "chemical castration" to describe occurrence of TOs or ovarian tissue in the testis of male frogs exposed to environmental chemicals (Hayes 2004). In clinical terminology, castration is defined as the removal of the gonads or their destruction by an external influence, resulting in a nonfertile organism. However, Hayes (2004) did not investigate any possible effects on the fertility of the test animals and thus did not know if these animals were truly castrated. Similarly, terms such as intersex, hermaphrodite, and sex reversal have been used in ways that appear inappropriate with regard to their clinical meaning in a series of different studies with fish or frogs (see previous sections for a detailed discussion).

To ensure the appropriate use of certain terminology in a field as controversial and complex as the study of endocrine disruption, we have attempted, in this chapter, to harmonize the terminology used to describe changes in gonadal development of vertebrates such as fish and amphibians, especially frogs (see Table 3). Where appropriate, the terminology suggested was adopted directly from the clinical terminology. However, as outlined here there are substantial differences between the developmental biology of oviparous vertebrates and mammals, and especially humans, that necessitate modification of the definitions of some of the clinical terms. Where appropriate, therefore, the terminology proposed in this manuscript was redefined based on the biological meanings of the terms used in clinical diagnosis. Considering the large increase in research in the area of reproductive endocrine disruption over the past decades, the authors see an increasing need for a harmonization of terms to be used to describe effects observed in the investigated species. Agreement on a common terminology will allow scientists to better communicate and compare their work, and will enable risk assessors to conduct large-scale evaluations of environmental endocrine disruption by fitting the information from individual studies into a synthesis of normal and abnormal conditions of gonadal tissues.

Acknowledgments

We are grateful to Dr. L. Karbe and M. Hoffman University of Hamburg, Hamburg, Germany for providing the photomicrographs of the cross sections of bream. Thanks go to A. Hosmer for many helpful comments. We also thank S. Williamson for her logistic support during the writing of this manuscript. This manuscript was developed under the supervision of the Atrazine Endocrine Ecological Risk Assessment Panel, Ecorisk, Inc., Ferndale, WA, with a grant from Syngenta Crop Protection, Inc.

References

Allen Y, Matthiessen P, Scott AP, Haworth S, Feist S, Thain JE (1999) The extent of oestrogenic contamination in the UK estuarine and marine environments: further surveys of flounder. Sci Total Environ 233:5–20.

Ankley G, Mihaich E, Stahl R, Tillitt D, Colborn T, McMaster S, Miller R, Bantle J, Campbell P, Denslow N, Dickerson R, Folmar L, Fry M, Giesy JP, Gray LE, Guiney P, Hutchinson T, Kennedy S, Kramer V, LeBlanc G, Mayes M, Nimrod A, Patino R, Peterson R, Purdy R, Ringer R, Thomas P, Touart L, Van Der Kraak G, Zachrewski T (1998) Overview of a workshop on screening methods for detecting potential (anti-) estrogenic/androgenic chemicals in wildlife. Environ Toxicol Chem 17:68–87.

Blazer VS (2002) Histopathological assessment of gonadal tissue in wild fishes. Fish Physiol Biochem 26:85–101.

Bögi C, Schwaiger J, Ferling H, Mallow U, Steineck C, Sinowatz F, Kalbfus W, Negele RD, Lutz I, Kloas W (2003) Endocrine effects of environmental pollution on *Xenopus laevis* and *Rana temporaria*. Environ Res 93:195–201

Carr JA, Gentles A, Smith EE, Goleman WL, Urquidi LJ, Thuett K, Kendall RJ, Giesy JP, Gross TS, Solomon KR, Van Der Kraak G (2003) Response of larval *Xenopus laevis* to atrazine: assessment of gonadal and laryngeal morphology. Environ Toxicol Chem 22:396–405.

Chang CY, Witschi E (1956) Genic control and hormonal reversal of sex differentiation in *Xenopus*. Proc Soc Exp Biol Med 93:140–144.

Cheng T-H (1929) Intersexuality in *Rana cantabrigensis*. J Morphol Physiol 48:354–369.

Christensen K (1929) Hermaphroditism in *Rana pipiens*. Anat Rec 43:345–358.

Clemens WA (1921) A case of complete hermaphroditism in a bullfrog (*Rana catesbeiana*). Anat Rec 22:179–181.

Coady KK, Murphy MB, Villeneuve DL, Hecker M, Jones PD, Carr JA, Solomon KR, Smith EE, Van Der Kraak G, Kendall RJ, Giesy JP (2004) Effects of atrazine on metamorphosis, growth, and gonadal development in the green frog (*Rana clamitans*). J Toxicol Environ Health A 67:941–957.

Coady KK, Murphy MB, Villeneuve DL, Hecker M, Carr JA, Solomon KR, Smith EE, Van Der Kraak G, Kendall RJ, Giesy JP (2005) Effects of atrazine on metamorphosis, growth, and gonadal and laryngeal development in *Xenopus laevis*. Ecotoxicol Environ Saf 62: doi:10.1016/j.ecoenv.2004.10.010.

Cody RP, Bortone SA (1997) Maculinization of mosquitofish as an indicator of exposure to kraft mill effluent. Bull Environ Contam Toxicol 58:429–436.

Crew FAE (1921) Sex-reversal in frogs and toads. A review of the recorded cases of abnormality of the reproductive system and an account of a breeding experiment. J Genet 11:141–181.
Crain DA, Spiteri ID, Guillette LJ Jr (1999) The functional and structural observations of the neonatal reproductive system of alligators exposed in ovo to atrazine, 2,4-D, or estradiol. Toxicol Ind Health 15:180–185.
Devlin RH, Nagahama Y (2002) Sex determination and sex differentiation in fish: an overview of genetic, physiological, and environmental influences. Aquaculture 208:191–364.
Dumont JN (1972) Oogenesis in *Xenopus laevis* (Daudin). I. Stages of oocyte development in laboratory maintained animals. J Morphol 136:153–179.
Du Preez LH, Jansen van Rensburg PJ, Joostea AM, Carr JA, Giesy JP, Gross TS, Kendall RJ, Smith EE, Van Der Kraak G, Solomon KR (2005) Seasonal exposures to triazine and other pesticides in surface waters in the western Highveld corn-production region in South Africa. Environ Pollut 135:131–141.
Eggert B (1929) Der Hermaphroditismus der Tiere. 1. Beitrag zur Intersexualitaet der der Anuren. Zeit Wissen Zool 133:562–585.
Evans TC (1931) Sex reversal in *Rana pipiens*. Anat Rec 48:47–53.
Foote CL, Witschi E (1939) Effect of sex hormones on the gonads of frog larvae (*Rana clamitans*): Sex inversion in females; stability in males. Anat Rec 75:75–83.
Gallien L (1974) Intersexuality. In: Loft B (ed) Physiology of the Amphibia, vol 2. Academic Press, New York, pp 523–549.
Gamapurohit NP, Shanbhag BA, Saidapur SK (2000) Pattern of gonadal sex differentiation, development, and onset of steroidogenesis in the frog, *Rana curtipes*. Gen Comp Endocrinol 119:256–264.
Getsfrid W, Thiyagarajah A, Hartley WR (2004) Ovotestis in a Japanese medaka. J Aquat Anim Health 16:164–168.
Gimeno S, Komen H, Gerritsen AGM, Bowmer T (1998a) Feminization of young males of the common carp, *Cyprinus carpio*, exposed to 4-*tert*-pentylphenol during sexual differentiation. Aquat Toxicol 43:77–92.
Gimeno S, Komen H, Jobling S, Sumpter J, Bowmer T (1998b) Demasculinization of sexually mature male common carp, *Cyprinus carpio*, exposed to 4-*tert*-pentylphenol during spermatogenesis. Aquat Toxicol 43:93–109.
Harries JE, Janbakshs A, Jobling S, Matthiessen P, Sumpter JP, Tyler CR (1999) Estrogenic potency of effluent from two sewage treatment works in the United Kingdom. Environ Toxicol Chem 18:932–937.
Hayes TB (2004) There is no denying this: defusing the confusion about atrazine. BioScience 54:1138–1149.
Hayes TB, Collins A, Lee M, Mendoza M, Noriega N, Stuart AA, Vonk A (2002) Hermaphroditic, demasculinized frogs after exposure to the herbicide atrazine at low ecologically relevant doses. Proc Natl Acad Sci USA 99:5476–5480.
Hayes TB, Haston K, Tsui M, Hoang A, Haeffele C, Vonk A (2003) Atrazine-induced hermaphroditism at 0.1 ppb in American leopard frogs (*Rana pipens*): laboratory and field evidence. Environ Health Perspect 111:568–575.
Hecker M, Tyler CR, Hoffmann M, Maddix S, Karbe L (2002) Plasma biomarkers in fish provide evidence for endocrine modulation in the Elbe River, Germany. Environ Sci Technol 36:2311–2321.

Hecker M, Coady KK, Villeneuve DL, Murphy MB, Jones PD, Giesy JP (2003) Response of *Xenopus Laevis* to atrazine exposure: Assessment of the mechanism of action of atrazine MSU-04. ECORISK, Greensboro, NC, USA.
Hecker M, Giesy JP, Jones PD, Jooste AM, Carr JA, Solomon KR, Smith EE, Van der Kraak G, Kendall RJ, Du Preez L (2004) Plasma sex steroid concentrations and gonadal aromatase activities in African clawed frogs (*Xenopus laevis*) from South Africa. Environ Toxicol Chem 23:1996–2007.
Hoffmann M (2005) Sexual differentiation and gonadal development of male and female bream (*Abramis brama* [L.]) of the Elbe river: Natural variability or evidence of endocrine modulation. Reports of the Center for Marine and Atmospheric Science, Series E. University of Hamburg, Hamburg, Germany.
Hsu CY, Liang HM (1971) Sex races of *Rana catesbeiana* in Taiwan. Herpetologica 26:214–221.
Jobling S, Nolan M, Tyler CR, Brighty G, Sumpter JP (1998) Widespread sexual disruption in wild fish. Environ Sci Technol 32:2498–2506.
Jooste AM, Du Preez LH, Carr JA, Giesy JP, Gross TS, Kendall RJ, Smith EE, Van Der Kraak GL, Solomon KR (2005) Gonadal responses of *Xenopus laevis* larvae exposed to atrazine in microcosms. Environ Toxicol Chem 39:5255–5261.
Kang IJ, Yokota H, Oshima Y, Tsuruda Y, Hano T, Maeda M, Imada N, Tadokoro H, Honjo T (2003) Effects of 4-nonylphenol on reproduction of Japanese medaka (*Oryzias latipes*). Environ Toxicol Chem 22:2438–2445.
Kavlock RT, Daston GP, De Rosa C, Fenner-Crisp P, Gray LE, Kaattari S, Lucier G, Luster M, Mac MJ, Maczka C, Miller R, Moore J, Rolland R, Scott G, Sheehan DM, Sinks T, Tilson HA (1996) Research needs for the risk assessment of health and environmental effects of endocrine disruptors: a report of the US EPA sponsored workshop. Environ Health Perspect 104:715–740.
Kinnberg K, Toft G (2003) Effects of estrogenic and antiandrogenic compounds on the testis structure of the adult guppy (*Poecilia reticulate*). Ecotoxicol Environ Saf 54:16–24.
Kloas W, Lutz I, Einspanier R (1999) Amphibians as a model to study endocrine disruptors: II. Estrogenic activity of environmental chemicals in vitro and in vivo. Sci Total Environ 225:59–68.
Laenge R, Hutchinson TH, Croudace CP, Siegmund F, Schweinfurth H, Hampe P, Panter GH, Sumpter JP (2001) Effects of synthetic estrogen 17α-ethinylestradiol on the life-cycle of the fathead minnow (*Pimephales promelas*). Environ Toxicol Chem 20:1216–1227.
Lee PA (1969) Histology of the oviduct of the leopard frog, *Rana pipiens*. Anat Rec 165:493–502.
Leino RL, Jensen KM, Ankley GT (2004) Gonadal histology and characteristic histopathology associated with endocrine disruption in the adult fathead minnow (*Pimephales promelas*). Environ Toxicol Pharmacol 19:85–98.
Levy G, Lutz I, Krueger A, Kloas W (2004) Bisphenol A induces feminization in *Xenopus laevis* tadpoles. Environ Res 94:102–111.
Lloyd JH (1929) Hermaphroditism in the common frog (*Rana temporaria*). Am Nat 63:130–138.
Mac Lean BL (1929) Abnormal reproductive organs in a specimen of *Rana pipiens*. Anat Rec 43:53–63.

Mackenzie CA, Berril M, Metcalfe C, Pauli PD (2003) Gonadal differentiation in frogs exposed to estrogenic and antiestrogenic compounds. Environ Toxicol Chem 22:2466–2475.

Marshek WJ, Kraychy S, Muir RD (1972) Microbial degradation of sterols. Micobiol 23:72–77.

Matthiessen P, Allen Y, Bamber S, Craft J, Hurst M, Hutchinson T, Feist S, Katsiadaki I, Kirby M, Robinson C, Scott S, Thain J, Thomas K (2002) The impact of oestrogenic and androgenic contamination on marine organisms in the United Kingdom: summary of the EDMAR programme. Mar Environ Res 54:645–649.

Mayer LP, Dyer CA, Propper CR (2003) Exposure to 4-*tert*-octylphenol accelerates sexual differentiation and disrupts expression of steroidogenic factor 1 in developing bullfrogs. Environ Health Perspect 111:557–561.

Metcalfe TL, Metcalfe CD, Kiparissis Y, Niimi AJ, Foran CM, Benson WH (2000) Gonadal development and endrocrine responses in Japanese medaka (*Oryzias latipes*) exposed to o,p′-DDT in water or through maternal transfer. Environ Toxicol Chem 19:1893–1900.

Metcalfe CD, Metcalfe TL, Kiparissis Y, Koenig BG, Khan C, Hughes RJ, Croley TR, March RE, Potter T (2001) Estrogenic potency of chemicals detected in sewage treatment plant effluents as determined by *in vivo* assays with Japanese medaka (*Oryzias latipes*). Environ Toxical Chem 20:297–308.

Mikaelian I, de Lafontaine Y, Harshbarger JC, Lee LLJ, Martineau D (2002) Health of lake whitefish (*Coregonus clupeaformis*) with elevated tissue levels of environmental contaminants. Environ Toxicol Chem 21:532–541.

Miles-Richardson SR, Kramer VJ, Fitzgerald SD, Render JA, Yamini B, Barbee SJ, Giesy JP (1999a) Effects of waterborne exposure to 17 β-estradiol on secondary sex characteristics and gonads of fathead minnows (*Pimephales promelas*). Aquat Toxicol 47:129–145.

Miles-Richardson SR, Pierens SL, Nichols KM, Kramer VJ, Snyder EM, Snyder SA, Render JA, Fitzgerald SD, Giesy JP (1999b) Effects of waterborne exposure to 4-nonylphenol on secondary sex characteristics and gonads of fathead minnows (*Pimephales promelas*). Environ Res Sect A 80:S122–S137.

Minier C, Caltot F, Leboulanger F, Hill EM (2000) An investigation of the incidence of intersex fish in Seine-Maritime and Sussex regions. Analysis 28:801–806.

Miyata S, Kubo T (2000) In vitro effects of estradiol and aromatase inhibitor treatment on sex differentiation in *Xenopus laevis* gonads. Gen Comp Endocrinol 119:105–110.

Miyata S, Koike S, Kubo T (1999) Hormonal reversal and the genetic control of sex differentiation in *Xenopus*. Zool Sci 16:335–340.

Murphy MB, Hecker M, Coady KK, Tompsett AR, Jones PD, DuPreez LH, Everson GJ, Solomon KR, Carr JA, Smith EE, Kendall RJ, van der Kraak G, Giesy JP (2005) Atrazine concentrations, gonadal gross morphology and histology in ranid frogs collected in Michigan agricultural areas. Aquat Toxicol (in press).

Neal HV (1924) A case of unilateral or true gynandromorphism in a leopard frog. Anat Rec 29:113.

Pancak-Roessler MK, Norris DO (1991) The effects of orchidectomy and gonadotropins on steroidogenesis and oogenesis in Bidder's organs of the toad *Bufo woodhousii*. J Exp Zool 260:323–336.

Papoulias DM, Villalobos SA, Meadows J, Noltie DB, Giesy JP, Tillit DE (2003) *In ovo* exposure to o,p′-DDE affects sexual development but not sexual differentiation in Japanese medaka (*Oryzias latipes*). Environ Health Perspect 111:29–32.

Pickford DB, Morris ID (2003) Inhibition of gonadotropin-induced oviposition and ovarian steroidogenesis in the African clawed frog (*Xenopus laevis*) by the pesticide methoxychlor. Aquat Toxicol 62:79–94.

Pickford DB, Hetheridge MJ, Caunter JE, Hall AT, Hutchinson TH (2003) Assessing chronic toxicity of bisphenol A to larvae of the African clawed frog (*Xenopus laevis*) in a flow-through exposure system. Chemosphere 53:223–235.

Piferrer F, Donaldson EM (1992) The comparative effectiveness of the natural and a synthetic estrogen for the direct feminization of Chinook salmon (*Oncorhynchus tschawytscha*). Aquaculture 106:183–193.

Purdom CE, Hardiman PA, Bye VJ, Eno NC, Tyler CR, Sumpter JP (1994) Estrogenic effects of effluents from sewage treatment works. Chem Ecol 8:275–285.

Qin Z-F, Zhou J-M, Chu S-G, Xu X-B (2003) Effects of Chinese domestic polychlorinated biphenyls (PCBs) on gonadal differentiation in *Xenopus laevis*. Environ Health Perspect 111:553–556.

Reeder AL, Foley GL, Nichols DK, Hansen LG, Wikoff B, Faeh S, Eisold J, Wheeler MB, Warner R, Murphy JE, Beasley VR (1998) Forms and prevalence of intersexuality and effects of environmental contaminants on sexuality in cricket frogs (*Acris crepitans*). Environ Health Perspect 106:261–266.

Seki M, Yokota H, Matsubara H, Tsuruda Y, Maeda M, Tadokoro H, Kobayashi K (2002) Effect of ethinylestradiol on the reproduction and induction of vitellogenin and testis-ova in medaka (*Oryzias latipes*). Environ Toxicol Chem 21:1692–1698.

Skakkebaek NE, Rajpert-De Meyts E, Main KM (2001) Testicular dysgenesis syndrome: an increasingly common developmental disorder with environmental aspects. Hum Reprod 5:972–978.

Smith EE, Du Preez L, Solomon KR (2003) Gonadal and laryngeal responses to field exposure of *Xenopus laevis* to atrazine in areas of corn production in South Africa. SA-OIC ECORISK, Greensboro, NC, USA.

Smith EE, Du Preez LH, Gentles A, Solomon KR, Tandler B, Carr JA, Van Der Kraak GL, Kendall RJ, Giesy JP, Gross T (2005) Assessment of laryngeal muscle and testicular cell types in *Xenopus laevis* (Anura Pipidae) inhabiting maize and non-maize growing areas of South Africa. Afr J Herpetol 54:69–76.

Sumner FB (1894) Hermaphroditism in *Rana virescens*. Anat Anz 9:694–695.

Sumpter JP, Johnson AC (2005) Lessions from endocrine disruption and their application to other issues concerning trace organics in the aquatic environment. Environ Sci Technol 39:4321–4332.

Sumpter JP, Jobling S, Tyler CR (1996) Oestrogenic substances in the aquatic environment and their potential impact on animals, particulary fish. In: Taylor EW (ed) Toxicology of Aquatic Pollution. Cambridge University Press, New York, pp 205–224.

Van Aerle R, Nolan M, Jobling S, Christiansen LB, Sumpter JP, Tyler CR (2001) Sexual disruption in a second species of wild cyprinid fish (the gudgeon, *Gobio gobio*) in United Kingdom freshwaters. Environ Toxicol Chem 20:2841–2847.

Van Tienhoven A (1983) Reproductive Physiology of Vertebrates, 2nd Ed. Cornell University Press, Ithaca, NY.

Vethaak DA, Lahr J, Kuiper RV, Grinwis GCM, Rouhani Rankouhid T, Giesy JP, Gerritsen A (2002) Estrogenic effects in fish in The Netherlands: some preliminary results. Toxicology 181–182:147–150.

Villalpando I, Merchant-Larios H (1990) Determination of the sensitive stages for gonadal sex-reversal in *Xenopus laevis* tadpoles. Int J Dev Biol 34:281–285.

Weber LP, Kiparissis Y, Hwang GS, Niimi AJ, Janz DM, Metcalfe CD (2002) Increased cellular apoptosis after chronic exposure to nonylphenol and quercetin in adult medaka (*Oryzias latipes*). Comp Biochem Physiol C 131:51–59.

Whitman-Elia JF, Queenan JT (2002) Ovotestis. eMedicine, http://www.emedicine.com/med/topic1702.htm.

WHO (2002) Global assessment of the state-of-the-science of endocrine disruptors (Damstra T, Barlow S, Bergman A, Kavlock R, Van Der Kraak G, eds). WHO/PCS/EDC/02.2.

Witschi E (1921) Development of gonads and transformation of sex in the frog. Am Nat 55:529–538.

Yokota H, Tsuruda Y, Maeda M, Oshima Y, Tadokoro H, Nakazono A, Honjo T, Kobayashi K (2000) Effect of bisphenol A on the early life stage in Japanese medaka (*Oryzias latipes*). Environ Toxicol Chem 19:1925–1930.

Manuscript received July 26; accepted August 8, 2005.

Rev Environ Contam Toxicol 187:133–251

Behavior of Pesticides in Water–Sediment Systems

Toshiyuki Katagi

Contents

I. Introduction

Many kinds of pesticides and their metabolites have been detected in various water bodies and bottom sediments even under normal agricultural practices (USEPA 1997; Gilliom 2001; Martin et al. 2003). Pesticides can potentially enter surface water by several routes and be partitioned to

---

Communicated by George W. Ware

T. Katagi
Sumitomo Chemical Co., Ltd. Environmental Health Science Laboratory, 2–1 Takatsukasa 4-Chome, Takarazuka, Hyogo 665-8555, Japan

bottom sediments even if they are appropriately used for crop protection in accordance with good agricultural practices. Spray drift, surface runoff, and field drainage are relevant routes of exposure, and contamination via groundwater discharge may occur. The direct application of pesticides to water occures either for rice protection in a paddy field or for control of undesirable emergent vegetation of weeds and algae.

To assess either ecotoxicological impacts on aquatic species living in either the water column or bottom sediment or contamination of drinking water, the investigation of pesticide behavior in a laboratory water–sediment system would give valuable information when their fate in the real environment is predicted. Recently, this experimental approach has been taken for contamination of water bodies by pharmaceuticals (Löffler et al. 2005) and some organic contaminants such as nonylphenol (Lalah et al. 2003).

There are many excellent reviews dealing with pesticide behavior in laboratory and natural water–sediment systems (Bennett 1990; Wolfe et al. 1990; Muir 1991; Groenendijk et al. 1994; Warren et al. 2003). The relative importance of each entry route is determined by balance among physicochemical properties and transformation rates of pesticides, method of their application, soil or sediment properties, climate and season for consideration, and location and dimensions of the water body near the applied fields. First, there are many factors controlling distribution of pesticides between water and sediment phases, as schematically shown in Fig.1a. Natural water contains suspended matter, dissolved organic and inorganic matter, and

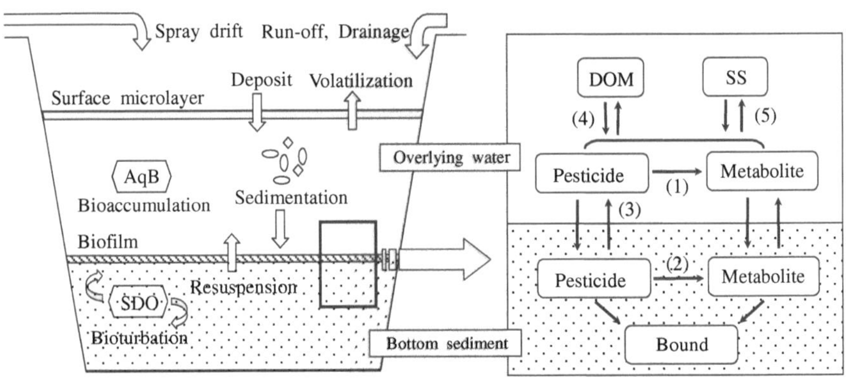

(a) Natural aquatic environment or larger-scale water-sediment system.

(b) Laboratory water-sediment system.

Fig. 1. Transport, distribution and transformation processes of pesticides in water–sediment system. *DOM*, dissolved organic matters; *SS*, suspended solids; *AqB*, aquatic biota such as fish, invertebrates, plankton, and macrophytes; *SDO*, sediment-dwelling organisms; (*1*) hydrolysis, photolysis, redox reactions, and biodegradation; (*2*) (1) except photolysis; (*3*) adsorption, desorption, and diffusion; (*4*) solubilization, complex formation, and catalysis; (*5*) adsorption, desorption, and catalysis.

many kinds of aquatic biota. These fractions interact with a pesticide molecule depending on its physicochemical properties such as hydrophobicity (log P_{ow}) and determine the pesticide distribution. The freely dissolved fraction of pesticide is reduced by association with these substances and/or bioaccumulated. The interstitial porewater in bottom sediment is a different water phase from overlying water and is deeply involved in the partitioning of pesticide to sediment. The sediment is a very complex phase consisting of clay minerals, organic matter, and living microorganisms. Therefore, their adsorptivity of a pesticide molecule, redox potential, and mass of microbes are considered to greatly affect pesticide behavior.

The processes controlling degradation of pesticide in a water–sediment system can be conveniently classified into abiotic and biotic (Wolfe et al. 1990; Waren et al. 2003). Water chemistry such as pH and concentrations of dissolved and suspended matter, as well as temperature, is the most important factor controlling the abiotic processes such as hydrolysis, redox reactions, and photolysis. The presence of two interfaces, air–water and water–sediment, is characteristic to a water–sediment system. The deposition and volatilization of a pesticide proceed via the former interface, where an existence of surface microlayer may alter the photolytic profiles of the pesticide. Redox reactions depend on redox potential in the neighborhood of a water–sediment interface as well as the amount of settled metal (hydr)oxide species by sedimentation. Biodegradation by microbes and algae is also important in considering the fate of pesticides, especially in this interface where biofilm is formed by their growth and where bioturbation by sediment dwellers such as chironomids and oligochaetes may modify their distribution and degradation profiles. Although its definition is somewhat ambiguous, the water–sediment system is simply considered to consist of two heterogeneous phases. The stagnant system where sediment is settled to the bottom would be a model of a larger water body such as ponds and lakes (Fig. 1b), whereas a suspended or mixing situation is considered to be a riverine model with turbulence of sediment. Therefore, the appropriate study design should consider the target natural system to be assessed as well as the processes to be examined (Cripe and Pritchard 1990). To estimate the pesticide behavior in a wide variety of natural aquatic environments, the kinetic and statistical analyses of experimental data being utilized as input data for computer simulation become more important (Groenendijk et al. 1994), together with knowledge about the differences between laboratory water–sediment systems and small to large micro/mesocosms.

On the basis of these considerations, this review covers the dominant processes determining pesticide distribution and degradation in water–sediment systems, study designs, and analytical methods. The existing results on pesticides in various water–sediment systems are reviewed by taking account of differences in pesticide behavior between laboratory and larger-scale systems.

II. Factors Controlling Distribution of Pesticides

A. Water Phase

A natural water body contains many kinds of dissolved and suspended species, such as organic compounds with a low molecular weight, humic substances, metal oxides, and clay particles originating from many kinds of biota, soil, and sediment (Katagi 2002). Therefore, pesticides in the water–sediment system would most likely interact with each fraction. Thurman (1985) extensively reviewed the amount of organic carbon in various natural waters and reported that the concentration of dissolved organic carbon (DOC) varies with type of water from approximately 0.5 ppm for groundwater to more than 30 ppm for colored water from swamps. The concentration of DOC in streams and rivers depends on the size of the water body, climate, season, and vegetation within the basin and is usually 1–4 mg/L in small streams and 2–10 mg/L in larger rivers. Particulate organic carbon (POC) amounts to approximately 10% of DOC. The majority of organic carbon in lakes is in a dissolved form and increases with trophic status. The largest amounts of DOC are usually observed in wetland waters of swamp, marsh, and bog. The typical composition of DOC in freshwater is summarized in Table 1, according to Thurman (1985). Humic substances consisting of fulvic, hydrophilic, and humic acids are the main component, with nonvolatile polymers with a molecular weight of approximately 500–5000 having carboxylic acid, phenolic hydroxyl, carbonyl, and hydroxyl as the major functional groups. Interactions of a pesticide with these fractions are known to cause an increase of its apparent water solubility and sometimes to retard or catalytically accelerate its hydrolysis via adsorption or reaction with functional groups therein (Katagi 2002).

Suspended sediment particles are the main components besides dissolved organic matter (DOM) in natural waters. They usually form a complex matrix and tend to exist as larger flocculated particles. Droppo and Ongley (1992) developed a direct observation method using a digitizer for photographs taken in optical microscopy observation to examine particle-size distribution and geometry of particles for suspended sediments. Particle size of the stable suspended sediments ranged from 3 to 40 μm with a median of 7 μm. Although flocculent particles were only 10%–27% of the total number of particles, they represented 92%–97% of the total volume of suspended sediments. They found, by transmission electron microscopy and scanning confocal laser microscopy on riverine and lacustrine sediment flocs, that many of the macropores are filled with a complex network of cells and fibrils, which would be extracellular polymeric substances secreted by an active microbial community within flocs (Droppo et al. 1997). As compared with overlying water, the components in porewater depend highly on the aerobicity of a sediment phase, and much larger amounts of DOM were reported for anaerobic porewaters (10–390 mg/L) than oxygenated porewaters (4–20 mg/L) (Thurman 1985).

Table 1. Typical composition of dissolved organic carbon (DOC) in freshwater.

Chemical class	Dissolved organic carbon (DOC) (%)	Concentration (μg/L)	Properties and typical chemical species
Humic substances	~80	500–30,000	50% C, 4%–5% H, 35%–40% O, 1%–2% N, <1% P + S
Fulvic acids	~40		Mol wt <2000
Humic acids	~10		Mol wt >2000–5000
Hydrophilic acids	~30		
Carbohydrates	5–12	500	
Polysaccharide	~5		Cellulose, amylose, hyaluronic acid, aliginic acid
Humic saccharide	~2		Saccharides bound to humic substances
Monosaccharide	~1		Glucose, arabinose, xylose, mannose, galactose, fructose
Carboxylic acids	5–8		
Nonvolatile fatty	~4	100–600	C_{14}–C_{20}; palmitic, stearic, myristic, palmitoleic, linoleic
Volatile fatty	~2	40–125	Acetic, formic, propionic, butyric
Hydroxy and keto	~1	10–250	Lactic, pyruvic, glycolic
Dicarboxylic	<1	10–50	Oxalic, malonic, succinic, adipic, fumaric, maleic
Aromatic	<1	5–25	C_6–C_{11} acids; salicylic, gallic, vanillic, coumaric
Phenol	<1	<1	Phenol, cresol, alkylphenols, naphthols
Amino acids	1–3	100–600	Glycine, glutamic acid, alanine, aspartic acid, serine
Hydrocarbons	<1		CH_4, C_{15}–C_{30} alkanes, triterpenoids
Trace species	<1	<1	Simple alcohols, aldehydes and ketones, sterols, mononucleotides, phosphatidyl compounds, sugar phosphates, H_2S, CS_2, $(CH_3)_2S$

From Thurman (1985).

B. Sediment Phase

Sediments are characterized mainly in their physical properties by bulk density, water content, porosity, and particle-size distribution (Percival and Lindsay 1997). Particle-size classification is not strictly fixed, but according to Wentworth's definition sediments are classified into gravel (>2 mm), sand (0.063 μm–2 mm), silt (63 μm–2 μm), and clay (<2 μm), determined by either the classical sieve method or several instrumental analyses (Ongley 1996; Percival and Lindsay 1997). Silt and clay fractions are known to be more reactive because of their higher surface area and adsorption capacity, and thus a strong correlation has been reported between concentration of pesticides and these fractions in many studies. The other important factor in considering the fate of a pesticide in the water–sediment system is redox potential E_h, as defined here (Bohn 1971; Patrick et al. 1996):

$$O_x + mH^+ + ne^- \rightarrow R_{ed}$$

$$E_h\ (mV) = E^0 - (59/n) \log ([R_{ed}] / [O_x]) + 59\ (m/n)\ pH$$

where O_x and R_{ed} are oxidized and reduced species, respectively. E_h is a pH-dependent value and is usually treated as a potential at equilibrium where a mixture of redox couples reacts until the net donation and acceptance of electrons become zero. However, measurement using an electrode is always associated with disturbance of a sediment sample that results in release or adsorption of gases, particularly O_2 and H_2S, as well as reactions at the liquid junction of a reference electrode (Brassard 1997). Therefore, the accurate and stable measurement of E_h seems difficult, and it would be reproducible when a small number of well-poised redox reactions control the equilibrium. It usually takes about 1–2 wk for the sediment E_h to reach an equilibrium under aerobic incubation (Gambrell et al. 1984; OECD 2002). There are no definitive criteria classifying a sediment aerobicity, but Wolfe et al. (1990) have conveniently proposed the following classification from the aspect of the E_h value at pH 7: strongly oxidizing (800–400 mV), moderately oxidizing (400–200 mV), moderately reducing (200 to –50 mV), reducing (–50 to –200 mV) and strongly reducing (–200 to –400 mV).

The characteristics of bottom sediments may depend on their origin and settling places. However, Suedel and Rodgers (1991) have reported insignificant differences in their characteristics, either between freshwater and saltwater sediments or among physiographic provinces, through analysis of sediments collected from various rivers, streams, lakes, bayous, ocean beaches, and bays in the United States. E_h values in freshwater and saltwater sediments were –409 to –23 mV, showing tendency to a reduced condition. The median percent of sand, silt, and clay were 80.6%, 18.3%, and 1.2% for freshwater sediments and 92.4%, 5.9%, and 1.6% for saltwater sediments. These trends are also observed through monitoring of sediments in the U.S. east coast (Hastings et al. 2001), but more clay and silt fractions have been detected for the sediments from San Francisco estuary (SFERM

2004) and Indian wetlands (Handoo 1986). Characteristics of bottom sediments are known to also vary with depth. By using an experimental fluvarium channel constructed with river water and sediment, Allan et al. (2004) have shown the depth dependency of content of organic matter, sediment porosity, and surface area. The sediment porosity showed an exponential decay against depth, whereas both the organic matter content and surface area significantly decreased within the top several millimeters and fluctuated thereafter.

Both electron and optical microscopic observation of settled sediments collected from lacustine and shallow-water coastal environments showed complex structures, including aggregates of fine-grained minerals bridged by organic materials and diatoms with many pores (Hulbert et al. 2002). The main organic component is humic substances, which are derived from primary production within aquatic environments as autochthonous sources and also allochthonously from terrestrial biota via transport of leached and eroded materials. Similar organic components observed in natural water would be present in sediments, as Cranwell (1976) has reported for representative chemical components in lake sediments. In contrast to the water column, greater amounts of carbohydrates are detected because of microbial and algal activity. Amino acids, amino sugars, purines, and pyrimidines originating from microbes and plankton are also detected as minor components. The variety of hydrocarbons, fatty acids, alcohols, and sterols have their origin from terrestrial plants, algae, microbes, and plankton through breakdown processes.

C. Interfaces

Through the air–water interface, molecular oxygen enters the water phase to maintain system aerobicity, whereas volatile compounds including a parent pesticide and its degradates escape from the system together with carbon dioxide or methane, generated mainly via biodegradation. Volatilization from water can be conveniently estimated by using the two-layer model based on the basic physicochemical properties of a pesticide such as water solubility and vapor pressure (Thomas 1990). The higher biological activity in this interface is characteristic of natural aquatic environments, and a surface microlayer consisting of lipids and other organic chemicals gives a unique reaction environment for a pesticide, as described in Section VI.A. However, the much smaller area of overlying water in laboratory systems makes its contribution insignificant. The apparent water–sediment interface can be practically described by the area separating overlying water and bottom sediment, but its definition becomes unclear for natural systems where a large volume of detritus precipitates over the bottom sediment (Meyers and Schelske 2000). Less diffusion of molecular oxygen into a deeper sediment layer and existence of several redox couples such as Fe^{2+}/Fe^{3+} and SO_4^{2-}/HS^- resulted in the gradient of redox potential

(E_h) at this interface, and thus an upper sediment layer a few millimeters thick is aerobic but the deeper layer has reduced conditions (Sethunathan 1973; Wolfe et al. 1990; DeVitre et al. 1994).

The real interface exists between sediment particles and interstitial porewater. Therefore, the partition of a pesticide and its degradates between water and sediment should be considered not only by the apparent adsorption–desorption processes in equilibrium but also by depth-dependent distribution involving their diffusion. Furthermore, this interface gives a productive environment for various types of microbes, algae, plankton, and sediment-dwelling organisms and also becomes a sink for reactive inorganic species such as iron hydroxides by precipitation (DeVitre et al. 1994).

Biofilm. The surface of particles at the top layer of a bottom sediment is partially covered with microbes such as the hyphae of water molds, as demonstrated by electron microscopy (Hulbert et al. 2002). In a shallow water body, sunlight exposure would enhance algal activity at the water–sediment interface, resulting in formation of a biofilm. Woodruff et al. (1999a,b) have examined the effect of diatom biofilm on water chemistry and transport of chemicals through the interface using an experimental fluvarium channel. The pH and dissolved oxygen concentration near the surface increased by algal photosynthesis activity under illumination with concomitant formation of calcite. The E_h value of approximately 100 mV at the surface abruptly decreased in the top 4-mm layer to –150 to –200 mV. The concentration of silicon in overlying water gradually decreased with photosynthetic activity of diatoms because the biofilm acted as a barrier to the diffusion of silicon from sediment porewater to the overlying water. The chemical composition of biofilm has been recently analyzed by pyrolysis gas chromatography-mass spectrometry (GC-MS) and Fourier transform infrared spectroscopy (FTIR) in a diffuse reflectance mode (Gallé et al. 2004). Polysaccharides are the main component, followed by aliphatics, lignin, aromatic phenols, and polyphenolics with fatty acids and lipids as minor ones.

Both adsorption and degradation capacity of biofilm have been investigated using a bioreactor. Headley et al. (1998) have examined the partition of eight pesticides in different chemical classes to biofilm prepared from Elbe River water. The estimated adsorption rate constants at equilibrium reached within several hours were found to weakly correlate with log P_{ow} values of these pesticides. The interaction with extracellular polymeric substances in biofilm that are produced by microbes and mainly consist of hydrophilic polysaccharides may account for this weak dependency. The biodegradation in biofilm prepared from pond and river waters was briefly investigated for the butoxyethyl ester of 2,4-D (29 in Table 6) and *p*-cresol (Kollig et al. 1987). The biodegradability was found to be dependent on a water origin, probably because of the involvement of different microbial consortia. Bacterial biodegradation of methyl parathion (12 in Table 5) was found to be dominant in biofilm prepared from a stream water (Lewis and Holm 1981). Insignificant contribution of algae, fungi, and protozoa was demonstrated by the unchanged reaction rates in darkness and by addition of candicidin, and the rate was found proportional to the cell weight. Lawrence et al. (2001) have recently reported the biodegradation of atrazine (110 in Table 15) and diclofop methyl (methyl 2-[4-(2,4-dichlorophenoxy)phenoxy]-propanoate) in biofilm. Atrazine was degraded via hydrolytic dechlorination followed by successive N-dealkylation finally to cyanuric acid, and the latter herbicide underwent ester hydrolysis, decarboxylation, and ether cleavage.

Adsorption-Desorption. Adsorption forces originate from physical, electrostatic, and chemical interactions and depend on the physicochemical properties of a pesticide and the surface of an adsorbent, for which several adsorption models have been proposed (Voice and Weber 1983; Site 2001). Linear adsorption is the simplest one, expressed by $q_e = K_p C_e$, where q_e, C_e, and K_p are the amount of a pesticide per unit weight of an adsorbent such as sediment, concentration of a pesticide in water, and adsorption coefficient, respectively. This relationship generally occurs at very low concentrations of a solute and low loading of an adsorbent in any model. Although Langmuir, BET, and Gibbs adsorption models have been developed, the empirically derived Freundlich equation defined here has been frequently utilized with satisfactory description of an adsorption isotherm in water–soil and water–sediment systems:

$$ln\ q_e = ln\ K_F + (1/n)\ ln\ C_e$$

where K_F and $1/n$ are the Freundlich adsorption coefficient and a joint measure of both the relative magnitude and diversity of energies associated with a particular adsorption process. As reported for adsorption of nonionic organic molecules to soil (Lotse et al. 1968; Chiou et al. 1979), organic matter content of soil and sediment is of great importance to determine their adsorption capacity and, therefore, the K_F value is usually normalized to K_{oc} against a fraction of organic carbon (foc); $K_{oc} = K_F/\text{foc}$. The K_{oc} value was demonstrated to be almost constant for nonpolar tetrachloromethane and 1,2-dichlorobenzene by using 36 U.S. and Chinese bed sediments (Kile et al. 1995). The hydrophobic interaction with organic matter is dominant for nonpolar pesticides such as pyrethroids, and the adsorption to clay minerals as model suspended matter in water column is less important (Zhou et al. 1995). Therefore, the K_{oc} value becomes one of the most convenient indices to evaluate the distribution of a pesticide in water–sediment systems. These values of pesticides (ARS Pesticide Properties Database) are plotted as shown in Fig. 2, which shows some trends of adsorption to sediment for each chemical class. As it happens, the influence of biota such as algae is necessary to be considered in natural larger-scale water body (Gobas and Maclean 2003).

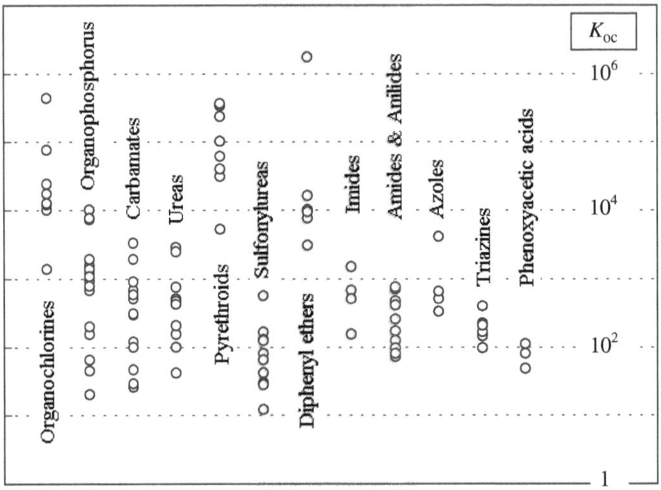

An adsorption coefficient generally decreases with an increasing temperature, but its effect is normally low for hydrophobic and polar organic compounds (Helweg et al. 2003). Sediment consists of various sizes of particles having different surface areas and a content of organic matter. Karickhoff et al. (1979) have shown the bell-shaped dependence of K_{oc} on particle size of sediments for pyrene and methoxychlor (7). The K_{oc} values for the sand fraction were smallest whereas medium to fine silt showed larger values. Although this trend was not so clear, the larger K_{oc} values for finer fractions as well as gravel having a larger content of organic matter were shown for several pesticides including atrazine (110) (Gao et al. 1997, 1998a).

Based on the correlation of K_F (K_{oc}) with organic matter content, several methods using a HPLC retention time, log P_{ow}, and water solubility (Site 2001), as well as a computational method (Meylan et al. 1992), have been developed to conveniently estimate these values. However, the batch equilibrium method prescribed in OECD guideline 106 is usually applied to obtain the K_F or K_{oc} value by measurement of equilibrated concentrations of a pesticide in each phase of a water–sediment (soil) system after shaking. The stagnant condition was found to reduce the adsorption of a pesticide to bottom sediment, possibly because of less contact with adsorption sites (Ying and Williams 2000). The presence of both unsettled sediment particles and macromolecules are considered to increase the apparent concentration of a pesticide in a water phase, which results in underestimation of the coefficient.

Gschwend and Wu (1985) examined this effect on adsorption of polychlorinated biphenyls and showed that the underestimation occurs either at higher concentration of sediment without prewashing or when sediment particles are separated through insufficient centrifugation. In contrast, more partitioning of a pesticide in a sediment phase can be supposed by its association with DOM in the porewater. Gao et al. (1998b) reported this effect for several pesticides by using DOM prepared from a pond porewater. Much higher concentration of cypermethrin [racemic mixtures of (35)] was observed in porewater than that in the overlying water (Maund et al. 2002). However, because the volume of porewater is usually much less than that of the overlying water, this effect would be much less compared with that caused by unsettled particles and macromolecules. Eadie et al. (1990) developed a sequential separation method through glass fiber filters and a Sep-pak column to fractionate dissolved DOM- and sediment-associated pesticides, but there is a concern that the sorption or partition equilibrium may be shifted during filtration. Alternatively, Mayer et al. (2000) successfully applied the solid-phase microextraction (SPME) method to quantify the extremely low concentration of hydrophobic chemicals in sediment porewater by using a glass fiber coated with poly(dimethylsiloxane). Liu et al. (2004) applied the SPME method to quantify the concentrations of pyrethroids in a freely dissolved form in stream water.

Adsorption coefficients of several pesticides increased with incubation period, showing the existence of a slow adsorption process (Bondarenko and Gan 2004). The fraction of the slower adsorption process was estimated to be 0.1–0.9, with the period for equilibration much longer (10–300 d) than that of the faster adsorption (1–7 d) (Pignatello and Xing 1996). Furthermore, the adsorption and desorption processes are ideally reversible, but hysteresis is observed for natural sediments because of either chemical and microbial degradation or specific interactions of a pesticide with organic matter and/or clay. The involvement of degradation was clearly demonstrated in a desorption study for atrazine (110) (Mersie et al. 1998b; Seybold et al. 1999).

Recently, both hysteresis in desorption and nonlinearity in an adsorption isotherm have been explained by heterogeneity of organic matter in soils and sediments (Huang et al. 2003). The organic matter is a mixture of humic substances, kerogen, and black carbon and is not in a homogeneous gel-like phase. Kerogen and black carbon have a three-dimensional structure with aromatic nuclei cross-linked by aliphatic chainlike bridges and backbones of stacked aromatic sheets, respectively, which gives spaces for small hydrophobic organic solutes to be trapped. To explain the adsorption kinetics of polychlorinated benzenes, Wu and Gschwend (1986) have applied the radial diffusive penetration model modified by a retardation factor reflecting microscale partitioning of an adsorbate between intraaggregate pore fluids and solids making up aggregate grains. They could reproduce slower adsorption when a chemical becomes more hydrophobic and the size of adsorbent particles becomes larger. The effect of DOC in porewater on adsorption has also been analyzed theoretically. A very weak correlation between adsorption coefficient and log P_{ow} was found for polychlorinated chemicals, but a higher correlation was obtained when the coefficient was normalized against the effective diffusivity in sediments (Capel and Eisenreich 1989). It is believed that the porewater colloids, associating with a hydrophobic chemical more tightly, can compete with the sediment particles in the adsorption process. The effect of DOC in the overlying water and porewater on adsorption was also demonstrated by Deane et al. (1999).

Diffusion. In a system prepared from sieved pond sediment and water, lindane (1) was distributed mostly at the top of the sediment layer (Rodgers et al. 1983). The dominant distribution of a water-spiked chemical in the sediment surface was also reported for DDT (6) (Gambrell et al. 1981) and atrazine (110) (Mersie et al. 2000). In the case of permethrin (34), radioanalysis of sediment showed that 40%–50% of the applied ^{14}C, mostly consisting of (34), was found in the top 1-cm layer (Sharom and Solomon 1981). Radioactivity in the second lower layer amounted to less than 20% and half of that originated from the corresponding chrysanthemic acid, showing more diffusion of less-hydrophobic chemical in sediment. More detailed observation of pesticide distribution in sediment was conducted for fenthion (16) (O'Neill et al. 1989). After application to the overlying water, (16) was distributed mainly at the sediment surface with its amount decreasing in an exponential-like fashion with sediment depth. In larger-scale systems, a similar trend was also reported. When the top 10-cm layer of the sediment in a paddy model ecosystem was treated with carbofuran (161), its vertical migration to the 10- to 20-cm layer was found to be significant with much less release to the overlying water (Jinhe et al. 1989). The vertical diffusion rate in bottom sediments was estimated to be 0.4–1.1 cm/d for toxaphene (reaction mixture of chlorinated camphenes containing 67%–69% chlorine) applied to the water surface (Veith and Lee 1971). Daniels et al. (2000) reported the vertical movement of nine chemicals including five pesticides in two natural river sediments where the stable polyaromatic hydrocarbons tended to be distributed on the sediment surface.

The diffusion of a pesticide in a water–sediment system has been discussed by using a mathematical model. Van Rees et al. (1991) examined three types of diffusion of ^3H_2O as a model chemical from water to sediment, from sediment to water, and within sediment by using a half-cell method. Although the adsorption process was insignificant, the higher coefficients were estimated for diffusion from water to sediment than the other two cases. Formica et al. (1988) applied Fick's second law to the exponential-like distribution of PCB congeners in aroclor 1242 and found that the effective diffusion coefficient varied with depth of sediment as well as incubation period. Similar mathematical approaches including a diffusion process have been undertaken by many researchers to explain the depth-dependent distribution of a chemical in sediment (Pritchard et al. 1986; Ding and Wu 1993; Koelmans et al. 2000; Allan et al. 2004). The effect of the physicochemical properties of a pesticide on its vertical distribution in sediment has been examined by Daniels et al. (1998) for simazine [*N*,*N*-diethyl analogue of atrazine (110)] and lindane (1). More-polar simazine was considered to be partitioned to sediment in a lesser extent but less adsorptive activity with a higher diffusion than (1) would result in deeper penetration. Valsaraj and Sojitra (1997) demonstrated the importance of colloidal matter for diffusion of a chemical in porewater to explain the depth-dependent distribution of pyrene in a water–sediment system.

III. Processes Controlling Degradation of Pesticides

Hydrolysis is a dominant abiotic process, and sometimes redox or photochemical reactions play an important role in pesticide degradation, whereas microbial degradation via oxidation, reduction, and conjugation mainly proceed as biotic processes. It is usually difficult to differentiate abiotic and biotic processes, but either sterilization of the system or the usage of antibiotics and enzyme-blocking agents together with an appearance of lag time in transformation is helpful to understand which process predominates (Wolfe and Macalady 1992). Typical reactions observed in the water–sediment system are conveniently summarized in Table 2.

A. Abiotic Processes

Hydrolysis. Hydrolysis at low concentrations of pesticide in water mostly obeys first-order kinetics wherein the rate of disappearance is proportional to its concentration, [P], where k_{obs} is the observed hydrolysis rate constant and the half-life ($T_{1/2}$) can be easily calculated (Katagi 2002):

$$\text{Rate} = -d[\text{P}]/dt = k_{obs}\,[\text{P}]$$

$$T_{1/2} = 0.693/k_{obs}$$

Because H^+ or OH^- catalyzes hydrolysis, usually with less importance of general acid or base catalysis, the k_{obs} value is really a pseudo-first-order rate constant and can be expressed as follows:

$$k_{obs} = k_{H}\,[\text{H}^+] + k_{O} + k_{OH}\,[\text{OH}^-]$$

where subscripts H, O, and OH represent the specific acid-catalyzed, neutral, and specific base-catalyzed hydrolysis, respectively. In considering the effect of temperature on hydrolysis, the Arrhenius equation should be taken into account:

$$k = \text{A}\exp(-E_a/\text{RT})$$

where E_a is the activation energy in kcal/mol, R is the gas constant (1.987 cal/K·mol), T is the temperature in Kelvin, and A is the frequency

Table 2. Typical transformation reactions observed in water–sediment systems.

Reaction	Type	Reaction scheme
Oxidation		
Alkyl oxidation	O1	$R(Ar){-}CH_3 \rightarrow R(Ar){-}CH_2OH \rightarrow R(Ar){-}CHO \rightarrow R(Ar)COOH$
β-Oxidation	O2	$R{-}OCH_2COOH \rightarrow R{-}OH$
N-Dealk(ox)ylation	O3	$-N{-}(O)CH_3 \rightarrow -N{-}(O)CH_2OH \rightarrow [-N{-}(O)CHO] \rightarrow -NH$
O-Dealkylation	O4	$-O{-}CH_3 \rightarrow [-O{-}CH_2OH] \rightarrow -OH$
Ring hydroxylation	O5	$Ar{-}H \rightarrow Ar{-}OH$
S-Oxidation	O6	$R{-}S{-}R' \rightarrow R{-}S(O){-}R' \rightarrow R{-}S(O_2){-}R'$
Desulfuration	O7	$P{=}S \rightarrow P{=}O$
Others	O8	Epoxidation, *N*-oxidation, etc.
Reduction		
Dehalogenation	R1	$R(Ar){-}X \rightarrow R(Ar){-}H + X^-$, $CX{-}CX \rightarrow C{=}C + X_2$ [X = halogen]
Dehydrohalogenation	R2	$CH{-}CX \rightarrow C{=}C + HX$ [X = halogen]
Multiple bond reduction	R3	$C{\equiv}C \rightarrow CH{=}CH \rightarrow CH_2CH_2$, $C{\equiv}N \rightarrow CH{=}NH \rightarrow CH_2NH_2$
Nitro reduction	R4	$-NO_2 \rightarrow [-NO] \rightarrow -NHOH \rightarrow -NH_2$
Sulfone, sulfoxide	R5	$R{-}S(O_2){-}R' \rightarrow R{-}S(O){-}R' \rightarrow R{-}S{-}R'$

Table 2. *Continued*

Reaction	Type	Reaction scheme
Hydrolysis		
Hydrolytic dehalogenation	H1	—CHRX → —CHROH + X^-, Ar—X → Ar—OH + X^- [X = halogen]
Nitrile	H2	—C≡N → —C(=O)NH_2 → COOH + NH_3
Carboxyl ester	H3	—C(=O)OR(Ar) → —COOH + R(Ar)—OH
Amide	H4	—C(=O)NR(Ar)— → —COOH + —NHR(Ar)
Carbamate	H5	—NC(=O)O(or S)R(Ar) → —NH + R(Ar)O(or S)H
[Sulfonyl]urea	H6	—[SO_2]NHC(=O)NR(Ar)— → —[SO_2]NH_2 + R(Ar)NH—
Phosphoryl ester	H7	—P(=X)—YR(Ar) → —P(=X)—OH + R(Ar)YH [X = O, S; Y = S, O, NH]
Other bond cleavage	H8	Oxime, sulfate ester, etc.
Conjugation		
Alkylation	C1(X)	—XH → —$XCH_3(C_2H_5)$ [X = O, S, N]
N-Acylation	C2	—NH_2 → —NHCHO, —NHC(=O)CH_3
Others	C3	Sulfonation, via glutathione *S*-transferase, etc.
Miscellaneous		
Isomerization	M1	*cis* ↔ *trans*, *E* ↔ *Z*, epimerization
Photo-induced cleavage	M2	Decarboxylation, bond cleavage, etc.
Ring opening	M3	Triazinyl ring, phenyl ring, imide, etc.
Cyclization	M4	Benzimidazole formation, etc.
Rearrangement	M5	P(=S)OR → P(=O)SR, bridge contraction, etc.

factor. In general, the change of temperature by 1° and 10°C causes a 10% and a factor of 2.5 changes in a rate constant. Because most pesticides are not freely solubilized in water, the application of a pesticide to a water–sediment system is usually conducted by dissolving in a minimum volume of cosolvent such as acetonitrile (OECD 2002). A volume of less than 1% is considered not to affect the rate of abiotic hydrolysis as well as biotic reactions.

Suspended clay and clay minerals are usually coated with organic matter, which reduces the importance of a clay surface in transformation reactions and adsorption becomes more important. The faster dissipation of the ester of silvex [2-(2,4,5-trichlorophenoxy)propanoic acid] was observed in a pond water–sediment system with a higher content of organic matter in sediment (Bailey et al. 1970), indicating the importance of adsorption. Assuming that the adsorption–desorption process is much faster than hydrolysis with no reaction in an adsorbed form under well-mixed conditions, Wolfe et al. (1977) have shown the possible effect of adsorption on the half-life ($T_{1/2}$) of hydrolysis by using a water–sediment ratio (S/W) and adsorption coefficient (K) as follows: $T_{1/2} = 0.693\ /\ [k_{obs}/(K * (S/W) + 1)]$, where k_{obs} is a hydrolysis rate constant in the absence of sediment; this means the possibility of different dissipation rates for pesticides having similar hydrolytic rates in water but showing different adsorption. Furthermore, the possible alteration of a hydrolysis rate in an adsorbed state has been demonstrated for several chemicals including organophosphorus pesticides (Macalady and Wolfe 1984, 1985). Although the system was shaken, hydrolytic degradation of diuron [3-(3,4-dichlorophenyl)-1,1-dimethylurea] was greatly enhanced by the presence of sediment (Salvestrini et al. 2004). The catalysis by buffer and DOM was unlikely to have been caused by their insignificant amounts in the aqueous phase, and the surface charge of sediment was considered to stabilize the zwitterionic step in hydrolysis. Therefore, the effect of adsorption should be considered not only from the adsorption coefficient but also from the hydrolytic mechanism. Incidentally, the association of a pesticide with humic substances causes a decrease of pesticide fraction being dissolved in water; the dissipation in total is considered to be slowed as reported for some organic esters and organophosphorus compounds (Katagi 2002). In contrast, triazine herbicides are known to undergo acid-catalyzed hydrolysis at the triazine ring, leading to formation of the hydroxylated derivative, and hence the humic substances possessing acidic moieties are likely to enhance this reaction.

Redox Reactions. Because humic substances are known to have a radical character, possibly originating from quinone substructures (Katagi 2004), and because there are various transition metals and their (hydr)oxides in a natural water–sediment system, one-electron oxidation or reduction is likely to proceed. The oxidation reaction has scarcely been reported for pesticides in darkness, but some metal–clay complexes are known to oxidize an aromatic ring, leading to formation of products with a higher molecular weight (Mortland and Halloran 1976). In contrast, a sediment phase whose interior is anaerobic even for an aerobic water–sediment system is likely to give a suitable environment where reductive dehalogenation and reduction of nitro, azo, sulfoxide, and sulfone groups easily proceed.

There seem to be reactive and unreactive sites for reduction of a chemical in sediments, as demonstrated in comparative reduction of nitrobenzene derivatives having the substituted alkyl chain with a different length (Zepp and Wolfe 1987). To examine the reductive process of a pesticide more clearly, the reactor system potentiometrically controlling an E_h value has been utilized. DDT (6) showed insignificant degradation in the sediment suspension at the E_h value of 50–450 mV, whereas its rapid degradation was reported in the presence of iron and sulfide in sediment (Gambrell et al. 1981). The similar rapid reduction of the nitro group in methyl parathion (12) was observed at $E_h \leq 50$ mV, showing the microbial process to be unlikely (Gambrell et al. 1984). However, the dissipation of (6) was inhibited by heat or chemical sterilization (Muir and Yarechewski 1984). In six sediment systems with different E_h at pH 6–7, the pseudo-first-order degradation rate constant of (12) in a logarithmic unit was found to negatively correlate with E_h ($r^2 = 0.99$) via formation of the amino derivative; however, the rate was significantly reduced by sterilization (Wolfe et al. 1986). These results show the partial involvement of abiotic processes.

The most abundant natural reductants in anaerobic conditions are Fe (II/III) oxides, Fe(II) sulfide, and hydrogen sulfide. DeVitre et al. (1994) indicated that almost half of iron species in freshwater–sediment systems is associated with clay minerals, but the remaining 10% and 40% are present in the reactive dissolved form and as slowly reactive iron-oxyhydroxide coatings on clay minerals. The redox states of the latter are considered to change in a water–sediment system with oxidation by molecular oxygen or reduction by DOM. The involvement of iron species in reduction has been demonstrated by Oyamada and Kuwatsuka (1979) for CNP (120), whose nitro group was more reduced in flooded soil having a larger amount of iron with a lower E_h value. In general, the reactions with these reactive species are rather slow, but the presence of both goethite (α-FeOOH) and Fe^{2+} at neutral pH efficiently transformed trifluralin (126), mainly via successive reduction of nitro groups before or after *N*-dealkylation (Klupinski and Chin 2003). The reaction rate was found to increase with pH in the range of 6.5–7.8 where Fe^{2+} is more favorably sorbed to goethite at a higher pH, showing that surface-mediated reactions are dominan. Similar degradation pathways were also reported using an E_h-controlled reactor at 50 mV by Willis et al. (1974). Klausen et al. (1995) examined extensively this type of reduction for 10 mono-substituted nitrobenzenes in the presence of seven minerals. They demonstrated the necessity of both Fe^{2+} and iron-containing mineral for the efficient surface-mediated reaction and that the relative reaction rate against nitrobenzene is controlled by a reduction potential of the nitrobenzene derivative. Dunnivant et al. (1992) reported the possible mediation of electrons by humic substances having a different origin in reduction of nitrobenzene derivatives by hydrogen sulfide. The reaction was found to be controlled by the reduction potential using a linear free-energy relationship analysis. Based on the reduction using natural quinones such as juglone and lawsone as model electron mediators, they proposed that hydroquinone moieties in humic substances would play a pivotal role in the mediation of electron transfer reactions involving organic pollutants.

Photolysis. A laboratory water–sediment study is mostly conducted in darkness, but photolysis becomes more important in outdoor studies, especially using a shallow water body with little macrophyte coverage. Even if a main photoproduct is further photodegraded in a laboratory system, the diurnal change of sunlight intensity would increase its persistence in a real aquatic environment (Higashi and Crosby 1987). Therefore, interpretation of laboratory results by taking account of accumulated solar irradiance would become important. In the case of a shallow and clear water body, sediment surface is considered to be also exposed to sunlight. However, it is usually difficult to distinguish reactions occurring on the sediment surface from those on adjacent suspended particles and in overlying water (Crosby 1994).

The extent of photolysis is highly dependent on UV absorption profiles of the pesticide, the surrounding medium, and emission spectrum of a light source. By passing through the atmosphere including ozone, the wavelength of sunlight available becomes >290–295 nm. There are two types of photochemical reactions, known as direct and indirect photolysis (Katagi 2004). Direct photolysis means the photoreaction proceeding by absorbing light energy whereas indirect photolysis is defined as reactions of a ground-state molecule with the other excited molecule or photochemically produced reactive species. The former indirect photolysis is called photosensitization or quenching, and the latter is a photo-induced reaction with reactive oxygen species. The *cis*/*trans* (or *E*/*Z*) geometric isomerization or *R*/*S* optical isomerization and photo-induced homolytic bond cleavage are the most typical reactions in direct photolysis, as reported for fluoxastrobin (84) (Borchers and Stupp 2004) and an experimental fungicide (Reeves 1999). Fipronil (148) underwent photo-induced desulfinylation in an aquatic ecosystem together with redox reactions at the S=O moiety and hydrolysis of the cyano group (Aajoud et al. 2003). In the case of photosensitization, a triplet energy transfer from humic substances is the most important type of energy transfer involved in photolysis of pesticides. The most important reactive oxygen species in the water–sediment system would be singlet oxygen ($^1O_2$) and hydroxyl radical (OH·), which are generated under illumination from humic substances and NO_3^-. In addition, illumination of the system including dissolved iron species may enhance a photo-induced

Fenton reaction in a water column (Katagi 2004). Reactions with these species result in the various types of oxidation of a pesticide. Although cyromazine (112) is known to be resistant to direct photolysis, it rapidly dissipated in the pond system by sunlight exposure (Hein et al. 2003). Furthermore, fluridone (152) was rapidly degraded in an experimental pond with more *o*- or *p*-hydroxylated derivatives than observed in the usual water–sediment study in darkness (Muir and Grift 1982), also indicating involvement of an indirect photolysis process.

Similar to hydrolysis, adsorption of a pesticide to suspended particles would alter its photolytic behavior in a water–sediment system. Plane et al. (1985) showed that approximately 80% of a chemical having a log P_{ow} value of 5 would exist in an adsorbed state, and as a result the photolytic half-life greatly increases if an insignificant photolysis is assumed for this state. Miller and Zepp (1979b) reported the reduced photolysis rate constants of DDE [dehydrochlorinated derivative of DDT (6)] in the presence of suspended sediments with increased formation of dechlorinated derivative, indicating that the adsorbed states are good hydrogen donors. Zepp and Schlotzhauer (1981) kinetically analyzed the dissipation of DDE and concluded that its diffusion to photochemically reactive adsorption sites would become a rate-determining step. The shielding effect by suspended sediments and clays was also reported for methoxychlor (7) by Oliver et al. (1979). Miller and Zepp (1979a) also examined the effect of light scattering in water containing suspended particles by using malachite green leucocyanide as a probe. There are many types of algae in natural aqueous environments such as lakes, and these can act as suspended particles against photolysis of a pesticide. Zepp and Schlotzhauer (1983) showed that photodegradation of polyaromatics, organophosphorus pesticides, and anilines is accelerated by green and blue-green algae and that this effect was minimal for a chemical undergoing rapid direct photolysis.

B. Biotic Processes

Microbial metabolism of a pesticide is also the primary force in its transformation or degradation in a water–sediment system, and bacteria and fungi are the two major groups among microorganisms in pesticide degradation. MacRae (1989) concisely summarized active microbial populations for each chemical class. There are many excellent reviews describing metabolic pathways and mechanisms for each class of pesticide (Sethunathan 1973; Kaufman 1974; Lal 1982; MacRae 1989; Bollag and Liu 1990; Häggblom 1992; Hoagland and Zablotowicz 2000; Zablotowicz et al. 2000), and Alexander (1981) has reported the classification of microbial reaction types. As reported by Paris et al. (1981), microorganisms in natural water can play a role in degrading a pesticide, but a sediment phase, especially in the neighborhood of the sediment–water interface, would be more important when microbial degradation is considered. In some cases pesticides are metabolized as an energy source for microbial growth (biodegradation) and in others transformed without usage as an energy by microorganisms (cometabolism). In the former case, a chemical will be finally mineralized to carbon dioxide and inorganic components, while different microorganisms transform a pesticide molecule in the latter by sequential cometabolic attacks. As listed in Table 2, the major reactions observed in microbial transformation of a pesticide consist of oxidative, reductive, and hydrolytic reactions, and some metabolites are known to be further conjugated.

Oxidation. Various types of oxidation have been reported for pesticides in water–sediment systems (see Table 2). Many kinds of enzymes, such as monooxygenases, dioxygenases, peroxidases, laccases, and mixed-function oxidases, are considered to be involved (Bollag and Liu 1990; Häggblom 1992; Spain 1995). The alkyl carbon is usually oxidized successively via alcohol and aldehyde to carboxylic acid, which is also the mechanism of *N*- and *O*-dealkylation in carbamate, urea, and traizine pesticides (Kaufman 1974; Sørensen et al. 2003). As a special case, β-oxidation results in reduction of an alkyl chain by two carbon atoms. In the case of unsaturated bonds, epoxidation would occur. Oxidation of the sulfur atom produces sulfoxide and sulfone derivatives and the release of sulfur proceeds in oxidation of the P=S moiety of phosphorothioates (Lal 1982). Mono- and dioxygenases can transfer one or two oxygen atoms into a phenyl ring, leading to formation of the corresponding phenol and catechol. The oxygen atom in the hydroxyl group usually originates from molecular oxygen but sometimes from water (Häggblom 1992). The catechol derivatives are further degraded via two possible mechanisms (Fig. 3). Ortho cleavage means the bond scission between the carbon atoms being attached to hydroxyl groups, which

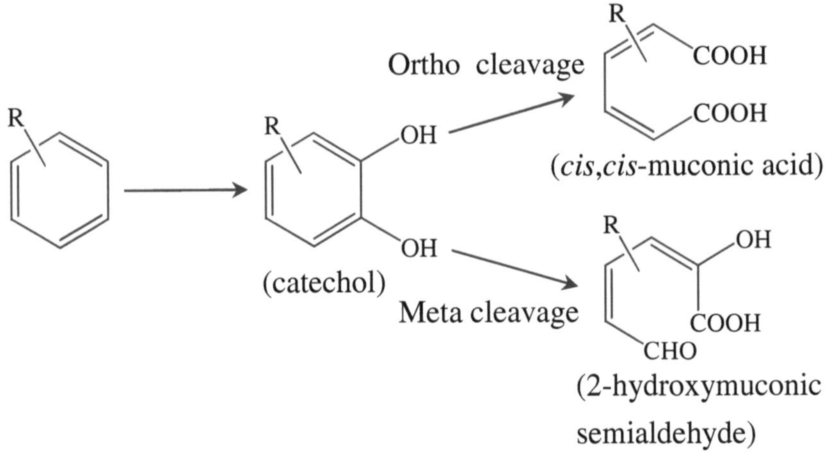

Fig. 3. Oxidative opening of a phenyl ring.

results in formation of *cis*, *cis*-muconic acid, whereas the *meta*-form gives 2-hydroxymuconic semialdehyde (Ou 2000). The microbial degradation of 2,4-D (29) has been extensively investigated, and seven enzymes are involved in its transformation ultimately to succinic acid. The first step is β-oxidation by 2,4-dioxygenase to the corresponding phenol, which is further transformed by hydroxylase, dioxygenase, isomerase, and hydrolase via ortho cleavage of the phenyl ring. The oxidative coupling of the resultant 2,4-dichlorophenol was reported to proceed in the presence of laccases or peroxidases (Bollag and Liu 1990). Similar routes in ring transformation have been reported for many types of nitrogen heterocyclics in their microbial degradation (Kaiser et al. 1996).

Reduction. Reductive dechlorination and dehydrochlorination are well known for DDT (6). Many microorganisms have the ability for dehydrochlorination, which is easily affected by bases and iron (Kaufman 1974). However, the anaerobic reduction of halogenated aromatics is less well understood as compared with the aerobic route. Microbial consortia under methanogenic and sulfate-reducing conditions can reduce halogenated aromatics by dehalogenation as a dominant mechanism (Häggblom 1992). In the case of urea herbicides, stereospecificity of dechlorination has been reported. Furthermore, dehalogenase encoded on plasmids can also liberate chloride ion from chlorinated aromatics under aerobic conditions. Either multiple bonds ($C{\equiv}C$, $C{=}C$, and $C{\equiv}N$) or *S*-oxides in pesticides are known to be reduced under anaerobic conditions in water–sediment systems, but the mechanism is unclear. Biotic reduction of the nitro group is known to proceed either by one- or two-electron mechanisms, the former of which is oxygen sensitive under anaerobic conditions (Spain 1995). Sulfate-reducing bacteria, clostridia, and fungi can reduce the nitro group, and hydrogenase with ferredoxin in *Clostridium kluyveri* is responsible for one-electron reduction of nitroaromatics. The nitroso group formed via these pathways is further subjected to successive reduction to hydroxyamino and amino groups by nitroreductases, which are flavoproteins using NADH or NADPH as an electron source in the aid of FMN or FAD as cofactors (Zablotowicz et al. 2000). Reactions among these chemical species are known to result in formation of azo- and azoxy derivatives. Mono- or dioxygenases in bacteria can metabolize nitroaromatics by hydroxylation at the *ipso* position of the nitro group under aerobic conditions with release of NO_2^-, but *Pseudomonas pseudoalcaligenes* can partially reduce nitrobenzene to hydroxyaminobenzene by nitroreductases even under aerobic conditions, which is further transformed to 2-aminophenol by mutase.

Hydrolysis. Pesticides having amide, carbamate, and urea bonds or carboxyl and phosphoryl ester linkages are subjected to enzymatic hydrolysis with the aid of esterases and amidases. Many microbial hydrolytic enzymes are known to be extracellular and can operate either under aerobic or anaerobic conditions because molecular oxygen is not involved in their reactions (Hoagland and Zablotowicz 2000). Many esterases from microbes have been cloned, and prokaryotic and eukaryotic esterases have the serine motif (Gly-X-Ser-X-Gly) in their active sites where the hydroxyl group of the serine residue can act as a nucleophile against the ester moiety. Chrysanthemic acid and cephalosporin esterases have a slightly different serine motif. Many bacteria-degrading pyrethroids were isolated from sediments, and gamma-proteobacteria were found most dominant (Lee et al. 2004). As reported for microbial degradation of pyrethroids by *Bacillus cereus*, *Pseudomonas fluorescens*, and *Archromobacter* sp., the degradation would proccced via cleavage of an ester linkage (Maloney et al. 1988). Arylacylamidases widely detected in bacteria and fungi can cleave the amide linkage together with some carbamates and urea at an optimal pH range of 7–8, and those from bacteria have a hydrophobic conserved motif [Gly-Gly-Ser-Ser (amidase signature)]. Carbamate hydrolases in bacteria were reported to show specificity for phenyl and methyl carbamates. Partially purified enzymes from *Achromobacter* sp. degrading carbofuran (161) could also hydrolyze other *N*-methyl carbamates (Karns et al. 1987). Some bacterial strains have been shown to directly hydrolyze a urea linkage with some specificity (Sørensen et al. 2003). In the case of organophosphorus pesticides, many microorganisms are known to show hydrolytic acitivity. Parathion hydrolases and phosphotriesterases are known to be involved in enzymatic hydrolysis. The former enzymes isolated from a mixed microbial culture adapted for growth on parathion (11) can hydrolyze other organophosphorus pesticides (Lal 1982; Karns et al. 1987). Although chemical hydrolysis partly proceeded during decrease of pH, sulfonylurea herbicides were found to be microbially degraded by *Pseudomonas fluorescens* via breakdown of a sulfonylurea linkage (Zanardini et al. 2002; Boschin et al. 2003). In the case of atrazine (110), the gene relating to each degradation path has been revealed and is known to be plasmid borne (Radosevich and Tuovinen 2004). The gene controlling the primary hydrolytic dechlorination (atzA) is widespread in the environment, and opening the triazine ring of cyanuric acid is catalyzed by amidohydrolase encoded by the gene atzD.

Conjugation. Pesticides and their metabolites are linked together with other endogeneous substrate(s), leading to formation of methylated and acetylated derivatives, and sometimes reactions with glycosides or amino acids are known. Methylation of (C=O)OH, SH and NH_2 groups, *N*-acetylation, and *N*-formylation are the most frequently observed reactions, but the relating enzymes seem to be nonspecific. The unique conjugation reaction mediated by glutathione-*S*-transferase is known for acetanilide herbicides (Stamper and Tuovinen 1998). The primary glutathione conjugate formed via reaction at the 2-chloroethyl group of acetanilides such as alachlor (54) and metolachlor (56) is stepwise transformed by carboxypeptidases, *gamma*-glutamyl transpeptidases, and cysteine *beta*-lyases, and finally oxidized to the corresponding ethanesulfonic and oxalic acid derivatives.

IV. Experimental Design and Kinetic Analysis

A. Experimental Design

A simple laboratory water–sediment system, not intentionally introducing aquatic biota such as invertebrates, algae, macrophytes, and fish, is considered here. Many researchers have developed various types of apparatus to examine the behavior of a pesticide, but the system basically consists of an incubation vessel in darkness containing water and sediment equipped with volatile traps (Guth 1981; Scholz et al. 1988; Cripe and Pritchard 1990). Although there is some variation in system size from a glass vial (Zhong et al. 1998) to a 1- to 2-L flask (Guth 1981; Scholz et al. 1988), the apparatus shown in Fig. 4 with a volume of approximately 200–500 mL is typically utilized for metabolic study of a pesticide. In general, appropriate traps including organic solvent, alkaline solution, or resin follow the incubation vessel to collect volatiles by scrubbing with CO_2-free air (Katagi 2004). An appropriate surface area, water–sediment ratio, and depth of each phase should be adjusted to obtain realistic fate profiles of a pesticide under the conditions that undesirable stratification of a pesticide concentration in a water phase is to avoided after application (Mersie et al. 2000) without disturbance of the bottom sediment. Scholz et al. (1988) examined the direct mixing of a water phase by a freely suspended stirrer and moderate rota-

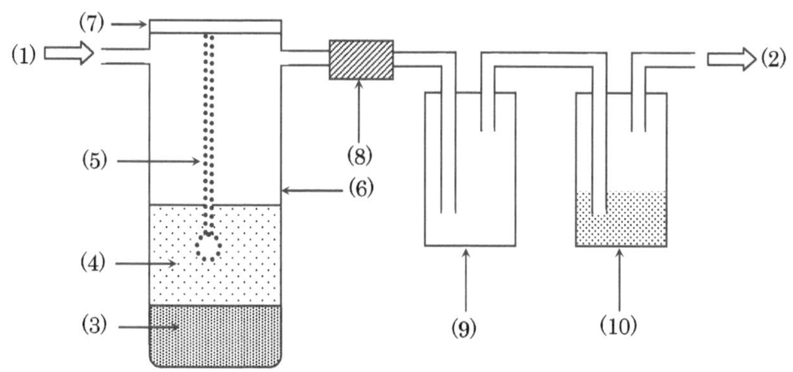

Fig. 4. Schematic diagram of typical continuous-flow water–sediment apparatus: (*1*) humidified CO_2-free air (in the case of anaerobic study, for example, N_2); (*2*) suction by pump; (*3*) sediment; (*4*) water; (*5*) stirrer; (*6*) glass vessel with or without rotatory shaker under the thermostated conditions; (*7*) glass cap (quartz glass when illuminated study is conducted); (*8*) volatile trap such as polyurethane foam; (*9*) vacant trap (if further trap is necessary, for example, filled with ethylene glycol); (*10*) alkaline trap.

tion of an incubation vessel by a mechanical shaker to determine if differences in distribution and mineralization of a pesticide showed more than 80% of the parallel results for six pesticides varying by less than 5% between the two methods.

On the basis of these considerations, several experimental conditions have been proposed by the several guidelines for pesticide registration (Adriaanse et al. 2002), summarized in Table 3 from the point of view of the factors controlling distribution and degradation of a pesticide. Concerning water–sediment ratio, van der Kolk and Crum (1993) reported the possible difference of degradation kinetics for chlorpyrifos (155) by varying water depth in a static water–sediment system. Kodaka et al. (2002) briefly examined the effect of an aeration method on distribution and degradation of fenitrothion (13). Air scrubbing of the upper layer of the water column resulted in slightly less formation of the reduced product, but with no differences in products, and caused only insignificant differences in degradation rate compared to air being passed over the water column.

Concerning the water phase, either continuous-flow or static conditions can be used. The former condition may focus on realistic dissipation of the contaminant from a moving water column such as a stream or ditch and may be used to investigate transport, distribution, and biodegradation of pesticides (Rodgers et al. 1983; Pritchard et al. 1979, 1986). The latter would be representative of a stagnant aquatic system such as pond or lake and also convenient for obtaining information on material balance together with its degradation pathway.

When a sediment phase is considered, the experimental system can be classified into two types, suspension and stagnant water–sediment. In the case of suspension, the sediment–water ratio is rather low (<1 g/L) and the potential biodegradability as well as an abiotic degradation process for a pesticide can be conveniently assessed (Walker 1984; Walker et al. 1988). The stagnant system is usually undertaken to investigate environmental behavior of a pesticide and the intact undisturbed core (sometimes referred to as an eco-core) of sediment with its overlying water has been evaluated. More contaminant was degraded with mineralization, and its partitioning from water to sediment was slightly facilitated by bioturbation with sediment-dwelling organisms (Pritchard et al. 1979; Van Veld and Spain 1983; Houx and Dekker 1987). However, from the point of view of standardization, a more homogeneous reconstitution, achieved by settling of the water and sediment system after passing them through appropriate sieves, is the more popular method for a laboratory study.

To examine material balance and product distribution in detail, a radiolabeled pesticide in a minimum volume of organic solvent is usually applied to the system. However, pesticides are usually used as a formulation to realize better biological efficacy with a well-dispersed aqueous suspension. Because the apparent water solubility of a pesticide is known to increase in formulation and either hydrolytic or photolytic degradation profiles are

Table 3. Comparison of regulatory guidelines on design of laboratory water–sediment studies.

Guideline	OECD308	USEPA Subdivision N 162-4	SETAC, Germany BBA IV-5-1	The Netherlands Part G, 2.1.2	Canada T-1-255	Japan, 12 Nousan 8147, 2-5-1
Test duration	Max. 100 d[e]	Max. 30 d	100 d	Max. 3 months	Max. 1 yr	Max. 6 mo
Number of system	≥2	1	≥2	≥2	1	1
Test matrix[a]	Natural source[f]	Representative area	Natural source	Natural ditch	Natural water[h]	Fresh paddy soil
Sediment–water ratio (layer thickness)	1:3–1:4 (dry wt) Sediment, 2.5 ± 0.5 cm	Unspecified	1:4–1:10 (dry wt) water, 6 cm (BBA)	10% sediment, ≥2 cm thickness	Unspecified	Soil >5 cm Water >1 cm
Temperature	Approx. 20° ± 2°C	(18–30)° ± 1°C	20° ± 2°C	Approx. 20°C	3–8 and (20–30)° ± 2°C	25° ± 2°C
Aerobic conditions	Or stir water[g]	Unspecified	Or stir water[g]	Unspecified	Or static[g]	Unspecified[j]
Acclimation	1–2 wk (<4 wk)	Unspecified	Unspecified	6–8 wk	Unspecified	≥2 wk
Sediment E_h[b]	–80 to –190 mV	Unspecified	Unspecified	Unspecified	Unspecified	≤200 mV
Application rate[c]	Max. (1 m)	Max.	Max. (BBA, 30 cm)	Expected field conc.	Max.[i]	Max.[k]
Cosolvent	<1%	Unspecified	Max. 0.1%	Minimum	Min.	Minimum
Sampling[d]	≥6	Adequate number	6–8	Unspecified	16	≥6
Light regimen	Dark	Dark	Dark	Light (8–14 h)/dark	Light (16 h)/dark(8 h)	Dark
Reference	OECD (2002)	USEPA (1982)	SETAC (1995) BBA (1990)	Dutch Guideline (1995)	Agriculture Canada (1987)	JMAFF (2000)

[a]Microbial activity should be basically checked.
[b]Redox potential.
[c]Assumed depth of water phase in parentheses.
[d]Acceptable material balance is basically 90%–110% of the applied dose. Metabolites in >10% yield should usually be identified.
[e]Until degradation pathway and distribution pattern are established or when 90% active ingredient has been removed.
[f]Texture (organic C, clay + silt%): fine, 2.5%–7.5%, >50% and coarse, 0.5%–2.5%, <50%.
[g]Shaking flask, gentle bubbling of air, or passing air over water surface.
[h]Water–sediment system when sediment is a major sink.
[i]Expected concentration at runoff or spray drift.
[j]After application of active ingredient, the system is fully mixed by shaking.
[k]Based on a uniform distribution to 10-cm soil.

sometimes affected by components such as a surfactant in formulation (Katagi 2002, 2004), their effects on the behavior of a pesticide in a water–sediment system should be investigated, but relevant information is scarce. Bromilow et al. (2003) examined the distribution of eight pesticides having different hydrophobicity ($\log P_{ow} = 1.6–6.1$) between water and sediment phases by applying each active ingredient or the corresponding commercial formulation to the water. There were no significant differences in distribution caused by the formulation used, and hydrophobicity was found to be a key factor in controlling distribution. The effect of different formulations on degradation of fenitrothion (13) was briefly examined in a lake water–sediment system, and metabolism in aquatic plants was found to be affected (Krieger et al. 1989). Incidentally, a pesticide and its metabolites are usually extracted from water and sediment by either partition with organic solvents in the presence or absence of acid or solid-phase extraction and subjected to radio- and various instrumental analyses to quantify their amounts as well as their chemical identification. The bound residues are fractionated by alkaline extraction. Klaus et al. (1998) developed a stepwise extraction procedure of sediment by changing the extracting ability of solvent from water to trimethylsilyl chloride in dimethylformamide (DMF) under reflux and examined possible association mechanisms between a pesticide and sediment such as hydrogen bonding in the adsorption of anilazine (113) to sediment.

The sediment is sometimes sterilized to investigate the microbial contribution. Tuominen et al. (1994) examined the effectiveness of several sterilization methods to control microbial activity. Neither autoclaving of sediment at 120°C for 20min nor gamma-radiation at 25–50kGy could maintain sterility for up to 3wk because bacterial spores can survive and grow by uptake of soluble phosphorus and organic carbon released by these treatments. Addition of 0.02%–0.04% acid-free formaldehyde was found most effective to maintain sterility together with an unchanged sediment pH. Sediment may be stored for an experiment, but fresh sediment should be used because slow degradation of organic carbon was reported even when sediments were stored at 5°C in darkness for 6mon (Sijm et al. 1997).

There are two possible routes of exposure to a water–sediment system in natural environment, spray-drift and runoff events. A pesticide directly enters the water phase in the former case but through association with sediment in the latter. The water- and sediment spike methods respectively represent each route, and the lipophilicity of a pesticide would play an important role in its distribution and degradation after application to the water–sediment system. Kodaka et al. (2004) have shown that fenitrothion (13) dissipates from the water–sediment system at similar rates irrespective of the spiking method. However, radioactivity recovered from the upper layer of sediment was more than the lower layer in the water-spiked system and an opposite trend by the sediment spike was observed with more radioactivity in porewater. More hydrophobic chlorpyrifos-methyl (156) showed much less distribution in a water phase when applied to sediment (Setzo and Sundaram 1982). Lindane (1) was more mineralized in the sediment-spiked system, possibly by the aid of dehydrogenases (Kalsch et al. 1998). Similar differences dependent on the spike method have been reported for larger-scale systems. Significant binding with a slight desorption to water was observed for carbofuran (161) applied to the sediment of a model paddy ecosystem, while it dissipated slowly in water via hydrolysis and oxidation at the furan ring with a moderate partition to sediment when water spiked (Jinhe et al. 1989). In a pond microcosm study, esfenvalerate (41) was tightly bound to bottom sediment due to its higher lipophilicity and insignificantly released to the water phase when spiked to the sediment, whereas the significant fraction remained in the water phase and was subjected to abiotic degradation in the one water spiked (Samsøe-Petersen et al. 2001).

B. Kinetic Analysis

The behavior of a pesticide and its metabolites in a laboratory water–sediment system is a key issue, not only to know their persistency in natural aquatic environments but also to assess their aquatic ecotoxicology (EC 2002). The wide variation of natural aquatic environments makes it necessary to simulate their behavior in several representative situations, and for this purpose their degradation rates should be estimated as precisely as possible. The kinetic analysis of experimental data is sometimes difficult because of complexity of the system where abiotic and biotic degradation proceeds, as with volatilization from water and adsorption to and desorption from sediment and diffusion via porewater.

To analyze the dissipation of a parent pesticide, the simple first-order (SFO) equation (Eq. 1) should be first applied, assuming the homogeneity of a test system in a steady state. However, several factors originating from adsorption to sediment, time lag in microbial degradation, and heterogeneity of sediment usually result in a significant deviation from SFO. To estimate the apparent dissipation kinetics more appropriately, the several equations following have been proposed by many researchers (Dyson et al. 1999; Wolt et al. 2001; Tarr et al. 2002; Beulke et al. 2004).

$$C = C_0 * \exp(-a * t) \quad (1)$$

$$C = C_0 * [b * \exp(-a_1 * t) + (1 - b) * \exp(-a_2 * t)] \quad (2)$$

$$C = C_0 * \exp(-a_1 * t)\ [t < t_c],\ C = C_0 * \exp[-a_1 * t_c - a_2 * (t - t_c)]\ [t \geq t_c] \quad (3)$$

$$C = C_0 * (1 + b * t)^{-a} \quad (4)$$

where C is concentration at time t, C_0 is initial concentration; a, a_1, a_2, and b are constants, and t and t_c are time.

Equation 2 (biexponential model) and Eq. 3 (hockey-stick model) describe the nonconstant decay. The first-order multicomponent (FOMC) model described by Eq. 4 assumes the spatial variability in a dissipation rate. Beulke and Brown (2001) have shown a problem situation in kinetic analysis even when SFO is applied. Direct fitting of experimental data (C) to Eq. 1 with regression in a logarithmic scale sometimes shows significantly different results in a rate constant from those in a nonlogarithmic scale. Furthermore, it also results in different rates if the initial concentration (C_0) is fixed in regression.

Concerning formation and decline of metabolites, a similar approach can be undertaken (Tarr et al. 2002; Beigel et al. 2004). When a pesticide is degraded to two metabolites, the profiles of each metabolite can be described by a similar biexponential formula to Eq. 2, assuming the formation fraction of each metabolite (Beigel et al. 2004). Sometimes the growth of microbes involved in degradation of a pesticide is apparently taken into account in kinetic analysis. Kollig et al. (1987) have introduced the formulae $B(t) = \mathrm{K}/[1 + C*\exp(p*t)]$ (C and p, constants) to describe a relevant microbial concentration $B(t)$ and analyzed the biotransformation of 2,4-D (29) ester and p-cresol as a function of $B(t)$ and concentration of the corresponding chemical. Alternatively, Paris et al. (1981) utilized the modified Monod expression, rate = $(\mu_m / \mathrm{Y})\ [\mathrm{S}]\ [\mathrm{B}] / (\mathrm{K_s} + [\mathrm{S}])$ (μ_m, maximal growth rate; Y and $\mathrm{K_s}$, constant; [S], concentration of a chemical; [B], concentration of bacteria). In any case, the estimation should be carefully conducted by choosing an appropriate equation with a plausibility check (Erzgräber et al. 2002; Boesten et al. 2003).

The model to be used should scientifically reflect the observed degradation pathway of a pesticide by focusing on significant paths while avoiding overparametrization. The fitting of a selected equation should be statistically checked by using F and χ^2 tests as well as an ordinal coefficient of correlation, and a visual assessment of fitting is also valuable. In contrast to these kinetic approaches, the thermodynamical method using a fugacity concept has been developed separately. The fate of several pesticides including a process of volatilization loss from water to air has been examined by the quantitative water–air–sediment–film interactions (QWASFI) model (Southwood et al. 1999; Chi and Huang 2002). The AQUATOX model is also known to simulate the fate of a pesticide by using fugacity to assess its impact on aquatic organisms (Park et al. 1995).

A pesticide molecule is converted to many metabolites via various pathways and finally transformed to unextractable bound residues and volatiles, including carbon dioxide. Therefore, the simple expression as above cannot always account for all metabolites and fractions detected in a water–sediment study. Thus, the kinetic approach using a compartment model has been frequently utilized, and the formation and degradation rates of each metabolite are estimated by analytically solving relevant differential equations. The example of a compartment model is shown in Fig. 1b. In general, the system is divided into water and sediment phases and the compartments in each phase are appropriately set to represent parent, metabolites, bound residues, and volatiles, as reported by Carlton and Allen (1994). If a sediment suspension system is being considered, either reactive or unreactive sites in solid particles can be considered as compartments to be analyzed (Zepp and Wolfe 1987). van der Kolk and Crum (1993) have applied a simple compartment model to analyze the behavior of chlorpyrifos (155) in experimental ponds. The decline of (155) in water was well reproduced, whereas its concentration in sediment failed to be well estimated, possibly due to the presence of an algal film on the sediment surface.

Similar approaches were undertaken to explain the fate of methyl parathion (12) and fenitrothion (113) in pond microcosm studies, and their dissipation in the systems was well simulated (Marshall and Roberts 1977; Crossland et al. 1986; Krieger et al. 1989). Kodaka et al. (2002, 2003) applied a few types of compartment models to laboratory water–sediment studies of organophosphorus pesticides and succeeded in simulating the behavior of the parent, main metabolites, bound residues, and volatiles. Inao and Kitamura (1999) further applied the concept of a compartment model to a paddy field and succeeded in describing the behavior of carbamate insecticides and triazine herbicides in the field. The concentration of a pesticide or its metabolites in sediment is usually difficult to estimate well because their partition from water to sediment is mainly controlled by diffusion, which cannot be incorporated into a compartment model. Concentrations in sediment are experimentally obtained by analysis of the whole sediment as a homogeneous medium, and thus there would be a gap compared to their real distribution. Diffusion of several stable pesticides in sediment has been analyzed by applying Fick's second law with consideration of sediment porosity and adsorption to sediment particles, and their distribution as dependent on the sediment depth in the water- or sediment-spiked systems was found to be well described (Pritchard et al. 1986; Koelmans et al. 2000). The codiffusion of a pesticide with DOM was also analyzed by a diffusion theory (Ding and Wu 1993). Recently, more detailed mathematical analysis has been conducted to simulate the fate of pesticides in an experimental fluvarium channel as a model of riverine environment (Daniels et al. 1998; Allan et al. 2004).

As previously demonstrated by Armbrust (1999), computer simulation such as EXAMS using both physicochemical properties and degradative profiles of a pesticide was found to be very useful to estimate its concentration in a natural aquatic environment. However, information relating to crops, field, and climate also control the distribution and fate of a pesticide. Thus, a series of simulations for surface water have been developed by the European FOCUS group (Linders et al. 2003), and the concentrations obtained are subjected to ecotoxicological assessment in the European regulation of pesticides (EC 2002). Depending on the complexity of a model and input data, four scenarios (Steps 1–4) are given, and in the higher steps the TOXSWA program (Adriaanse et al. 2002) is utilized. It has been demonstrated that TOXSWA can also be applied to a small laboratory water–sediment systems by adjusting the scale of the target environment. Optimization of degradation and soil adsorption parameters was necessary to give a satisfactory simulation (Adriaanse 2004). Kodaka et al. (2004) applied the TOXSWA program to laboratory water–sediment studies of fenitrothion (13) and diethofencarb (81). Optimized half-lives in either water or sediment phase were in good agreement with those separately reported by their hydrolysis and soil metabolism.

V. Fate of Pesticides in Laboratory Water–Sediments Systems

The degradation profiles of pesticides in each chemical class are summarized in Tables 4–16. The main routes of degradation are tabulated as reaction types defined in Table 2, together with the estimated dissipation half-lives either in the water phase or total system. Brief descriptions of the experimental conditions such as temperature, aerobicity, and duration of incubation are also summarized.

A. Organochlorines

Organochlorine pesticides are generally hydrophobic and have high soil adsorption coefficients (K_{oc}), greater than 1000 (see Fig. 2). Upon introduction to an aqueous phase, they are rapidly partitioned to sediment. The main routes of degradation are reductive dechlorination and dehydrochlorination, followed by oxidative cleavage of a ring especially for aromatic compounds (Häggblom 1992). When pesticides have additionally more reactive functional groups such as nitrile and alkoxy groups, stepwise hydrolysis to the carboxyl group and oxidative *O*-dealkylation proceed (Table 4). Regioselectivity of aromatic dechlorination was observed for picloram (4) (Ramanand et al. 1993), and stepwise dechlorination of pentachlorophenol (5) was demonstrated by GC-MS analysis of soil extracts (Ide et al. 1972). Peijnenburg et al. (1992a,b) examined the reductive transformation of various halogenated aromatic hydrocarbons in anaerobic systems and found that the electronic effect of substituents controlling an electron transfer to halogen atoms and their steric constraint determine the reductive reactivity. In the case of halophenols, the regioselectivity of $o > p > m$ in dehalogenation was observed. Organochlorine compounds having more chlorine atoms tend to be more resistant to degradation (Lee and Ryan 1979). Anaerobicity is considered essential for reductive dehalogenation, which was demonstrated by higher degradation rates of DDT (6) and methoxychlor (7) in the system with a lower redox potential (Gambrell et al. 1981; Muir and Yarechewski 1984). Susarla et al. (1997) have shown insignificant degradation of lindane (1), DDT (6), and hexachlorobenzene in autoclaved sediments, indicating that microbial processes are dominantly involved in these reactions (Kaufman 1974; Häggblom 1992).

Lag time due to microbial adaptation is observed in many cases, and methanogenic enrichment with bromoethane sulfonic acid has enhanced

Table 4. Degradation profiles of organochlorine (halogen) pesticides in laboratory water–sediment systems.

No.	Pesticide/structure	System information[a] DT_{50} (w/s)[b]	Route of degradation[c]	Reference
1	Lindane 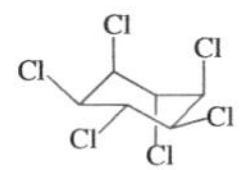	Rivers, 15°C, aerobic, 50 d N.R. / N.R.	M, B	Kalsch et al. (1998)
2	Endosulfan 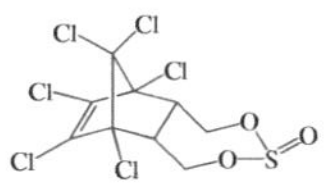	Estuary, 20°C, aerobic, 20 d N.R. / 8.3–22 d	H8 (S—O)	Cotham and Bidleman (1989)
3	Dicamba 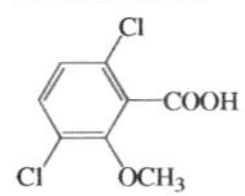	Pond, 10°–20°C, aerobic, 165 d N.R. / 231 d	N.R.	Scifres et al. (1973)
4	Picloram 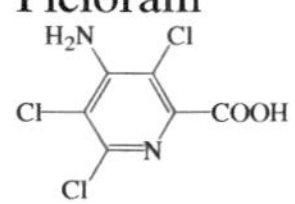	US pond, r.t., anaerobic, 200 d N.R. / 44 d	R1 (p-position to COOH)	Ramanand et al. (1993)
5	Pentachlorophenol 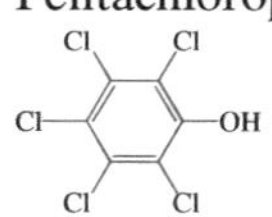	Two paddy soils, 28°C, aerobic, 4 wk N.R. / N.R.	R1	Ide et al. (1972)
6	DDT 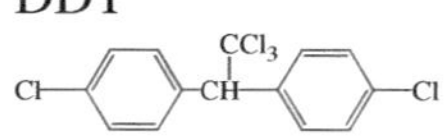	Lake and pond, 22.5°C, aerobic, 448 d N.R. / 123–131 d	R1, R2	Muir and Yarechewski (1984)

Table 4. *Continued*

No.	Pesticide/structure	System information[a] DT_{50} (w/s)[b]	Route of degradation[c]	Reference
7	Methoxychlor 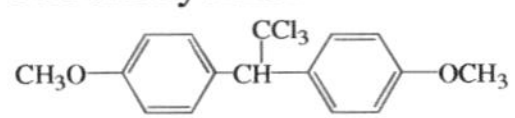	Lake and pond, 22.5°C, aerobic, 448 d N.R. / 116–206 d	O4, R1, M, B	Muir and Yarechewski (1984)
8	Dichlobenil 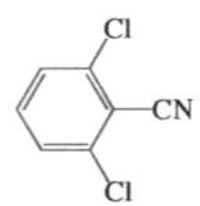	Lake, N.R., aerobic, N.R. N.R. / 2.4 d	H2, V	EPA (1998b)
		Farm pond, 25°C, aerobic, 29 d N.R. / N.R.	H2, V	Miyazaki et al. (1975)
9	Bromoxynil 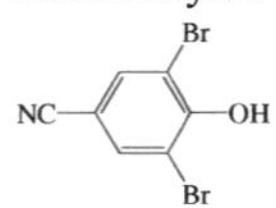	Five systems, 18.5°–25°C, aerobic, 30 d–26 wk <1 hr / <4 hr–3.7 d	H2, H3, R1	EU-Ex (2004c) EPA (1998a)
10	Ioxynil 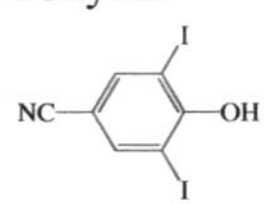	Two systems, 20°C, aerobic, N.R. 3.5–3.8 d / 4.6–4.9 d	H2	EU-Ex (2004e)

B, bound formation; M, mineralization; V, volatilization; r.t., room temperature; N.R., information not reported.
[a]Type of water and sediment (soil), temperature, aerobicity, period of incubation.
[b]Time of 50% disappearance of a pesticide in water layer (w) and water–sediment system (s).
[c]Reaction number listed in Table 2. Terms in parentheses mean a position of reaction or product.

anaerobic degradation of chloroaromatic compounds (Sharak Genthner et al. 1989). Furthermore, a sediment surface was found to be necessary for biodegradation of *p*-chlorophenol (Pritchard et al. 1987) and 2,3,6-trichlorophenylacetic acid (Rosenberg 1984), therefore providing an environment where bacteria can adhere and exhibit their metabolic activity. Dehydrochlorination of (1) was known to be one of the main routes (Kaufman 1974; Wolfe and Macalady 1992). Extensive degradation profiles were not available for (1) in the water–sediment system, but significant mineralization indicated the possible involvement of oxidative ring cleavage (Kalsch et al. 1998). In the case of endosulfan (2), the cleavage of the S—O bond to form the corresponding diol was a primary degradation pathway instead of dechlorination (Cotham and Bidleman 1989). One of the other reactions is *O*-demethylation reported for methoxychlor (7) (Muir and Yarechewski 1984) and epoxidation of the C═C bond of cyclodiene pesticides (Kaufman 1974). Subsequent hydrolysis of nitrile to amide and carboxylic acid is the main degradation route for dichlobenil (8), bromoxynil (9), and ioxynil (10). Although dissipation of (8) from the system was mainly controlled by volatilization, its hydrolysis to amide, partly abiotic, proceeded and the aerobic microbial processes were also demonstrated by *Arthrobacter* cell suspensions (Miyazaki et al. 1975). In contrast, (8) was stable under anaerobic conditions (EPA 1998b). Bromoxynil (9) and ioxynil (10) are usually used as octanoate esters, which are known to be rapidly hydrolyzed in aquatic systems (EPA 1998a; EU-Ex 2004c,e). Bromoxynil (9) underwent stepwise hydrolysis of nitrile, ultimately to carboxylic acid with concomitant mono- and di-debromination, whereas (10) gave only the corresponding amide derivative.

B. Organophosphorus Esters

These pesticides are generally susceptible to abiotic hydrolysis (Katagi 2002) and are also known to undergo biodegradation by various bacteria, algae, fungi, and protozoans (Lal 1982). The dissipation processes are complex depending on their distribution, because of the wide range of K_{oc}, from 10 to 10^4 (see Fig. 2) and degradation profiles (Table 5). Most pesticides dissipate moderately or rapidly from a water–sediment system via various degradation pathways such as hydrolysis of P-Oaryl linkage, *O*-dealkylation, oxidative desulfuration, redox reactions at SOx ($x = 0, 1, 2$) moiety, thiono-thiolo rearrangement, and reduction of a nitro group.

Phosphorothioates. Parathion (11) and its methyl analogue (12) rapidly dissipated via cleavage of the P-Oaryl linkage with microbial reduction of the nitro group in an aerobic nonsterile system (Graetz et al. 1970). Dissipation profiles were not strictly dependent on either origin of sediments or an acclimation period (Houx and Dekker 1987). The undisturbed sediment core gave a higher aerobic degradation rate for (12) than the reconstituted

Table 5. Degradation profiles of organophosphorus pesticides in laboratory water–sediment systems.

No.	Pesticide/Structure	System information[a] DT_{50} (w/s)[b]	Route of degradation[c]	Reference
11	Parathion 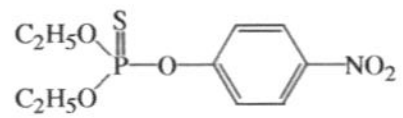	Two lakes, 23°C, aerobic, 92 d 8 d / N.R.	R4, H7	Graetz et al. (1970)
		Ditch and lake eco-cores, 22°C, aerobic, 14 d N.R. / N.R.	R4	Houx and Dekker (1987)
12	Methyl parathion 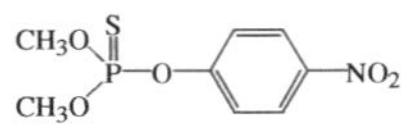	Salt marsh eco-core, 24°C, aerobic, 32 d N.R. / N.R.	R4, H7, B	Pritchard et al. (1979)
		Sandy loam soil, 25°C, aerobic, 30 d N.R. / 4.1 d	H7	EPA (2003)
		Sandy loam soil, 25°C, anaerobic, 1 yr N.R. / 12.2 hr	R4, H7, O7, B	EPA (2003)
		Seven U.S. sediments, 22°C, anaerobic, 6–7 hr N.R. / 3.3–15.6 hr	R4	Wolfe et al. (1986)
13	Fenitrothion 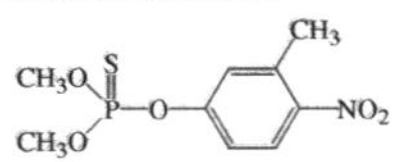	Lake and pond, 20°C, aerobic, 31 d 2.8–4.4 d / 4.9–7.9 d	R4, H7, C2, B	Kodaka et al. (2002)
14	Cyanofos 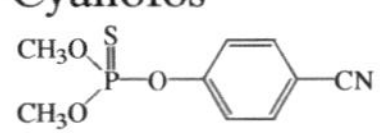	French lake, 20°C, aerobic, 29 d 5.0 d / 8.8 d	O4, O7, H2, H7, M, B	Kodaka et al. (2003)
15	Tolclofos-methyl 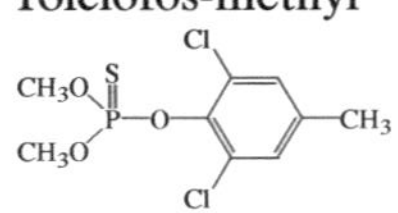	French lake, 20°C, aerobic, 31 d 7.7 d / 24.5 d	O4, O7, H7	Kodaka et al. (2003)
16	Fenthion 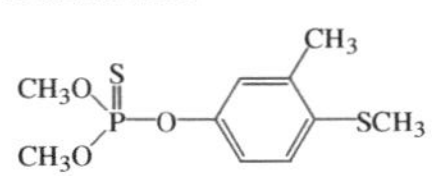	Kansas pond, 22°C, aerobic, 66 d N.R. / N.R.	O4, O6, O7, H7, M, B	Scholz et al. (1988)
		Three salt marshs, 25°C, aerobic, 500 hr N.R. / 3.8–14 d	Polar degradates	Cripe et al. (1989)

Table 5. *Continued*

No.	Pesticide/Structure	System information[a] DT$_{50}$ (w/s)[b]	Route of degradation[c]	Reference
17	Diazinon 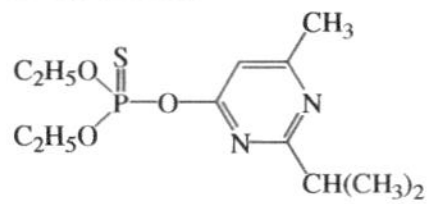	Two soils, 30°C, aerobic, 50 d N.R. / 9.0–9.9 d	H7, M	Sethurathan and MacRae (1969)
18	Tebupirimphos 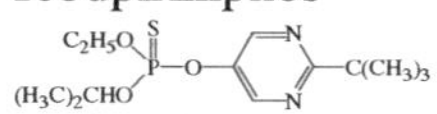	Pond, 22°C, anaerobic, 127 d N.R. / 194 d	O4, H7, M5 (thiono-thiolo)	Halarnkar et al. (1997)
19	Profenofos 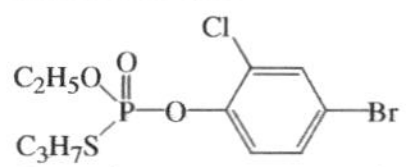	Soil–creek water, N.R., anaerobic, 1 yr N.R. / 3.2 d	H7, formation of cyclohexadienyl sulfate, C1 (O)	EPA (2000b)
20	Oxydemeton-methyl 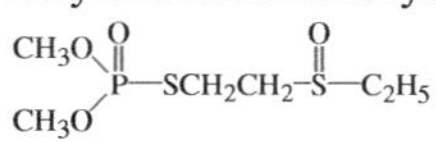	Pond, 25°C, anaerobic, 1 yr N.R. / 3.5 d	O6, R5, H7, C1(S),B	EPA (1999b)
21	Fenitrooxon 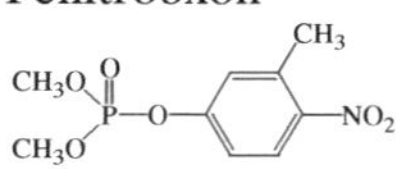	Lake and pond, 20°C, aerobic, 31 d 2.4 d / 2.5 d	H7, B	Kodaka et al. (2002)
22	Naled 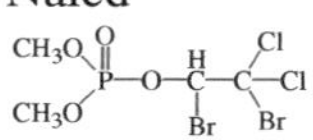	Soil–bog water, 25°C, anaerobic, 190 d N.R. / 0.2–0.5 d	O4, R1, H7, M	EPA (2002b)
23	Malathion 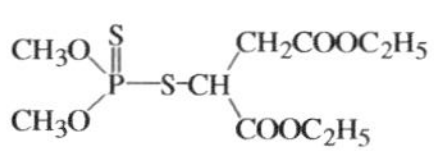	Sandy loam soil, N.R., aerobic, N.R. 1.09 d / N.R.	H3	EPA (2000a)
		Sandy loam soil, N.R., anaerobic, N.R. 2.5 d / N.R.	O4, H3	
		U.S. estuary, 20°C, aerobic, 20 d N.R. / 2 d	N.R.	Cotham and Bidleman (1989)

Table 5. *Continued*

No.	Pesticide/Structure	System information[a] DT_{50} (w/s)[b]	Route of degradation[c]	Reference
24	Acephate 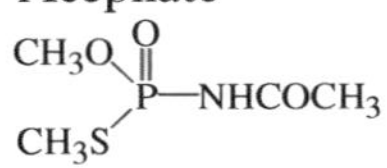	Pond and creek, 9°C, aerobic, N.R. N.R. / N.R.	H4	EPA (2001a)
		Clay sediment, N.R., anaerobic, 20 d N.R. / 6.6 d	O4, H4, H8 (P—N), MF	
		Pond and creek, 9°C, aerobic, 42–50 d N.R. / ~21 d	H4	Setzo et al. (1979)
25	Methamidophos 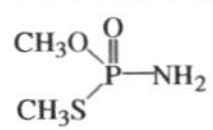	Sandy loam – pond water, N.R., anaerobic, N.R. N.R. / 41 d	V	EPA (2002a)
26	Butamifos 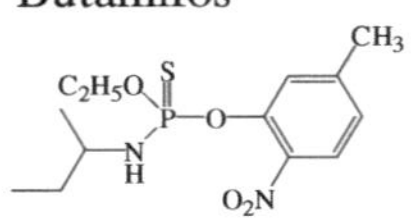	French lake, 20°C, aerobic, 29 d 8.7 d / 16.2 d	Minor unknowns, V, B	Kodaka et al. (2003)
27	Tribufos $(C_3H_7S)_3P{=}O$	Silty clay sediment, 25°C, anaerobic, 1 yr N.R. / 4–6 mon	O6, H7	EPA (2000c)
28	Glyphosate 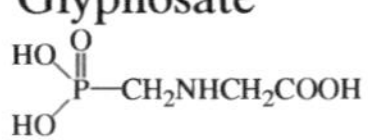	Three systems, 20°C, aerobic, 91–100 d 1–4 d / 27–146 d	O2, M, B	EU-Ex (2002g)

B, bound formation; M, mineralization; V, volatilization; MF, methane formation; N.R., information not reported.
[a]Type of water and sediment (soil), temperature, aerobicity, period of incubation.
[b]Time of 50% disappearance of a pesticide in water layer (w) and water–sediment system (s).
[c]Reaction number listed in Table 2. Terms in parentheses mean a position of reaction or product.

one, whereas sterilization of the system completely inhibited nitro reduction (Pritchard et al. 1979). In anaerobic flooded soil, a few minor metabolites formed via thiono-thiolo rearrangement and oxidative desulfuration were identified (EPA 2003). The nitro reduction was also the major degradation route for fenitrothion (13) in aerobic systems (Kodaka et al. 2002). Further transformation of the amino derivative was *N*-acetylation and *O*-demethylation, and these metabolites tended to exist in either overlying water or porewater (Kodaka et al. 2003). Cyanofos (14) and tolclofos-methyl (15) showed slightly slower dissipation than (13) via ester hydrolysis, *O*-demthylation, and stepwise hydrolysis of nitrile (Kodaka et al. 2003). In the higher sediment concentration, fenthion (16) was more rapidly degraded in the sediment slurry (Cripe et al. 1989), mainly with stepwise *S*-oxidation and *O*-demethylation (Scholz et al. 1988). Microbial processes accounted for degradation of diazinon (17), possibly through an ester cleavage followed by oxidative ring opening (Sethunathan and MacRae 1969). Tebupirimphos (18) mainly underwent ester hydrolysis to release the corresponding phenol together with *O*-deethylation or *O*-deisopropylation in the flooded soil, and additionally thiono-thiolo rearrangement to form S-ethyl isomer was confirmed similarly as (12) (Halarnkar et al. 1997).

Phosphorothiolates and Phosphates. These pesticides are less persistent than phosphorothioates because of their increased hydrolytic instability (Katagi 2002). Anaerobic metabolism of profenofos (19) has shown its rapid degradation via ester hydrolysis to the corresponding phenol, which is further *O*-ethylated and degraded to cyclohexadienyl sulfate (EPA 2000b). Similar rapid dissipation was reported for oxydemeton-methyl (20), mainly via hydrolytic cleavage of the P—S bond to form the corresponding sulfide, which was further *S*-methylated followed by *S*-oxidation (EPA 1999b). Fenitrooxon (21) was aerobically degraded more rapidly than (13) via ester cleavage with more bound residues and mineralization (Kodaka et al. 2002). Much more rapid degradation to dichlorvos [*O*-(2,2-dichlorovinyl) *O*,*O*-dimethyl phosphate] via debromination was reported for naled (22) in an anaerobic system (EPA 2002b).

Phosphorodithioates and Others. Because malathion (23) has a very low K_{oc} value (151–183) (EPA 2000a), its partition to sediment is slow. In a static seawater–sediment system, (23) was degraded at a rate approximately twice that in the sterile system (Cotham and Bidleman 1989). Aerobicity of the system seems not to significantly affect the dissipation profiles, and hydrolysis of the two carboxylic esters accompanied with *O*-demethylation was reported (EPA 2000a). Acephate (24) aerobically dissipated with a slight conversion to methamidophos (25), which was easily lost from the system by vaporization (EPA 2002a), while its slightly rapid degradation was observed under anaerobic conditions with significant formation of CH_4 and CO_2 (Setzo et al. 1979; EPA 2001a). Biodegradation was important for hydrolytically stable butamifos (26) (Kodaka et al. 2003). Different from (11), (12), and (13), the nitro group of (26) could not be reduced in the water–sediment system, which may result from its low reactivity caused by steric hindrance of the bulky phosphoramidothioate moiety at the *o*-position. Tribufos (27) was distributed mainly in the sediment phase due to its large K_{oc} value and biphasically degraded under anaerobic conditions to *n*-butylsulfonic acid (EPA 2000c). Glyphosate (28) is a very hydrophilic pesticide but is strongly adsorbed to sediment and underwent β-oxidation to form aminomethyl phosphonic acid (EU-Ex 2002g).

C. Carboxylic Acids and Esters

The esters of phenoxyalkanoic acids undergo moderate to very slow abiotic hydrolysis (Katagi 2002). The reaction also proceeds biotically but at a higher rate, as demonstrated for 2,4-DB (30) in various natural waters (Paris et al. 1981). The K_{oc} values of the corresponding acids are around 100 (see Fig. 2), indicating more favorable partition in a water phase. In contrast, pyrethroid insecticides are rather hydrolytically stable at neutral pH (Katagi 2002). They exhibit much larger K_{oc} values due to their high hydrophobicity (Laskowski 2002), resulting in their rapid partition to suspended particles and bottom sediment upon introduction to a water column (Zhou et al. 1995; Maund et al. 2002). The strobin analogues exhibit similar hydrolytic profiles to phenoxyalkanoic acid esters. Except for azoxystrobin (42) and kresoxim-methyl (44), this class has a higher K_{oc} value, greater than 1000, and tends to be distributed more in the sediment phase.

Phenoxyalkanoic Acids and Esters. The degradation profiles are listed in Table 6. 2,4-D (29) showed moderate degradation with its predominant distribution in water (EU-Ex 2001a). More 2,4-DB (30) was distributed in water than sediment with a slightly rapid degradation via β-oxidation to (29) (EU-Ex 2002d). Significant mineralization was reported for mecoprop-P (31), possibly through formation of the corresponding phenol (EU-Ex 2003b). Different from these, fluroxypyr (32) has a pyridyl ring, but this change seems not to affect its dissipation profiles and its meptyl ester was rapidly hydrolyzed to (32) followed by dechlorination at the *o*- and *p*-positions (Lehmann et al. 1993; EU-Ex 1999a). Cyhalofop-butyl (33) rapidly underwent ester cleavage with successive hydrolysis of a nitrile group, and the resultant acids and amide, having smaller K_{oc} values, showed rapid to moderate degradation (Jackson and Douglas 1999; EU-Ex 2002c).

Pyrethroids. Degradation in water–sediment systems is moderate, with half-lives of a week to a month (Laskowski 2002). Very rapid dissipation of several pyrethroids in water was mainly caused by their partition to sediment (Agnihotri et al. 1986, 1989; Lutnicka et al. 1999). Hydrolysis is the main degradation pathway for permethrin (34) (Table 7). Sharom and

Table 6. Degradation profiles of phenoxyacetic acids in laboratory water–sediment systems.

No.	Pesticide/structure	System information[a] DT_{50} (w/s)[b]	Route of degradation[c]	Reference
29	2,4-D 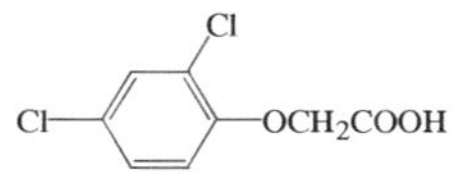	N.R., 20°C, aerobic, N.R. N.R. / 29 d	Minor metabolite	EU-Ex (2001a)
30	2,4-DB 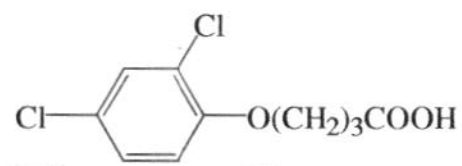	N.R., 20°C, aerobic, N.R. 12.2–12.6 d / 15.8–18.4 d	O2	EU-Ex (2002d)
31	Mecoprop-P 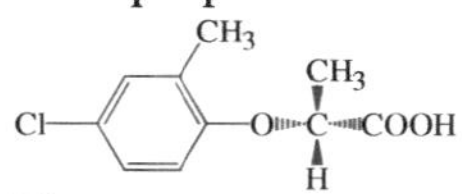	Stream and river, 20°C, aerobic, 100 d 24–49 d / 23–67 d	M	EU-Ex (2003b)
32	Fluroxypyr 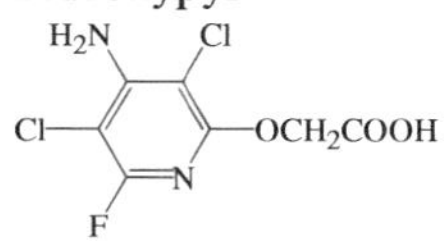	N.R., 25°C, aerobic, N.R. N.R. / 2 d (meptyl), 24 d	H3 (meptyl ester)	EU-Ex (1999a)
		Brewer Lake, 25°C, aerobic, 56 d N.R. / 1 wk	O2, *p*-R1, M, B	Lehmann et al. (1993)
		Brewer Lake, 25°C, anaerobic, 56 d N.R. / 0.5 wk	O2, *o,p*-R1, B	Lehmann et al. (1993)
33	Cyhalofop-butyl 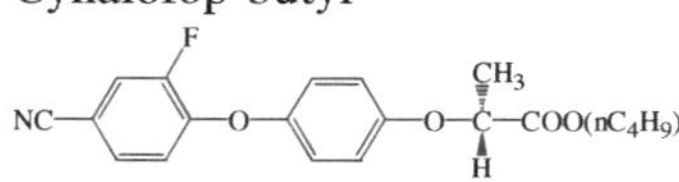	Two systems, 20°C, aerobic, 98 d 1.7–4.5 hr / 1.4–5.3 hr	H2, H3, M	EU-Ex (2002c) Jackson and Douglas (1999)

B, bound formation; M, mineralization; N.R., information not reported.
[a]Type of water and sediment (soil), temperature, aerobicity, period of incubation.
[b]Time of 50% disappearance of a pesticide in water layer (w) and water–sediment system (s).
[c]Reaction number listed in Table 2. Terms in parentheses mean a position of reaction or product.

Table 7. Degradation profiles of pyrethroids in laboratory water–sediment systems.

No.	Pesticide/structure	System information[a] DT_{50} (w/s)[b]	Route of degradation[c]	Reference
34	Permethrin 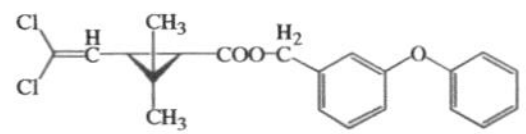	St. George lake, 21°C, aerobic, 12 wk N.R. / N.R.	H3 (*trans* > *cis*)	Sharom and Solomon (1981)
35	*Alpha*-Cypermethrin[d] 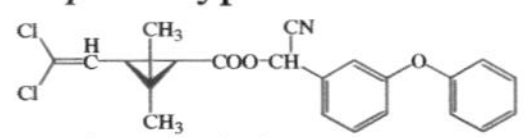	Two systems, 20°C, aerobic, N.R. 0.4–2.1 d / 6.4–35.4 d	H3	EU-Ex (2004a)
36	Deltamethrin 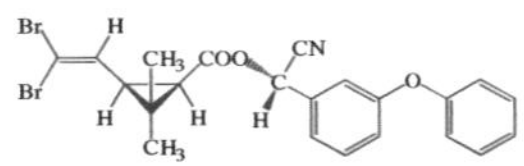	Two systems in duplicate, 20°C, aerobic, N.R. 17 hr / 40–90 d	M1 (benzyl C)	EU-Ex (2002e)
37	Fenpropathrin 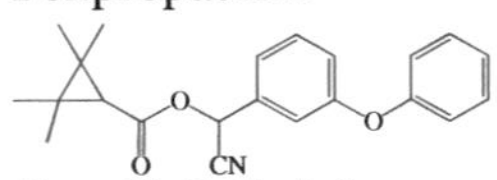	Pond, N.R., aerobic 3–7 d / N.R.	N.R.	Yin et al. (1994)
38	*Beta*-Cyfluthrin[e] 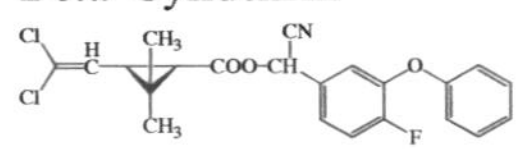	N.R., 20°C, aerobic, 100 d 2.4–3.8 hr / 0.22–3.5 d	H3, M, B	EU-Ex (2002a)

Table 7. *Continued*

No.	Pesticide/structure	System information[a] DT_{50} (w/s)[b]	Route of degradation[c]	Reference
39	*Lambda*-Cyhalothrin[f] 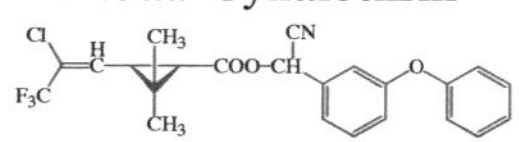	Two systems, 20°C, aerobic, 98 d 5–11 hr / 7–15 d	O5 (*p*-phenoxy), H3, M, B	EU-Ex (2001b)
40	Fenvalerate 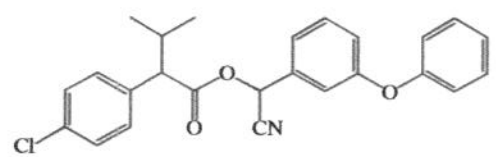	Estuary, 20°C, aerobic, 20 d N.R. / 12 d	N.R.	Cotham and Bidleman (1989)
41	Esfenvalerate 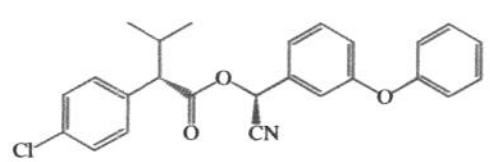	Pond and river (10°, 25°C), aerobic, 100 d N.R. / 54–80 d	H3	EU-Ex (2000b)
		Denmark pond, 15°C, aerobic, 90 d N.R. / 73–350 d	M, B	Samsøe-Petersen et al. (2001)

B, bound formation; M, mineralization; N.R., information not reported.
[a]Type of water and sediment (soil), temperature, aerobicity, period of incubation.
[b]Time of 50% disappearance of a pesticide in water layer (w) and water–sediment system(s).
[c]Reaction number listed in Table 2. Terms in parentheses mean a position of reaction or product.
[d]Mixtures of (1*R*)-*cis*-α*S* and (1*S*)-*cis*-α*R* isomers.
[e]Mixtures of (1*S*)-*cis*-α*R*, (1*S*)-*trans*-α*R*, (1*R*)-*cis*-α*S* and (1*R*)-*trans*-α*S* isomers.
[f]Mixtures of (Z)-(1*R*)-*cis*-α*S* and (Z)-(1*S*)-*cis*-α*R* isomers.

Solomon (1981) examined its metabolism in sterile and nonsterile lake water in the presence and absence of sediment. Hydrolysis in water was accelerated by microbes with a higher rate for the *trans*-isomer than the *cis*-isomer, and the rate of degradation of (34) was slower in the presence of sediment due to adsorption. Dichlorovinyl chrysanthemic acid (DVCA) and 3-phenoxybenzoic acid (PBacid) were formed via hydrolysis of *alpha*-cypermethrin (35) but were not persistent in the system (EU-Ex 2004a). Similar degradation was observed for deltamethrin (36) with the abiotic epimerization at the benzyl carbon to form its α-*R*-isomer (EU-Ex 2002e; Katagi 2002). *Beta*-cyfluthrin (38) underwent more rapid hydrolysis (EU-Ex 2002a). Dissipation of (38) as formulation showed hockey-stick profiles in water, which seem to originate from the rapid adsorption to sediment at an early stage followed by slower abiotic and biotic hydrolysis (Gupta and Gajbhiye 2005). The introduction of a trifluoromethyl group in the side chain of cyclopropyl ring did not significantly affect dissipation profiles of *lambda*-cyhalothrin (39) (EU-Ex 2001b). The reduction of an intermediate product, 3-phenoxybenzaldehyde (PBald), to the alcohol (PBalc) and ring hydroxylation at 4′-position of the 3-phenoxybenzyl moiety were observed as minor pathways (Bewick et al. 1984). Fenvalerate (40) and its (2*S*, α*S*) isomer esfenvalerate (41) were also quickly partitioned to sediment (Cotham and Bidleman 1989; EU-Ex 2000b). Hydrolysis was the main degradation pathway for (41) and the phenoxyphenyl moiety was mineralized more than the 4-chlorophenyl (Samsøe-Petersen et al. 2001). There were no specific changes in dissipation profiles for the other pyrethroids such as fluvalinate [(*RS*)-α-cyano-3-phenoxybenzyl *N*-(2-Chloro-α,α,α-trifluoro-*p*-tolyl)-DL-valinate] and flucythrinate [(*RS*)-α-cyano-3-phenoxybenzyl (*S*)-2-(4-difluoromethoxyphenyl)-3-methylbutyrate] (Othman et al. 1986; Agnihotri and Jain 1987).

Strobin Analogues. Degradation profiles of strobilurins are summarized in Table 8. Hydrolysis was a dominant pathway for azoxystrobin (42), kresoxim-methyl (44), and trifloxystrobin (45), resulting in significant formation of the corresponding acids, which persisted in water (EU-New 1998a,b, 2003l). Although pyraclostrobin (46) has the same methoxy carbamate structure as (44) and (45), demethoxylation was the main degradation route instead of ester hydrolysis (APVMA 2003). Fluoxastrobin (47) has a heterocyclic moiety in the position corresponding to a methoxyacrylate structure (Borchers and Stupp 2004). In contrast to the other strobin analogues, the rapid cleavage of ether bond at the 6-position of the pyrimidinyl ring in (47) was predominant under aerobic conditions, and the opening of 1,2,4-dioxazinyl ring to the corresponding acid or amide was found as a minor pathway (Fig. 5). In contrast, the anaerobic incubation under nitrogen gave this acid via ring opening as a major metabolite with the amide as a minor component.

Table 8. Degradation profiles of strobilurins in laboratory water–sediment systems.

No.	Pesticide/structure	System information[a] DT_{50} (w/s)[b]	Route of degradation[c]	Reference
42	Azoxystrobin 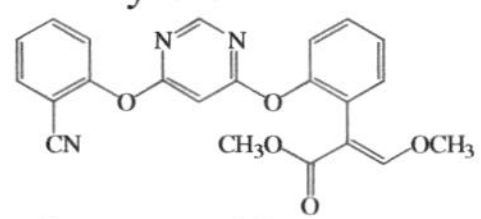	N.R., 20°C, aerobic, 152 d 34–57 d / 170–294 d	H3	EU-New (1998a)
43	Picoxystrobin 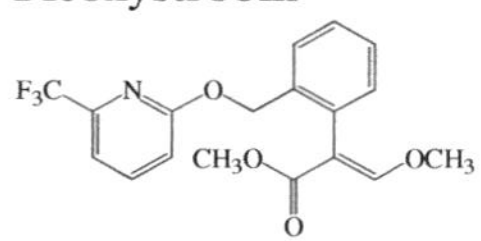	Two systems, 20°C, aerobic, 120 d 7.5–10.5 d / 44.5–67.4 d	Three metabolites	EU-New (2003i)
44	Kresoxim-methyl 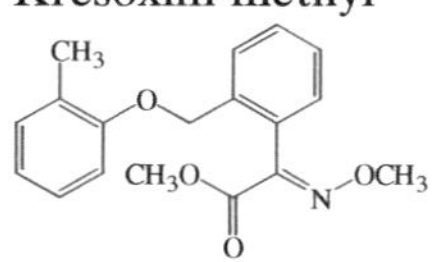	N.R., 20°C, aerobic, 100 d 0.8–0.9 d / 1.2–1.3 d	H3	EU-New (1998b)
45	Trifloxystrobin 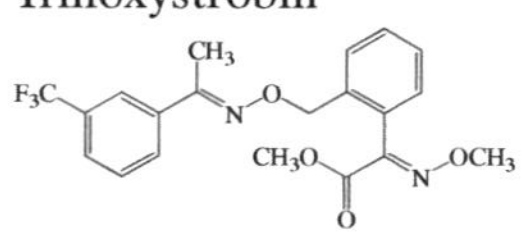	Four replicates, N.R., 20°C, aerobic, N.R. 1.1–1.2 d / 1.2–3.5 d	H3	EU-New (2003l)

Table 8. *Continued*

No.	Pesticide/structure	System information[a] DT_{50} (w/s)[b]	Route of degradation[c]	Reference
46	Pyraclostrobin 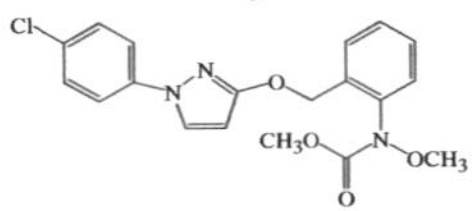	Pond and Rhine River, 20°C, aerobic, 100 d 3 d / 9–33 d	O3, B	APVMA (2003)
47	Fluoxastrobin 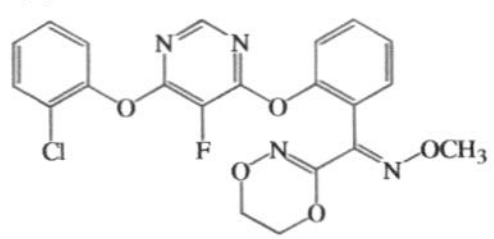	Gravel pit and pond, 20°C, aerobic, 122 d 2–5 d / N.R.	H8 (ether), M3 (dioxazine)	Borchers and Stupp (2004)

B, bound formation; N.R., information not reported.
[a]Type of water and sediment (soil), temperature, aerobicity, period of incubation.
[b]Time of 50% disappearance of a pesticide in water layer (w) and water–sediment system(s).
[c]Reaction number listed in Table 2. Terms in parentheses mean a position of reaction or product.

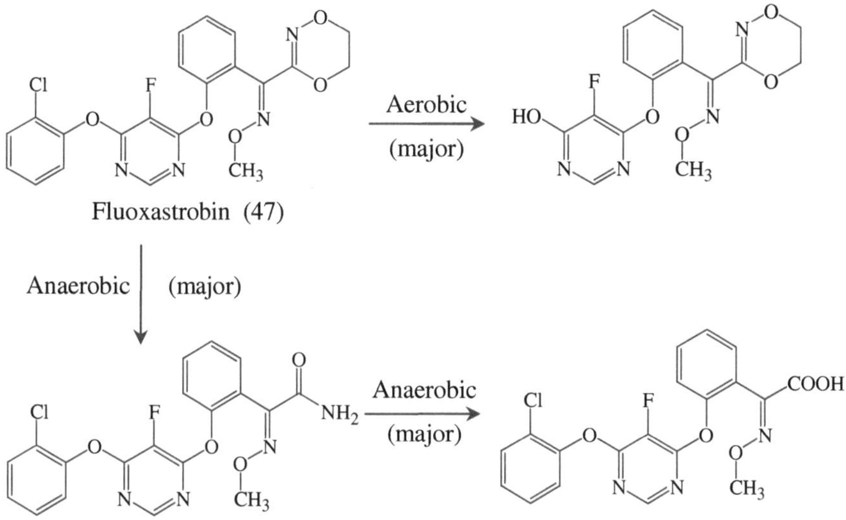

Fig. 5. Degradation pathways of fluoxastrobin (47).

D. Amides, Anilides, and Dicarboximides

At the pH relevant to natural aquatic environments, amides and anilides are considered to be resistant to abiotic hydrolysis whereas dicarboximides are easily hydrolyzed (Katagi 2002). These classes undergo various types of biodegradation including cleavage of the amide linkage, oxidation, reduction, and conjugation. As shown in Fig. 2, these pesticides exhibit low to high mobility with K_{oc} values of 100–1000 due to the existence of a hydrophilic NHC(=O) moiety in their molecules and thus are considered to be gradually partitioned to sediment when resistant to hydrolysis.

Amides and Anilides. The main degradation routes with their dissipation half-lives in the water–sediment system are listed in Table 9. Carpropamide (48) was resistant to hydrolysis; however, either the methyl group in the cyclopropyl moiety or *m*-position of the phenyl ring were hydroxylated in paddy soils (Kurogochi and Köster 1998). Unique cleavage of the Si—C bond was observed for silthiofam (49), but the resultant product seemed to be persistent in the system under aerobic conditions (EU-New 2003k). The reduction of the C≡C bond in propyzamide (50) proceeded to form 3,5-dichloro-*N*-(1,1-dimethylpropenyl)benzamide (EU-Ex 2003f). Fentrazamide (52) was rapidly partitioned to sediment with cleavage of the amide linkage to form cyclohexyl ethyl amine and 2-chlorophenyltetrazolinone (Hellpointner 2001).

Alachlor (53) was degraded via reductive dechlorination together with hydrolysis of the chloromethyl group (Bollag et al. 1986; Chesters et al. 1989). The anaerobic aquatic metabolism of metolachlor (55) gave a

Table 9. Degradation profiles of amide and anilide pesticides in laboratory water–sediment systems.

No.	Pesticide/structure	System information[a] DT_{50} (w/s)[b]	Route of degradation[c]	Reference
48	Carpropamid 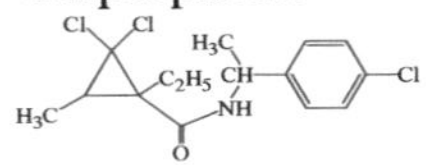	Two paddy soils, N.R., aerobic, N.R. N.R. / 120–220 d	O1 (cyclopropyl), M, B	Kurogochi and Köster (1998)
49	Silthiofam 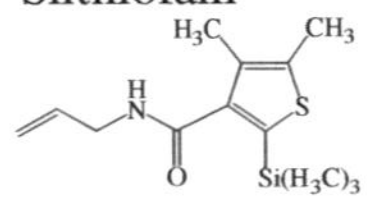	Pond and runoff systems, 20°C, aerobic, 100 d 5–52 d / 147–269 d	H8 (Si-C)	EU-New (2003k)
50	Propyzamide 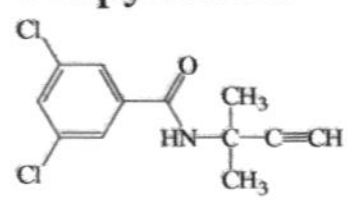	River and pond, 20°C, aerobic, 105 d 18–24 d / 69–118 d	R3	EU-Ex (2003f)
51	Zoxamide 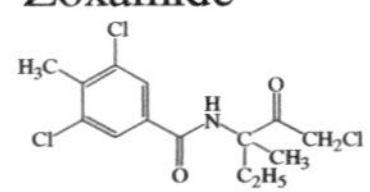	Pond and river, 10°–20°C, aerobic, N.R. N.R. / 6–21 d	Two minor metabolites	PMRA (2001c)
52	Fentrazamide 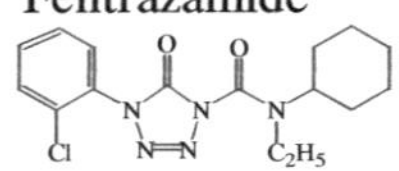	Two paddy soils, 28°C, aerobic, 105 d N.R. / <20 d	H4, C1(N) (1-N of tetrazole), M, B	Hellpointner (2001)
53	Alachlor 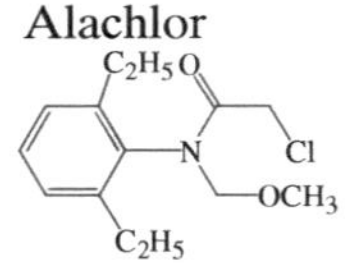	Creek, 28°C, anaerobic, 10 wk N.R. / N.R.	R1	Bollag et al. (1986)
		Flooded soil, N.R., N.R., N.R. N.R. / N.R.	R1, H1, C2	Chesters et al. (1989)

Table 9. *Continued*

No.	Pesticide/structure	System information[a] DT_{50} (w/s)[b]	Route of degradation[c]	Reference
54	Acetochlor 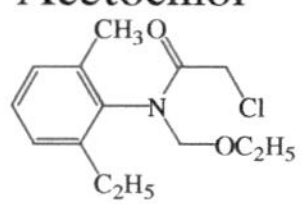	Clay loam soil, 25°C, anaerobic, 371 d N.R. / 16 d	MF, B	Loor-Vela et al. (2003)
55	Metolachlor 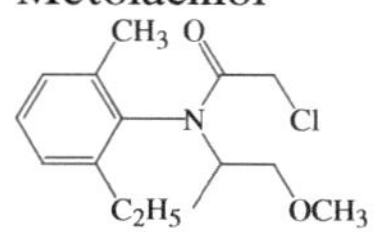	Lake, N.R., anaerobic, 56 d N.R. / N.R.	R1	Chesters et al. (1989)
		Soil–river water, 24°C, anaerobic, 112 d 8 d / N.R.	C3 (glutathione) to sulfonic and oxalic acids	Mersie et al. (2004)
		Iowa pond, 25°C, aerobic, 60 d N.R. / 8 d	M4 (morpholinone), H1, B	Rice et al. (2004)
56	Dimethenamid 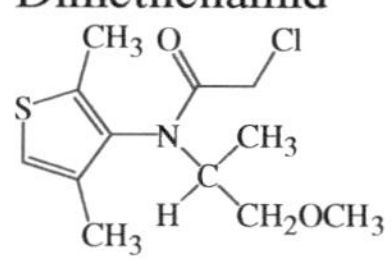	Stream, 25°C, anaerobic, 142 d N.R. / 31 d	B	Crawford et al. (2002)
57	Dimethenamid-P 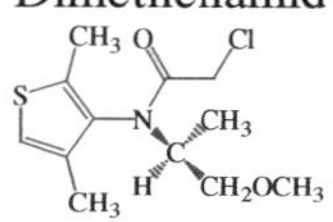	Pond and river, 20°C, aerobic, 105 d 20.3–27.7 d / 23.4–33.4 d	H1, O1	EU-New (2003b)
58	Metalaxyl 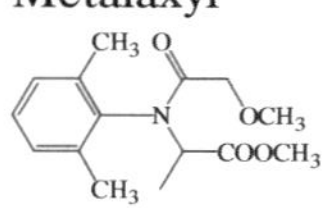	N.R., N.R., aerobic, 30 d 41 d / N.R.	H3	EPA (1994)
		N.R., N.R., anaerobic, 385 d 29.9 d / N.R.	H3, *m*-O5	

Table 9. *Continued*

No.	Pesticide/structure	System information[a] DT_{50} (w/s)[b]	Route of degradation[c]	Reference
59	Metalaxyl-M 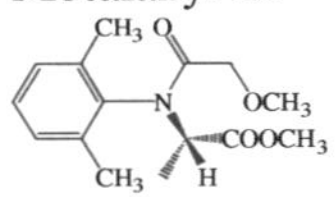	Two systems, 20°C, aerobic, 180 d 47.5 d / 47.5 d	O4, H3	EU-New (2002j)
60	Benalaxyl 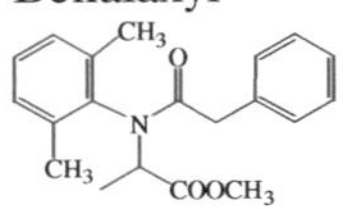	Pond and river, 20°C, aerobic, 100 d 17–58 d / 118–166 d	O1, H3	EU-Ex (2004b)
61	Flufenacet 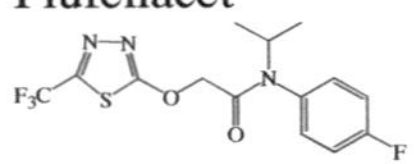	Two systems, 20°C, aerobic, N.R. 46.3–61.7 d / 20–84.6 d	H8 (O-C)	EU-New (2003d)
62	Naproanilide 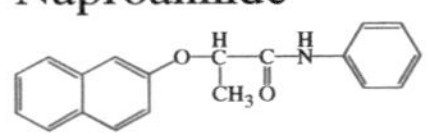	Three Japanese soils, 30°C, aerobic and anaerobic, 5 wk N.R. / <10 d	O5, H4, C1 (O), H8 (ether)	Oyamada et al. (1980)
63	Fenhexamid 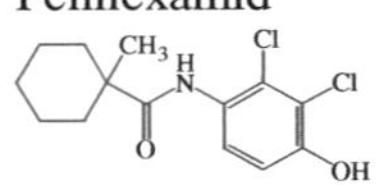	Two systems, N.R., aerobic, 100 d N.R. / 2–15 d	C1 (O), C3 (sulfonation)	APVMA (2001a) Anderson et al. (1999)
64	Niclosamide 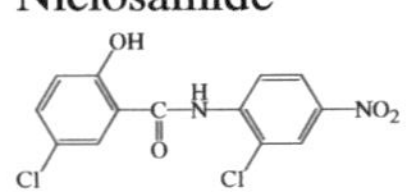	River and pond, 25°C, aerobic and anaerobic, 93 d 0.83–3.1 d / N.R.	R4, H4, M	Muir and Yarechewski (1982a)
		River, 25°C, (1) aerobic, 30 d and (2) anaerobic, 1 yr (1) N.R. / 4.9–5.4 d (2) N.R. / 0.65–2.8 d	 R4, B R4, B	Graebing et al. (2004)
65	Cyclanilide 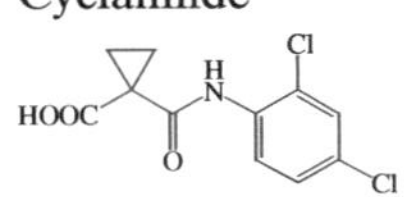	River and pond, 20°C, aerobic, 105 d 17–18 d / 56–63 d	H4, B	EU-New (2001a)

66	Picolinafen 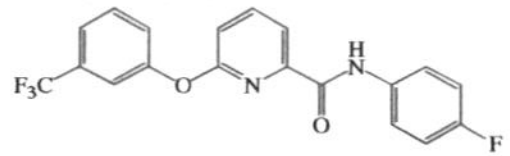	Two systems, 20°C, aerobic, 100 d 1.1–1.4 d / 6.2 d	H4, B	EU-New (2002m)
67	Boscalid 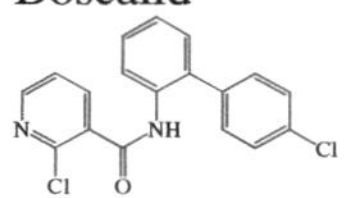	Pond and river, 20°C, aerobic, 100 d 3–9 d / 580–680 d	B	PMRA (2004b)
68	Cyazofamid 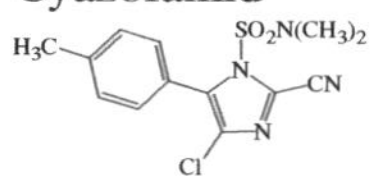	Two systems, 20°C, aerobic, 100 d 8.7–9.9 d / 10.8–16.5 d	H8 (N-S), H2, B	EU-Ex (2002b)
69	Florasulam 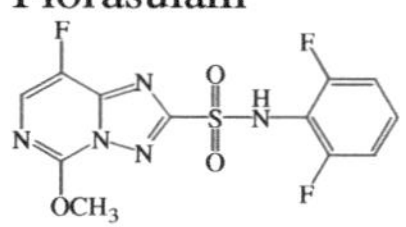	Two systems, 20°C, aerobic, 100 d 8.7–18 d / 8.7–18 d	O4	EU-New (2002e)
70	Cloransulam-methyl 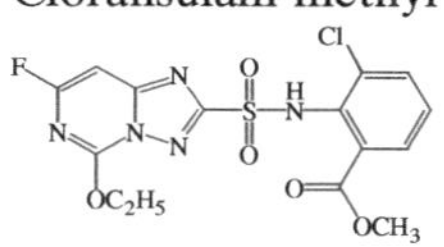	N.R., N.R., aerobic (1) and anaerobic (2), N.R. N.R. / 25.6 d (1), 16 d (2)	H3 (1 and 2), O4 (2)	PMRA (2001b)

B, bound formation; M, mineralization; MF, methane formation; N.R., information not reported.
[a]Type of water and sediment (soil), temperature, aerobicity, period of incubation.
[b]Time of 50% disappearance of a pesticide in water layer (w) and water–sediment system(s).
[c]Reaction number listed in Table 2. Terms in parentheses mean a position of reaction or product.

reduced product similar to (53) (Chesters et al. 1989); the presence of sediment enhanced its aerobic degradation mainly via hydrolytic dechlorination and formation of the morpholinone derivative (Rice et al. 2004), possibly via intramolecular nucleophilic attack of the methoxy oxygen at the carbonyl carbon (Katagi 2002). Mersie et al. (2004) reported significant formation of ethanesulfonic and oxalic acid derivatives where chlorine was substituted with SO_3H and COOH, respectively. The reactions were found to proceed in the presence of glutathione-*S*-transferase and oxidases when the sediment phase still remained aerobic and mildly anaerobic. The corresponding metabolites were not anaerobically formed for dimethenamid (56) (Crawford et al. 2002), while its optical isomer dimethenamid-P (57) showed formation of the corresponding oxanilic acid as a major metabolite in aerobic systems (EU-New 2003b). Metalaxyl (58) primarily underwent ester hydrolysis followed by ring hydroxylation at the *m*-position in the anaerobic condition (EPA 1994), but only ester cleavage was reported for its optical isomer metalaxyl-M (59) in aerobic conditions (EU-New 2002j). The aerobic aquatic metabolism of benalaxyl (60) favored ring opening of the benzyl ring ultimately to give the carboxymethyl moiety (EU-Ex 2004b). Either ester cleavage or oxidation of one of the arylmethyl groups to the carboxyl was found to be a minor degradation pathway.

Flufenacet (61) showed rather slow aerobic degradation by cleavage of the (1,3,4-thiadiazol-2-yl)oxy—acetylmethyl carbon bond (Fig. 6) (EU-New 2003d). Detection of the methylsulfide derivative implied involvement of glutathione-*S*-transferase in its biodegradation. Oyamada et al. (1980) examined the metabolism of naproanilide (62) under flooded conditions by changing a soil–water ratio and the depth of the overlying water. (62) was rapidly degraded irrespective of aerobicity via amide cleavage to 1-(2-naphthoxy)propionic acid, which was rapidly transformed via *O*-methylation and ether cleavage to 2-naphthol followed by further ring hydroxylation. The hydroxyl group of fenhexamid (63) was methylated or sulfonated in the aerobic system (Anderson et al. 1999; APVMA 2001a). Slower anaerobic degradation was reported with dechlorination and hydroxylation at the cyclohexyl ring. Niclosamide (64) has also a free hydroxyl group, but no conjugation has been reported (Muir and

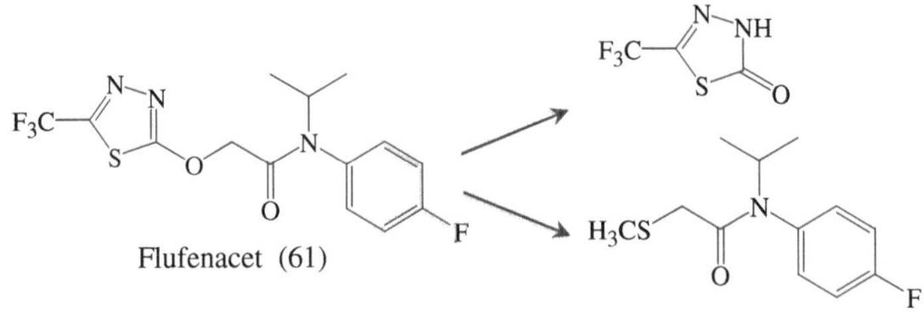

Fig. 6. Aerobic aquatic metabolism of flufenacet (61).

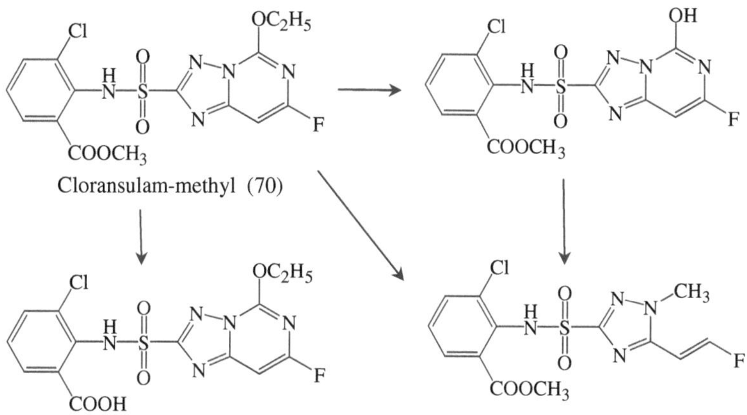

Fig. 7. Anaerobic aquatic metabolism of cloransulam-methyl (70).

Yarechewski 1982a; Graebing et al. 2004). The main degradation route was reduction of the nitro group with greater amounts under anaerobic than aerobic conditions, and as a minor aerobic pathway, amide cleavage was observed. Cyclanilide (65) was partitioned mainly to water due to its hydrophilicity and underwent amide hydrolysis at a moderate rate to 2,4-dichloroaniline (EU-New 2001a). Picolinafen (66) quickly dissipated by its partition to sediment with amide hydrolysis to form the corresponding acid (EU-New 2002m). Cyazofamid (68) underwent cleavage of the N(imidazole ring)–S bond followed by stepwise hydrolysis of nitrile to the corresponding amide and carboxylic acid with significant formation of bound residues (EU-Ex 2002b). *O*-Demethylation was predominant for florasulam (69) with the metabolite being more persistent (EU-New 2002e). Aerobic aquatic metabolism of cloransulam-methyl (70) gave the acid derivative formed by ester hydrolysis in water, while in anaerobic conditions *O*-deethylation of this acid and opening of the pyrimidinyl ring of [1,2,4]triazolo[1,5-*c*]pyrimidin moiety additionally proceeded (Fig. 7) at a slightly faster degradation rate (PMRA 2001b).

Dicarboximides. This class of pesticides rapidly dissipates in a water–sediment system (Table 10). Captan (71) showed a rapid conversion to the corresponding phthalimide THPI via cleavage of the N—S bond (EPA 1999a). THPI was further degraded by opening of the imide ring followed by amide hydrolysis together with epoxidation of the C=C ring bond. The imide hydrolysis was one of the main degradation pathways for flumioxazin (72) and cinidon-ethyl (73); the ester cleavage in the side chain also proceeded for the latter (EU-New 2002b,f). Famoxadone (74) rapidly dissipated from water by partition to sediment with opening of the 3-oxazolidine-2,4-dione ring (EU-New 2002d). Opening of the 2,4-

Table 10. Degradation profiles of dicarboximides in laboratory water–sediment systems.

No.	Pesticide/structure	System information[a] DT_{50} (w/s)[b]	Route of degradation[c]	Reference
71	Captan 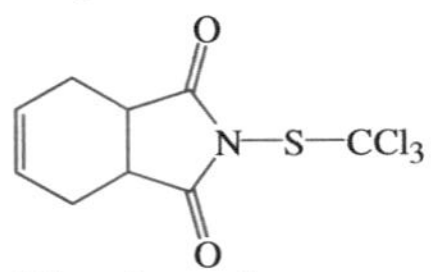	Two systems, N.R., aerobic, N.R. N.R. / <24 hr	O8 (epoxidation), H4, H8 (N-S), M3 (imide)	EPA (1999a)
72	Flumioxazin 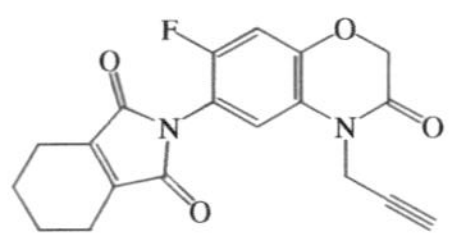	Two systems, 20°C, aerobic, 98 d <1.85 d / <1.85 d	H4, M3 (imide), M, B	EU-New (2002f)
73	Cinidon-ethyl 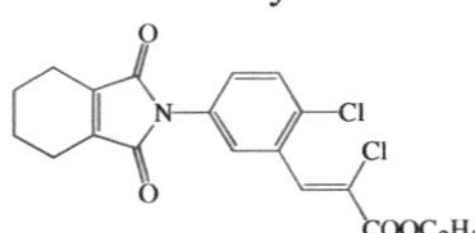	Two systems, 20°C, aerobic, 100 d 1.5–7 hr / 5 hr	H3, H4, M3 (imide), M, B	EU-New (2002b)
74	Famoxadone 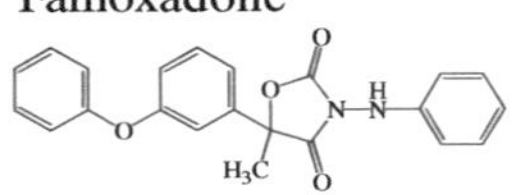	Two systems, 20°C, aerobic, 30 d 0.07–0.48 hr / 0.68–2.1 d	M3 (oxazolidinone), H4, M, B	EU-New (2002d)

Table 10. *Continued*

No.	Pesticide/structure	System information[a] DT_{50} (w/s)[b]	Route of degradation[c]	Reference
75	Iprodione 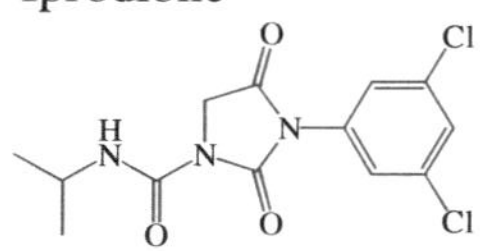	N.R., 20°C, aerobic, 100 d <6 hr / <30 d	M3 (imidazolinone), M4 (isomer)	EU-Ex (2002h)
76	Butafenacil 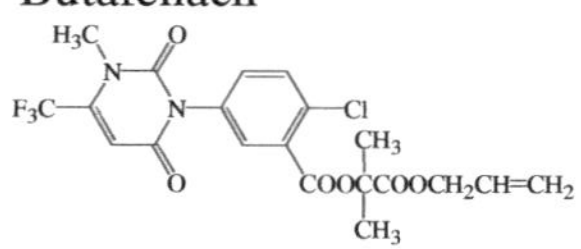	Pond and river, N.R., aerobic, N.R. N.R. / 3.7–6.1 d	H3, M3 (uracil)	APVMA (2002a)

B, bound formation; M, mineralization; N.R., information not reported.
[a]Type of water and sediment (soil), temperature, aerobicity, period of incubation.
[b]Time of 50% disappearance of a pesticide in water layer (w) and water–sediment system(s).
[c]Reaction number listed in Table 2. Terms in parentheses mean a position of reaction or product.

dioxoimidazolidinyl ring was similarly reported for iprodione (75) (EU-Ex 2002h). The product was again cyclized to form a different 2,4-dioxoimidazolidine isomer, as reported for its hydrolysis (Katagi 2002). Butafenacil (76) underwent either ester hydrolysis at the two positions in its side chain or opening of the 2,6-dioxopyrimidinyl ring to form the *N*-methylurea derivative (APVMA 2002a).

E. Carbamates

Carbamates are hydrolytically stable at environmentally relevant pH with the reaction mechanism depending on *N*-substituents and pK_a of a leaving group (Katagi 2002). Similarly as with amides and anilides, the presence of the NHC(=O)O moiety causes some variation in K_{oc} values from 10^1 to 10^4 (see Fig. 2). Therefore, the chemical structure of carbamates drastically affects their distribution and degradation profiles (Table 11).

The main degradation route of aldicarb (77) changes with system aerobicity (Kazumi and Capone 1995). The terminal thiomethyl sulfur of (77) was aerobically oxidized to sulfoxide and sulfone, whereas the carbamate linkage was first cleaved in anaerobic conditions with the generation of CH_4. Vink and van der Zee (1997) also reported formation of 2-methyl-2-methylthiopropionaldehyde and 2-methyl-2-methylthiopropionnitrile instead of these oxidized metabolites. The hydrolytic mechanism of (77) is known to depend on the medium pH (Katagi 2002), and the pH difference between sediments seemed to result in different degradations (Wolfe and Macalady 1992). Carbaryl (78) was gradually partitioned to sediment because of its hydrophobicity (Setzo et al. 1979) and microbially metabolized by successive oxidative *N*-demethylation, cleavage of carbamate linkage, and ring hydroxylation (Kaufman 1974). Thiobencarb (79) was preferably distributed in a sediment phase even in an early stage of incubation and likely to be persistent (EPA 1997b). Nakamura et al. (1977) examined its metabolism both in aerobic and anaerobic flooded conditions and identified many minor metabolites formed via oxidative *N*-deethylation, *S*-oxidation, hydroxylation at the *o*-position of the phenyl ring, and cleavage of the thiocarbamate linkage.

Chlorpropham (80) is aerobically degraded at a moderate rate by hydrolysis to form 3-chloroaniline (EU-Ex 2003a). In contrast, the corresponding aniline could not be detected for diethofencarb (81) except with very minor unknowns and significant bound residues (Kodaka et al. 2004). Desmedipham (82) and phenmedipham (83) were very susceptible to biodegradation via cleavage of the carbamate linkage connecting the phenyl rings (EPA 1996b; EU-Ex 2004d,f). The main metabolites of iprovalicarb (84) were *p*-methylphenethylamine via cleavage of the amide linkage and its *N*-acetyl derivative (Henneböle 1999; EU-New 2002i). Aerobic aquatic metabolism of asulam (85) gave four products including sulfanilamide (EPA 1995a). Distribution profiles of the few carbamates in

Table 11. Degradation profiles of carbamates in laboratory water–sediment systems.

No.	Pesticide/structure	System information[a] DT_{50} (w/s)[b]	Route of degradation[c]	Reference
77	Aldicarb 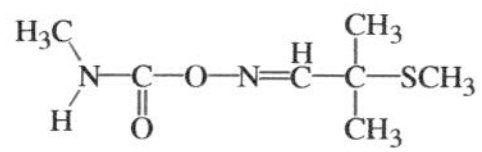	Three systems, 12°C, aerobic and anaerobic, 14–21 d N.R. / N.R.	O6, H5, M	Kazumi and Capone (1995)
		Markermeer Lake, N.R., aerobic, 70 d N.R. / <5 d	O6, H5	Vink and van der Zee (1997)
78	Carbaryl 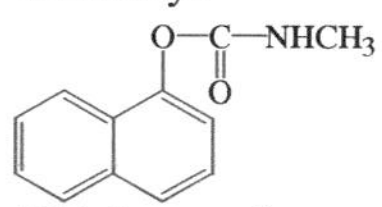	Pond and creek, 9°C, aerobic, 42–50 d N.R. / 14–21 d	N.R.	Setzo et al. (1979)
79	Thiobencarb 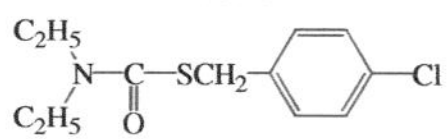	Sacramento River, N.R., anaerobic, 1 yr N.R. / 5.4 yr	O1, H5	EPA (1997b)
		Three Japanese soils, 30°C, aerobic and anaerobic, 5 wk N.R. / N.R.	O3, *o*-O5, O6, H5	Nakamura et al. (1977)
80	Chlorpropham 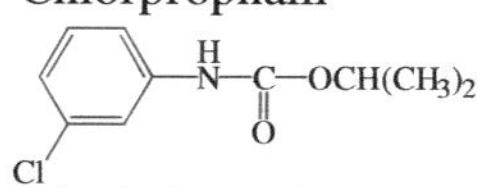	Six systems, N.R., aerobic, N.R. 10.2–21.2 d / 19.3–77 d	H5	EU-Ex (2003a)
81	Diethofencarb 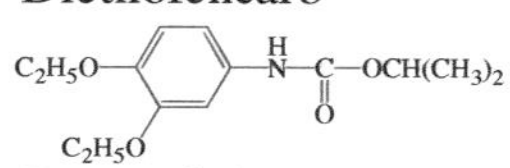	Lake and pond, 20°C, aerobic, 31 d 24.2 d / 45.3 d	B	Kodaka et al. (2004)
82	Desmedipham 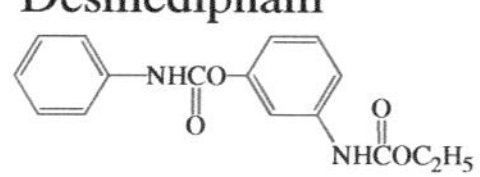	Two systems, 20°C, aerobic, 100 d 0.1–3.1 d / 2.2–4 d	H5	EU-Ex (2004d) EPA (1996b)

Table 11. *Continued*

No.	Pesticide/structure	System information[a] DT_{50} (w/s)[b]	Route of degradation[c]	Reference
83	Phenmedipham 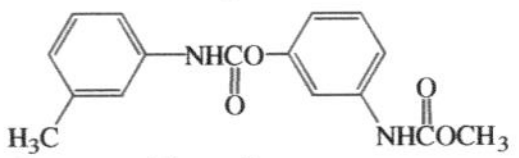	Three systems, 20°C, aerobic, 126 d 0.1–0.3 d / 0.11–0.18 d	H5	EU-Ex (2004f)
84	Iprovalicarb 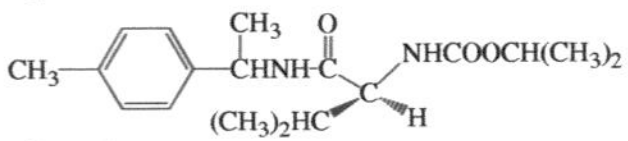	Two systems, 20°C, aerobic, 100 d 18.8–54.4 d / 25.3–55.7 d	O1, H5, C2, M, B	EU-New (2002i) Henneböle (1999)
85	Asulam 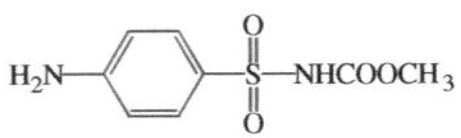	Sandy loam soil, 20°C, aerobic, N.R. N.R. / 105 d	H5	EPA (1995a)

B, bound formation; M, mineralization; N.R., information not reported.
[a]Type of water and sediment (soil), temperature, aerobicity, period of incubation. r.t.: room temperature.
[b]Time of 50% disappearance of a pesticide in water layer (w) and water–sediment system(s).
[c]Reaction number listed in Table 2. Terms in parentheses mean a position of reaction or product.

the formulation were briefly studied in aerobic water–sediment systems but with no information on metabolism (Agnihotri and Jain 1985; Garg and Agnihotri 1985).

F. Ureas and Sulfonylureas

In the case of sulfonylureas, the pK_a value of the SO_2NH moiety is about 4 and the dissociated form is considered to be predominant in natural aquatic environments (Katagi 2002); this is reflected in the K_{oc} values of sulfonylureas (10^1–10^3), lower than those of ureas (10^2–10^4) (see Fig. 2), and sulfonylureas would likely be partitioned in a water phase and possibly move downward in the soil column (Florip et al. 2003). Ureas are stable at neutral pH, and their biodegradation becomes important. Sulfonylureas undergo moderate to rapid abiotic hydrolysis, especially at acidic pHs (Katagi 2002). Berger and Wolfe (1996) studied the degradation rates of 12 sulfonylureas in stream and pond sediment systems; they partitioned to water and by QSAR analysis either hydrolysis or adsorption to sediment was found to be the main factors determining dissipation rates. Degradation profiles of urea, benzoylurea, and sulfonylurea pesticides are shown in Tables 12 and 13.

Ureas and Benzoylurea. Linuron (86) showed moderate aerobic degradation via successive *N*-demethoxylation and *N*-demethylation (Huber and Gémes 1981). Under anaerobic conditions, the main degradation pathways were *N*-demethoxylation and reductive dechlorination at the 3-position of the phenyl ring, with cleavage of a urea linkage becoming a minor reaction (EPA 1995b; EU-Ex 2002j). Similar degradation profiles, depending on system aerobicity, were reported for diuron (*N*,*N*-dimethyl derivative), but dechlorination proceeded specifically at the 4-position of the phenyl ring (Attaway et al. 1982a). Reductive dechlorination was significantly enhanced by pretreatment of sediment with diuron, indicating a microbial adaptation (Attaway et al. 1982b). In the case of isoproturon (87), the successive *N*-demethylation was the main aerobic pathway with the metabolites being distributed mainly in the water phase (Rönnefahrt et al. 1997; EU-Ex 2002i). Propineb (88) is very unstable in the environment and is quickly transformed to propylene thiourea (PTU). PTU was detected mainly in the water phase and aerobically degraded to propylene urea via oxidative desulfuration (EU-Ex 2003e). Concerning benzoylureas, anaerobic aquatic metabolism of diflubenzuron (89) was reported with the main degradation route being cleavage of the amide linkage liberating 2,6-difluorobenzoic acid (EPA 1997a).

Sulfonylureas. These herbicides can be conveniently classified from their structures at both molecular ends. The first class, (90)–(95), has a 4-methoxy-6-methyl-1,3,5-triazin-2-yl moiety and an aromatic ring, being connected

Table 12. Degradation profiles of urea and benzoylurea pesticides in laboratory water–sediment systems.

No.	Pesticide/structure	System information[a] DT_{50} (w/s)[b]	Route of degradation[c]	Reference
86	Linuron 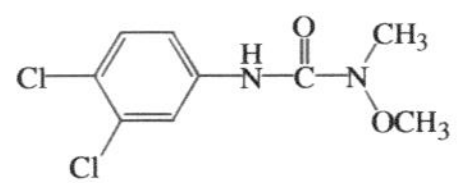	Silt loam and sand soils, 24°C, anaerobic, N.R. N.R. / <3 wk	O3, R1, H6	EU-Ex (2002j)
		Lake, N.R., aerobic, 60 d 10 wk / N.R.	O3	Huber and Gémes (1981)
87	Isoproturon 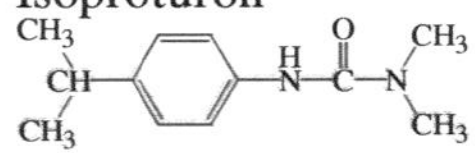	Six studies, N.R., aerobic, 120 d 20–61 d / 44–276 d	O3	EU-EX (2002i)
		One system, 20°C, aerobic, 200 d 137–216 d / 203–304 d	O3	Rönnefahrt et al. (1997)
88	Propineb 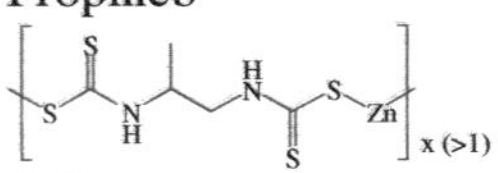	N.R., 20°C, aerobic, 100 d 4 d / <30 d	N.R.	EU-Ex (2003e)
89	Diflubenzuron 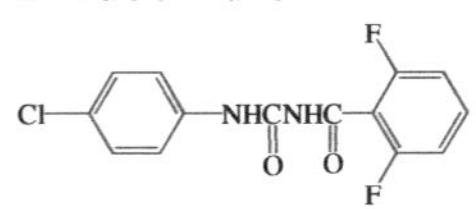	Silt loam soil, 24°C, anaerobic, N.R. N.R. / 34 d	H6	EPA (1997a)

N.R., information not reported.
[a]Type of water and sediment (soil), temperature, aerobicity, period of incubation.
[b]Time of 50% disappearance of a pesticide in water layer (w) and water–sediment system(s).
[c]Reaction number listed in Table 2. Terms in parentheses mean a position of reaction or product.

Table 13. Degradation profiles of sulfonylurea pesticides in laboratory water–sediment systems.

No.	Pesticide/structure	System information[a] DT_{50} (w/s)[b]	Route of degradation[c]	Reference
90	Chlorsulfuron 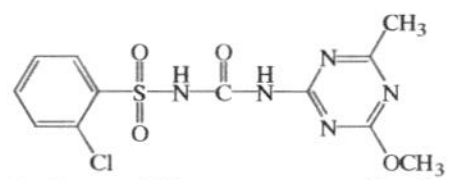	U.S. pond, 25°C, anaerobic, 365 d N.R. / >365 d	O4, H6, M3 (triazine)	Strek (1998)
91	Metsulfuron-methyl 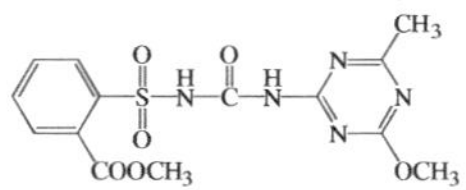	Two systems, 20°C, aerobic, N.R. 81–148 d / 105–175 d	O4, H3	EU-Ex (2000c)
92	Iodosulfuron 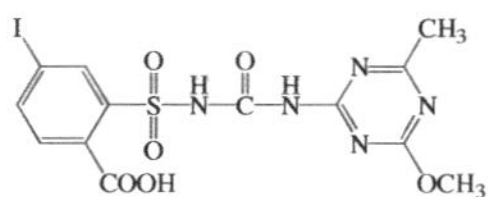	Two systems, 20°C, aerobic, 365 d 13–19 d / 13–25 d	O4, H6, R1	EU-New (2003f)
93	Prosulfuron 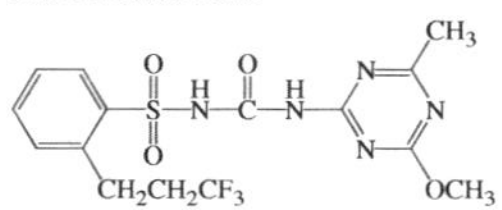	Two systems; 20°C, 90 d (aerobic; 1); 25°C, N.R. (anaerobic; 2) 1) 33–35 d / 120–191 d 2) 16 d / 16 d	 O4, H6, B O4, H6, M3 (triazine), B	EU-New (2002n)
94	Triasulfuron 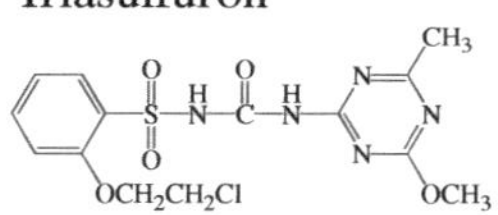	Sediment, 1% (1) and 50% (2), 20°C, aerobic, 10 wk and 30 d 1) 189–245 d / 189–245 d 2) 30–39 d / 38–44 d	 O4, H6 H6	EU-Ex (2000d)
95	Thifensulfuron-methyl 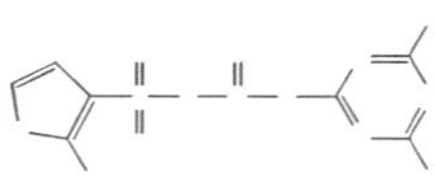	Four systems, 20°C, aerobic, 91–182 d 18–26 d / 15–27 d	O4, H6, H3	EU-Ex (2001e)

Table 13. *Continued*

No.	Pesticide/structure	System information[a] DT$_{50}$ (w/s)[b]	Route of degradation[c]	Reference
96	Flupyrsulfuron-methyl 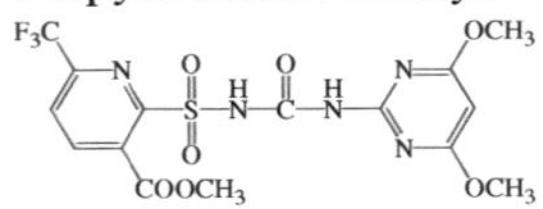	Two systems, 20°C, aerobic, 100 d 3–6 d / same	M4 (imide), M5 (bridge contraction)	EU-New (2001b)
		Two U.S. rivers, 20°C, aerobic, 120 d 3–6 d / 3–6 d	O4, M4 (imide), M5 (bridge contraction)	Singles et al. (1999)
97	Trifloxysulfuron sodium 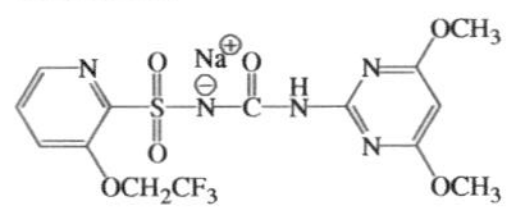	Pond and river, N.R., aerobic, N.R. 5.2–10.6 d / 23–26 d	O4, H6, M5 (bridge contraction)	APVMA (2002c)
98	LGC-42153 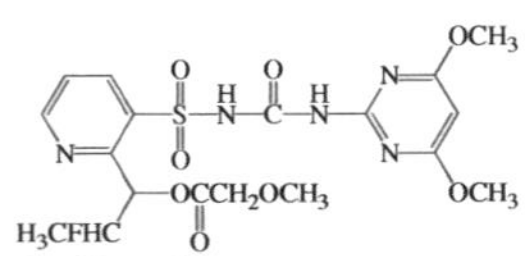	Korean loamy soil, 25°C, aerobic, 120 d N.R. / 2.7–3.2 d	H3, H6	Kim et al. (2003)
99	Sulfosulfuron 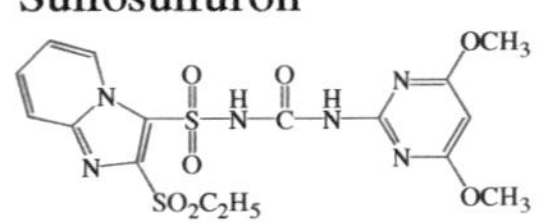	Two systems, 20°C, aerobic, 100 d 16.1–19.5 d / 19.8–32.2 d	O4, H6	EU-New (2002q)
100	Foramsulfuron 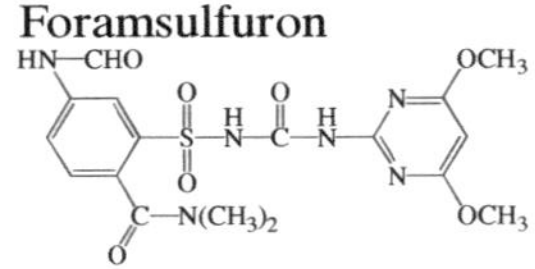	Two systems, 20°C, aerobic, 118–119 d 13–21 d / 34–55 d	Two metabolites	EU-New (2002g)

Table 13. *Continued*

No.	Pesticide/structure	System information[a] DT_{50} (w/s)[b]	Route of degradation[c]	Reference
101	Ethoxysulfuron 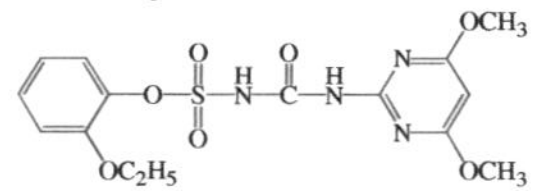	Two systems, 20°C, aerobic, 100 d 11–23 d / 24–35 d	Three metabolites, B	EU-New (2002c)
102	Oxasulfuron 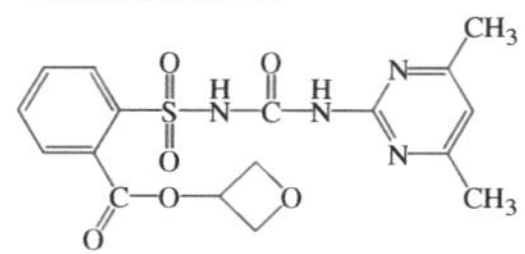	Pond and river, 20°C, aerobic, 180 d 10–12 d / 20–23 d	H6, M, B	EU-New (2002l)
103	Propoxycarbazone 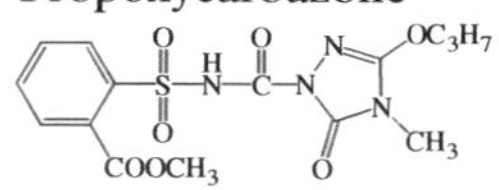	Two systems, 20°C, aerobic, 100 d 10.6–90.8 d / 12–189 d	H3, H6	EU-New (2003j)
104	Propoxycarbazone-sodium 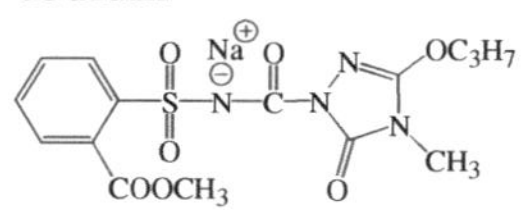	Two systems, N.R., aerobic, N.R. 12 d / 101 d	H3, H6	Malekani and Hellpointner (2002)

B, bound formation; M, mineralization; r.t., room temperature; N.R., information not reported.
[a]Type of water and sediment (soil), temperature, aerobicity, period of incubation.
[b]Time of 50% disappearance of a pesticide in water layer (w) and water–sediment system(s).
[c]Reaction number listed in Table 2. Terms in parentheses mean a position of reaction or product.

through a sulfonylurea bridge. The common degradation pathways are *O*-demethylation and cleavage of the sulfonylurea linkage. *O*-demethylation of chlorsulfuron (90), followed by opening of the triazine ring, gave acetyltriuret and carbamoyl guanidine derivatives (Strek 1998) (Fig. 8). The acetyltriuret derivative was anaerobically formed from prosulfuron (93) (EU-New 2002n). Ester hydrolysis in the side chain mainly proceeded for metsulfuron-methyl (91), together with *O*-demethylation (EU-Ex 2000c). In the case of iodosulfuron (92), reductive deiodination proceeded in addition to these reactions (EU-New 2003f). More rapid aerobic metabolism was observed for triasulfuron (94) in the presence of sediment (EU-Ex 2000d). Thifensulfuron-methyl (95) has a thiophene instead of a phenyl ring; however, their degradation profiles were not significantly changed (EU-Ex 2001e).

The second class (96)–(100) has a 4,6-dimethoxypyrimidin-2-yl ring at one molecular end. The unique contraction of a sulfonylurea bridge was reported for flupyrsulfuron-methyl (96) and trifloxysulfuron sodium (97) (Fig. 9). (96) further underwent intramolecular cyclization (Singles et al. 1999; EU-New 2001b), while the liberation of the carbamoyl moiety was observed for (97) (APVMA 2002c). Hydrolysis of the ester and sulfonylurea bridge were predominant pathways for an experimental herbicide LGC-42153 (98) and sulfosulfuron (99) (EU-New 2002q; Kim et al. 2003). Propoxycarbazone (103) and its sodium salt (104) have a triazolinone ring instead of pyrimidine and triazine rings; their degradation pathways were familiar ones, ester hydrolysis in the side chain and cleavage of the sulfonylurea bridge (Malekani and Hellpointner 2002; EU-New 2003j).

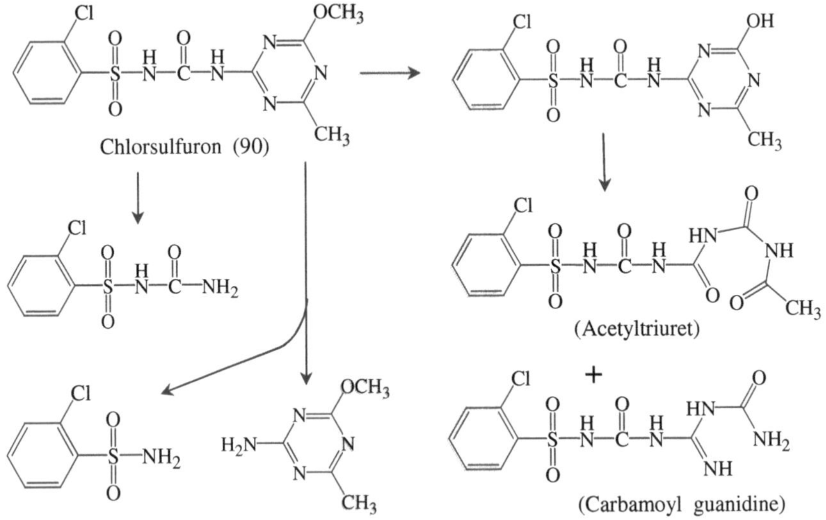

Fig. 8. Anaerobic aquatic metabolism of chlorsulfuron (90).

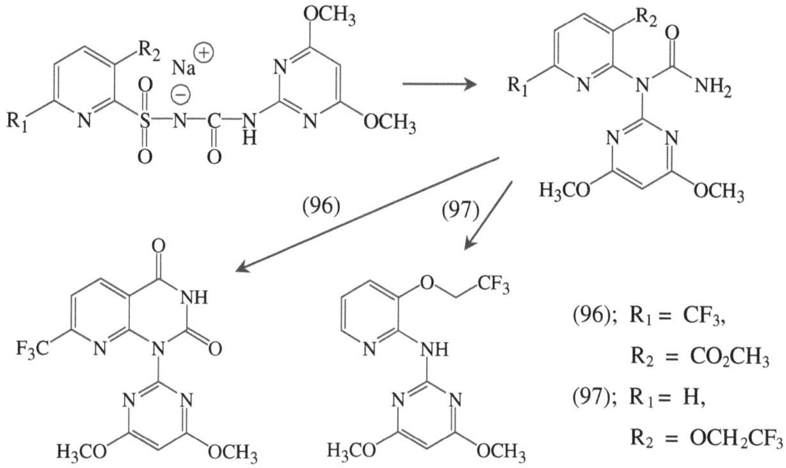

Fig. 9. Aerobic aquatic metabolism of flupyrsulfuron-methyl (96) and trifloxysulfuron-methyl (97).

G. Azoles and Triazines

Azole pesticides basically show moderate to high hydrophobicity but triazines are more hydrophilic with K_{oc} values around 10^2 (see Fig. 2). From the view point of groundwater contamination, both degradability in aquifer sediments and possible association with dissolved organic matter have been examined for atrazine (110) (Wijayaratne and Means 1984; Wood et al. 1991). Concerning abiotic degradation, azoles are usually stable, but triazines are known to undergo acid-catalyzed hydrolytic substitution of chlorine at the 6-position (Katagi 2002). Slow biodegradation is often reported for azoles, whereas oxidative *N*-dealkylation in addition to hydrolysis is a main degradative pathway for triazines (Kaufman 1974; Ou 2000). The degradation profiles of azoles and triazines are presented in Tables 14 and 15.

Azoles. Amitrole (105) is degraded to two minor unknowns in both aerobic and anaerobic conditions (EPA 1996a). The formation of amide linkage with dissolved organic matter at nitrogen atoms in the 1-position with the amino group was identified by solid-state ^{15}N-NMR (Spiteller et al. 2002). Propiconazole (106) was slowly degraded to four minor degradates including 1,2,4-triazole via opening of the 1,3-dioxolan-2-yl ring and cleavage of the C—N (triazole) linkage (EU-Ex 2003d). Strong adsorption to sediment was observed for fenbuconazole (107), but its biodegradation was found to be very slow, with formation of three minor degradates including the keto derivative via oxidation at the 4-chlorobenzyl carbon (PMRA 2003a). In contrast, moderate degradation was reported for bitertanol (108)

Table 14. Degradation profiles of azole pesticides in laboratory water–sediment systems.

No.	Pesticide/structure	System information[a] DT$_{50}$ (w/s)[b]	Route of degradation[c]	Reference
105	Amitrole 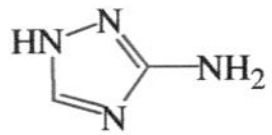	Sandy loam sediment, 21°–24°C, aerobic, 30 d N.R. / 57–74 d	N.R.	EPA (1996a)
		German river, N.R. aerobic, 100 d N.R. / N.R.	B	Spiteller et al. (2002)
106	Propiconazole 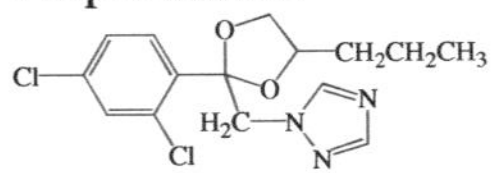	Rhine River and pond, 20°C, aerobic, 175 d 5.4–6.4 d / 485–636 d	M3 (dioxolan)	EU-Ex (2003d)
107	Fenbuconazole 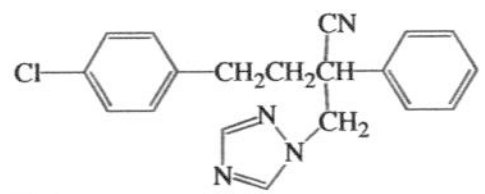	Pond and river, N.R., aerobic, 105 d 1.2–4.3 d / >1000 d	O1, H2, M4 (lactone)	PMRA (2003a)
108	Bitertanol 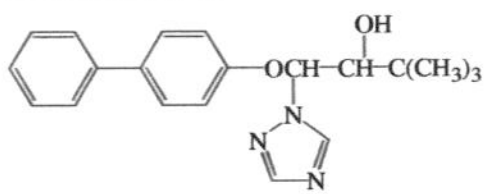	Gravel pit and ditch, 22°C, aerobic, 120 d 24–27 d / N.R.	O8 (ketone), M3 (phenyl), M, B	FAO (1999a)
109	Prothioconazole 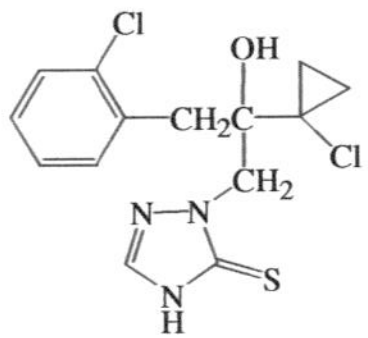	Two systems, 20°C, aerobic, N.R. <1 d / 2–3 d	H8 (C—S), C1 (S)	Hellpointner and Borchers (2004)

B, bound formation; M, mineralization; N.R., information is reported.
[a]Type of water and sediment (soil), temperature, aerobicity, period of incubation.
[b]Time of 50% disappearance of a pesticide in water layer (w) and water–sediment system(s).
[c]Reaction number listed in Table 2. Terms in parentheses mean a position of reaction or product.

Table 15. Degradation profiles of triazine pesticides in laboratory water–sediment systems.

No.	Pesticide/structure	System information[a] DT_{50} (w/s)[b]	Route of degradation[c]	Reference
110	Atrazine 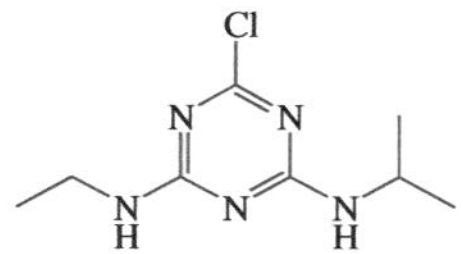	Chesapeake Bay, 12°–35°C, aerobic, 80 d 3–12 d / 13–16 d	O3, H1	Jones et al. (1982)
		Two soils–river water, 5°, 24°C, aerobic, 672 d N.R. / N.R.	O3, H1, B	Seybold et al. (1999)
		Five ditch, creek and estuarine, r.t., aerobic, 70 d 15–62 d / 21–68 d	O3	Aelion and Mathur (2001)
		Iowa pond, 25°C, aerobic, 60 d N.R. / 42 d	O3, H1, B	Rice et al. (2004)
111	Terbutryn 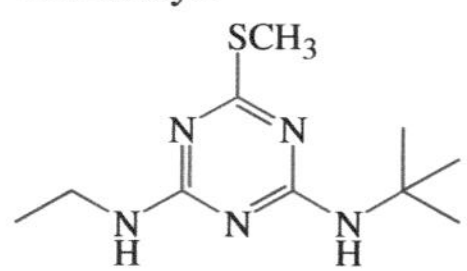	Manitoba pond and river, 25°C, aerobic, 515 d 180–240 d / N.R.	O3, O6, H8 (C—S), B	Muir and Yarechewski (1982b)
112	Cyromazine 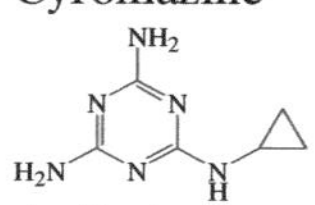	Pond, 20°C, aerobic, 30 d N.R. / 71 d	O3	Hein et al. (2003)
113	Anilazine 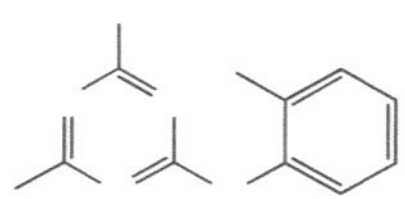	German river, N.R. aerobic, 100 d 16 h / N.R.	H1, B	Klaus et al. (1998) Spiteller et al. (2002)

B, bound formation; r.t., room temperature; N.R., information not reported.
[a]Type of water and sediment (soil), temperature, aerobicity, period of incubation.
[b]Time of 50% disappearance of a pesticide in water layer (w) and water–sediment system(s).
[c]Reaction number listed in Table 2. Terms in parentheses mean a position of reaction or product.

under aerobic conditions with formation of bound residues and mineralization (FAO 1999a). Two minor degradates were produced via oxidation of the alcohol moiety to ketone or oxidative opening of the terminal phenyl ring of the biphenyl moiety to 4-carboxyphenyl. Prothioconazole (109) underwent *S*-methylation irrespective of a system aerobicity; sulfur and 1,2,4-triazole were detected as major degradates under aerobic conditions (Hellpointner and Borchers 2004).

Triazines. Atrazine (110) in water was partitioned to sediment very slowly and rapidly released from (110)-treated sediment with its distribution controlled by temperature-dependent metabolism (Mersie et al. 1998a,b, 2000; Seybold et al. 1999). Hydroxylation was more favorable as compared with *N*-dealkylation. As with incubation, mineralization was usually minimal, but the bound fraction as alkaline extracted gradually increased (Smalling and Aelion 2004). The extent of *N*-dealkylation at either amino group was found to be dependent on the presence of sediment, but *N*-deisoprolylation seemed more probable when several coastal sediments were examined (Aelion and Mathur 2001). Rice et al. (2004) examined the aerobic degradation of (110) in detail and confirmed the several pathways including hydrolxylation and *N*-dealkylation, most of which proceeded microbially. In aerobic conditions, terbutryn (111) underwent hydrolysis at the 6-position (major) and *N*-deethylation (minor), with additionally *S*-oxidation to sulfoxide as a major pathway (Muir and Yarechewski 1982b). Cyromazine (112) has two free amino groups and a cyclopropylamino group in the molecule; melamine was identified as a main degradate in aerobic systems (Hein et al. 2003). Different from the above triazines, anilazine (113) rapidly dissipated from water to form the dihydroxylated derivative via hydrolysis (Klaus et al. 1998; Spiteller et al. 2002). The ^{13}C-NMR study also indicated that part of this product is attached to DOM via ether linkage and exists as bound residues.

H. Miscellaneous

There are many pesticides not simply classified by their functional groups, and in this section their degradation profiles are described (Table 16). Acrolein (114) has a conjugated C=C—C=O moiety that is subjected to various redox reactions. The corresponding alcohols and acids were formed, and among them glyceric acid and propanol were characteristic of aerobic and anaerobic conditions, respectively (Smith et al. 1995). Acibenzolar-*s*-methyl (115) underwent rapid hydrolysis to the corresponding persistent acid in the aerobic system (EU-New 2002a). Ester hydrolysis was also predominant for aerobic aquatic metabolism of trinexapac-ethyl (116) (PMRA 2001a). Very rapid hydrolysis, irrespective of aerobicity, was observed for spirodiclofen (117), with most of the aqueous radiocarbon due to the resultant enol (Babczinski 2002). The dominant aerobic reaction of propargite

Table 16. Degradation profiles of miscellaneous pesticides in laboratory water–sediment systems.

No.	Pesticide/structure	System information[a] DT_{50} (w/s)[b]	Route of degradation[c]	Reference
114	Acrolein $H_2C{=}\overset{H}{C}{-}CHO$	Canal, 25°C, (1) aerobic, 32 d and (2) anaerobic, 178 d (1) 9.5 d / N.R. (2) 10.3 d / N.R.	O8 (C═C & C═O), R3, M	Smith et al. (1995)
115	Acibenzolar-*S*-methyl 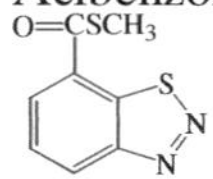	Pond and river, 20°C, aerobic, N.R. <1 d / <1 d	H3, M, B	EU-New (2002a)
116	Trinexapac-ethyl 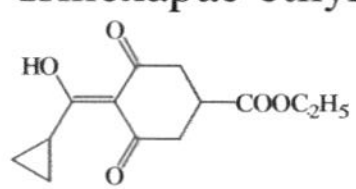	N.R., N.R., aerobic, N.R. N.R. / 10–25 d	H3	PMRA (2001a)
117	Spirodiclofen 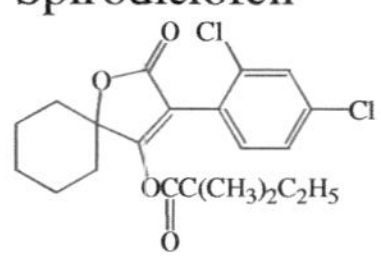	Two systems, N.R., aerobic, 110 d N.R. / 4–5 d	H3	Babczinski (2002)
118	Propargite 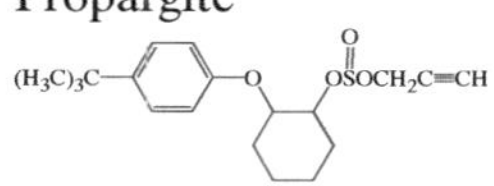	Lake, 25°C, aerobic, 30 d N.R. / 38 d	O1 (*t*-Bu), H8 (S—O, ether)	EPA (2001b)
		Sand hydrosoil, 25°C, anaerobic, 1 yr N.R. / 46.6 d	O1 (*t*-Bu), H8 (S—O, ether), B	
119	Pyridate 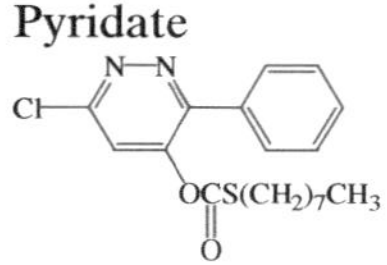	N.R., 20°C, aerobic, 84 d <0.4 d / <1 d	H8 (C—CO), C1 (O)	EU-Ex (2001c)

Table 16. *Continued*

No.	Pesticide/structure	System information[a] DT_{50} (w/s)[b]	Route of degradation[c]	Reference
120	CNP 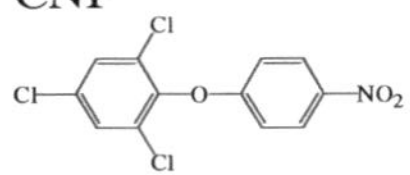	Two paddy systems, 30°C, aerobic, 20 d N.R. / 1–7 d	R4, H8 (ether), H8 (C—N), C2, B	Oyamada and Kuwatsuka (1979)
121	Bifenox 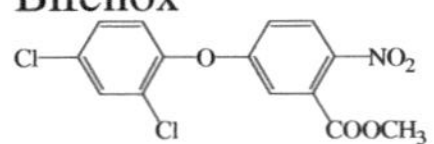	Four Japanese flooded soils, 30°C, aerobic, 60 d N.R. / 2–5 d	R4, H3, C2	Ohyama and Kuwatsuka (1978)
122	Fluazinam 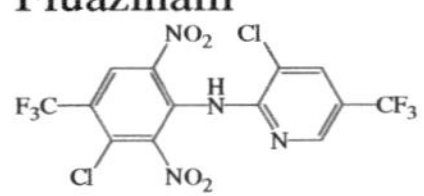	Three soils, aerobic (1) and anaerobic (2), 25°C, N.R. 1) N.R. / 20 hr–3.2 d 2) N.R. / <1 d	 R4, H1 (CF_3) R4	PMRA (2003b)
123	Pyraflufen-ethyl 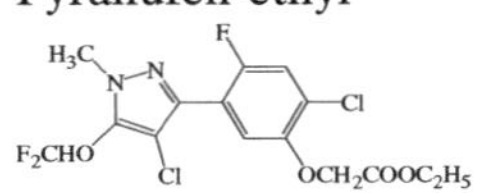	Two systems, 20°C, aerobic, 100 d 1–2 hr / 2 hr	O2, H3, C1(O)	EU-New (2002p)
124	Carfentrazone-ethyl 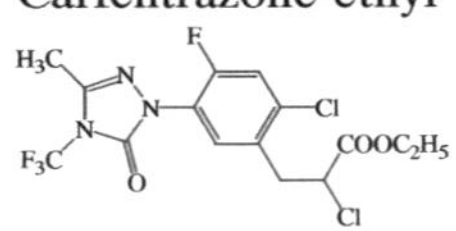	Two systems, 20°C, aerobic, 100 d <0.4 d / <0.4 d	O8, R1, R2, H3	EU-New (2003a)
		Two paddy systems, 25°C, aerobic, 30 d 1.04–1.5 d / 1.25–1.69 d	O8, R1, R2, H3	Elmarakby et al. (2001)
125	Oxadiargyl 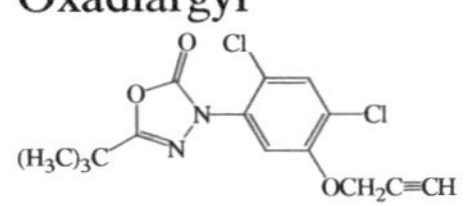	Two systems, 20°C, aerobic, 149 d 2.2 hr / 63 d	N.R.	EU-New (2002k)

126	Trifluralin 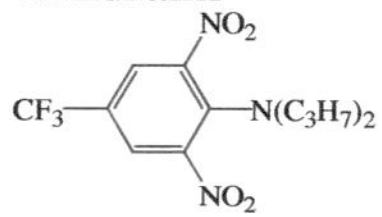	Clay soil, 25°–28°C, aerobic and anaerobic, 3 wk N.R. / 4–25 d	O3, R4	Willis et al. (1974)
		Flooded soil, N.R., N.R., 8 wk N.R. / N.R.	R4, M4 (benzimidazole)	Golab et al. (1979)
127	Pendimethalin 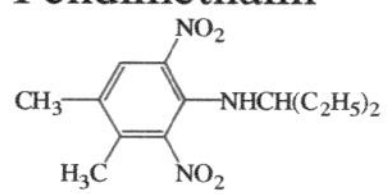	Four systems, 20°C, aerobic, 197 d N.R. / 4–28 d	B,V	EU-Ex (2003c)
128	Methoxyfenozide 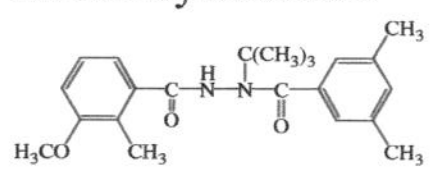	N.R., N.R., aerobic (1) and anaerobic (2), N.R. N.R. / 387–963 d (1), 654 d (2)	O1	PMRA (2004c)
129	Tebufenozide 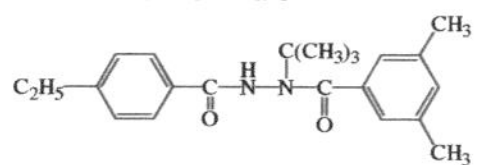	Two U.S. soils, N.R., aerobic, 1 yr N.R. / 99–101 d	O1, O2 (aryl ethyl)	PMRA (1996)
130	Acetamiprid 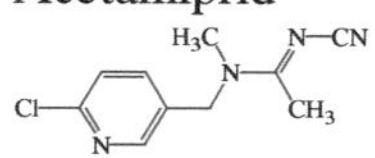	N.R., 25°C, aerobic, N.R. N.R. / 30 d	H2, H8 (N—(C═N))	PMRA (2002)
131	Thiacloprid 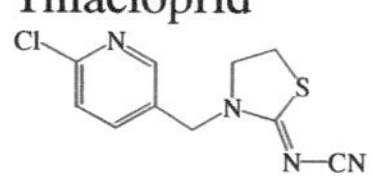	N.R., N.R., aerobic, 100 d N.R. / 12–20 d	H2, M3 (thiazolidine)	Krohn (2001)
132	Imidacloprid 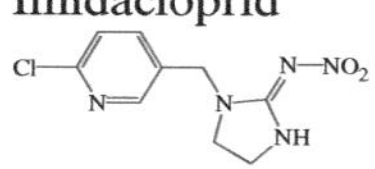	Two systems, N.R., aerobic, 120 d 30 d / N.R.	H8 (N—NO_2), B	Krohn and Hellpointner (2002)

133	Clothianidin 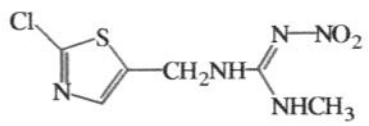	Two ponds, 20°C, aerobic, 100 d N.R. / 48–65 d	H8 (N—NO_2), B	Stupp and Fahl (2002)
134	Dinotefuran 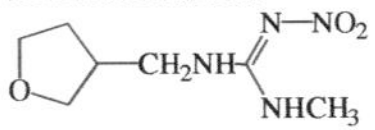	Two systems, N.R., aerobic, N.R. N.R. / 80.8 d	H8 (N—NO_2)	EPA (2004)
135	Thiamethoxam 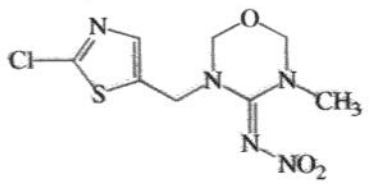	River and pond systems, N.R., aerobic, 100 d 5.8–13 d / 26–36 d	H8 (N—NO_2)	APVMA (2001b)
136	Isoxaflutole 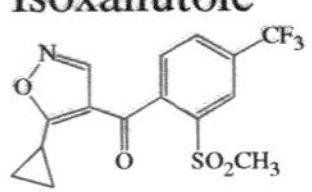	Two systems, 20°C, aerobic, 100 d N.R. / 0.34–0.54 d	M3 (isoxazole), R3, B	EU-New (2003g)
137	Flurtamone 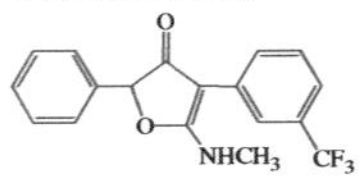	Two systems, 20°C, aerobic, 99 d 22–24 d / 59–>100 d	B	EU-New (2003e)
138	Mesotrione 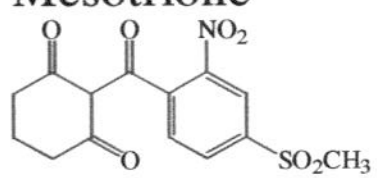	Two systems, 20°C, aerobic, N.R. 3.9–6.6 d / 3.9–6.5 d	H8 (C—CO), R4	EU-New (2003h)
139	Amitraz 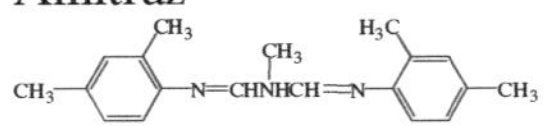	River, pond, ditch, 25°C, aerobic, 91 d N.R. / <1 d	H8 (oxime)	Allen and Arnold (1990)

Table 16. *Continued*

No.	Pesticide/structure	System information[a] DT_{50} (w/s)[b]	Route of degradation[c]	Reference
140	Fludioxinil 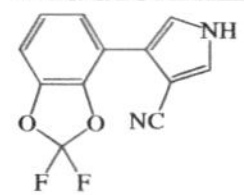	Pond and river, 25°C, aerobic, N.R. N.R. / 450–700 d	N.R.	APVMA (2002b)
141	Fenamidone 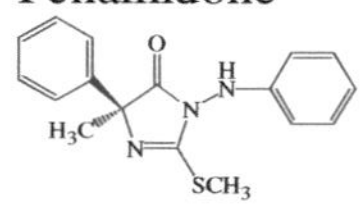	Two systems, 20°C, aerobic, N.R. 17.4–31 d / 67–127 d	H8 (N—N)	EU-New (2003c) Leake (2003)
142	Imazomox 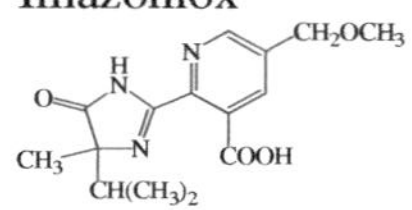	Two systems, 20°C, aerobic, 103 d 61 d / 129–154 d	Minor metabolites	EU-New (2002h)
143	Pyriproxyfen 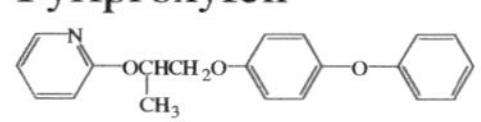	Lake, 25°C, aerobic, 28–31 d N.R. / 16–21 d	O5 (*p*-phenoxy-phenyl), O1, H8 (ether)	FAO (1999b)
144	Bentazone 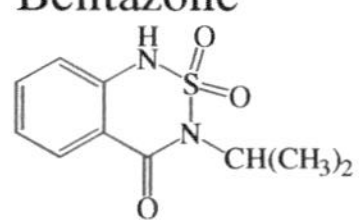	Two systems, 20°C, aerobic, N.R. 161 d / 523–908 d	C1 (N)	EU-Ex (2000a)
145	Thiabendazole 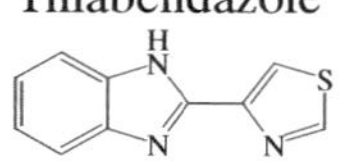	N.R., 20°C, aerobic, 181 d 1.6–2.3 d / 1.6–4.3 d	B	EU-Ex (2001d)

Table 16. *Continued*

No.	Pesticide/structure	System information[a] DT_{50} (w/s)[b]	Route of degradation[c]	Reference
146	Pymetrozine 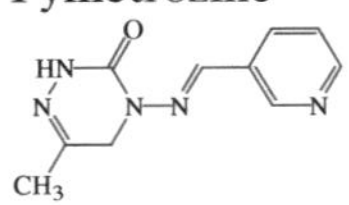	Pond, Rhine River, 20°C, aerobic, 120 d 5.2–6.6 d / 50–118 d	M, B	EU-New (2002o)
147	Clofentezine 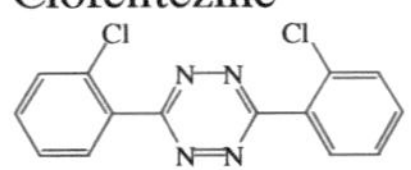	Two rivers, 20°C, aerobic, 180 d N.R. / <1 wk	M3 (tetrazine), M	Arnold et al. (1986)
148	Fipronil 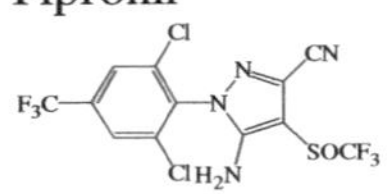	Sandy loam soil–pond water, N.R., aerobic, 1 yr N.R. / 14.5 d	R5, H2	CDPR (2001)
149	Spiroxamine 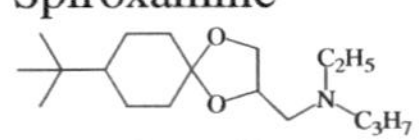	Two systems, 20°C, aerobic, 100 d 12–13 hr / 28–106 d	O8 (*N*-oxidation)	EU-Ex (1999b) Klein et al. (1997)
150	Tepraloxydim 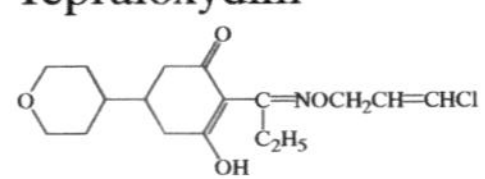	Two ponds, N.R., aerobic, N.R. 41–129 d / 49–171 d	H8 (oxime ether)	PMRA (2004a)
151	Ethofumesate 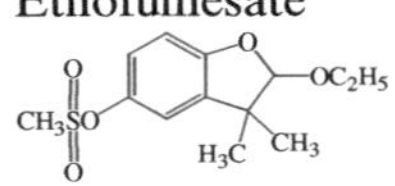	Three systems, 20°C, aerobic, 84–234 d 7–50 d / 105–285 d	B	EU-Ex (2002f)

Table 16. *Continued*

No.	Pesticide/structure	System information[a] DT_{50} (w/s)[b]	Route of degradation[c]	Reference
152	Fluridone 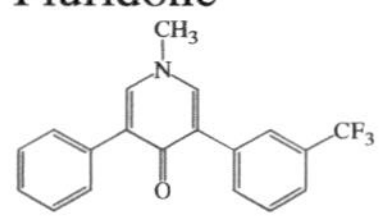	Three river systems, 22°–25°C, aerobic, 2 yr N.R. / N.R.	O5, M3	Muir and Grift (1982)
153	Spinosad 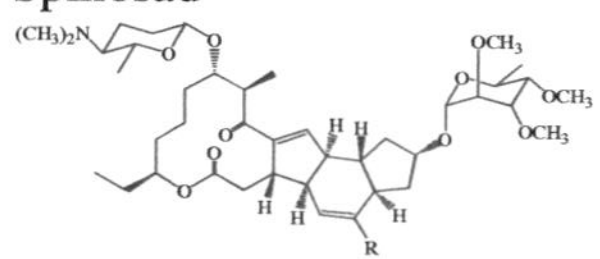 Spinosyn A: R=H, Spinosyn D: R=CH_3	Two Japanese soils, 25°C, aerobic, 100 d N.R. / 28–37 d	O3 (forosamine), H8 (C-sugars), O1, B	Cleveland et al. (2002)
		U.K. pond, 25°C, anaerobic, 365 d N.R. / 160–240 d	O3 (forosamine), H8 (C-rhamnose), O1, O4 (rhamnose)	Cleveland et al. (2002)
		Pond, 25°C, anaerobic, 365 d N.R. / 161–250 d	O3 (forosamine)	CDPR (2004)

B, bound formation; M, mineralization; V, volatilization; N.R., information not reported.
[a]Type of water and sediment (soil), temperature, aerobicity, period of incubation.
[b]Time of 50% disappearance of a pesticide in water layer (w) and water–sediment system(s).
[c]Reaction number listed in Table 2. Terms in parentheses mean a position of reaction or product.

(118) was hydrolysis of the sulfite moiety to the cyclohexyl alcohol, and ether cleavage and successive methyl oxidation at the *tert*-butyl group further proceeded but to a lesser extent (EPA 2001b). The thiocarbonate structure of pyridate (119) was also very susceptible to hydrolysis, with the phenol product gradually being partitioned to sediment and further *O*-methylated (EU-Ex 2001c). The main reactions of CNP (120), bifenox (121), and fluazinam (122) are reduction of the nitro group. The amino derivative of (120) was a transient species, and the following acetylation was very rapid (Oyamada and Kuwatsuka 1979). Ester hydrolysis also proceeded rapidly for (121), and the extent of reduction was correlated with either redox potential of sediment or free iron content (Ohyama and Kuwatsuka 1978). Conjugation at the amino group with formyl and acetyl was also detected as a minor pathway. Hydrolysis of the CF_3 group to carboxyl was also reported for (122) (PMRA 2003b). The dominant aerobic degradation pathway of pyraflufen-ethyl (123) was rapid hydrolysis to the acid, which was further subjected to β-oxidation at the side chain to form the phenol, followed by its *O*-methylation (EU-New 2002p). Similar aerobic reactions were observed for carfentrazone-ethyl (124) with reductive dechlorination and dehydrochlorination (EU-New 2003a). Hydroxylation of the methyl group at the 3-position of the triazolyl ring was also reported as a minor pathway (Elmarakby et al. 2001).

Trifluralin (126) aerobically underwent the successive reduction of the two nitro groups instead of *N*-depropylation being detected in upland soil (Golab et al. 1979). The effect of aerobicity on its degradation was investigated by Willis et al. (1974) by using a soil suspension under the electrical control of redox potential (E_h). The *N*-depropylated derivative was mainly produced at $E_h > 150\,mV$, while nitro reduction was a dominant process at $\leq 50\,mV$. The relation of the degradation rate to system anaerobicity was spectrophotometrically examined by monitoring the concentration of NO_3^- and Fe^{2+} (Tor et al. 2000). After depletion of NO_3^-, the concentration of Fe^{2+} gradually increased with concomitantly enhanced degradation of (126). The *N,N'*-dibenzoylhydrazide moiety in methoxyfenozide (128) and tebufenozide (129) was stable through aerobic and anaerobic aquatic metabolism. (128) was resistant to biodegradation but with a slight O-demethylation under aerobic conditions (PMRA 2004c). In contrast, the aerobic degradation of (129) was found moderate via successive oxidation of the ethyl group at the 4-position to give carboxy, methylcarbonyl, and carboxymethyl derivatives with significant mineralization and bound formation (PMRA 1996).

Hydrolysis of the nitrile group to amide was a common pathway for acetamiprid (130) and thiacloprid (131) (Fig. 10). More rapid degradation of (130) was observed in aerobic conditions via nitrile hydrolysis followed by cleavage of the acetamidine linkage to give *N*-(6-chloro-3-pyridyl)methyl-*N*-methylamine (PMRA 2002). In the case of (131), nitrile hydrolysis was followed by opening the 1,3-thiazolidinyl ring to the corre-

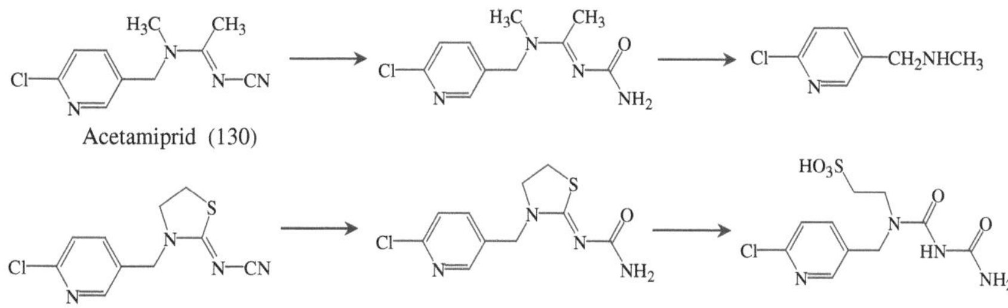

Fig. 10. Aerobic aquatic metabolism of acetamiprid (130) and thiacloprid (131).

sponding sulfonic acid (Krohn 2001). Imidacloprid (132), clothianidin (133), and thiamethoxam (135) underwent unique *N*-denitration (APVMA 2001b; Krohn and Hellpointner 2002; Stupp and Fahl 2003). In addition to *N*-denitration, hydrolytic conversion of the C=N moiety to the corresponding urea derivative and the C–N cleavage to give 1-methyl-2-nitroguanidine were observed in the aerobic system for dinotefuran (134) (EPA 2004). Isoxaflutole (136) was very rapidly degraded via hydrolytic opening of the isoxazolyl ring to the diketone derivative, followed by reduction of the nitrile group of this metabolite to the amine (EU-New 2003g). Because of its higher hydrophilicity, mesotrione (138) was distributed in the water phase and oxidative cleavage of the carbonyl linkage to give the carboxyl derivative predominantly proceeded in concomitant reduction of the nitro group (EU-New 2003h).

Amitraz (139) was unstable in the aerobic system via rapid hydrolysis to *N*-methyl-*N′*-2,4-xylylformamidine and formamide, which were then degraded to 2,4-dimethylaniline (Allen and Arnold 1990; Kewu et al. 1997). Fenamidone (141) showed moderate dissipation in the aerobic system with cleavage of the N–NHPh bond (EU-New 2003c; Leake 2003). The aerobic aquatic metabolism of pyriproxyfen (143) showed a rapid dissipation from the system via cleavage at either ether bond in the 4-phenoxyphenoxy moiety, and ring hydroxylation at the 4′-position was also detected as a minor pathway (FAO 1999b). 2-(Pyridin-2-yloxy)propionic acid was a major degradate in the water phase and also dominantly detected in anaerobic conditions. Bentazone (144) in the aerobic system gave the *N*-methyl derivative (EU-Ex 2000a). Rapid partition to sediment followed by hydrolysis of the tetrazine ring was observed for clofentezine (147), and the resultant 2-chlorobenzoic hydrazide was further degraded to carbon dioxide (Arnold et al. 1986). Fipronil (148) also showed a rapid adsorption to sediment where reduction of $SOCF_3$ to SCF_3 dominantly proceeded with hydrolysis of the nitrile group to the amide as a minor pathway (CDPR 2001). In the case of spiroxamine (149), rapid dissipation from water was followed by unique oxidation to form the corresponding *N*-oxide (Klein et

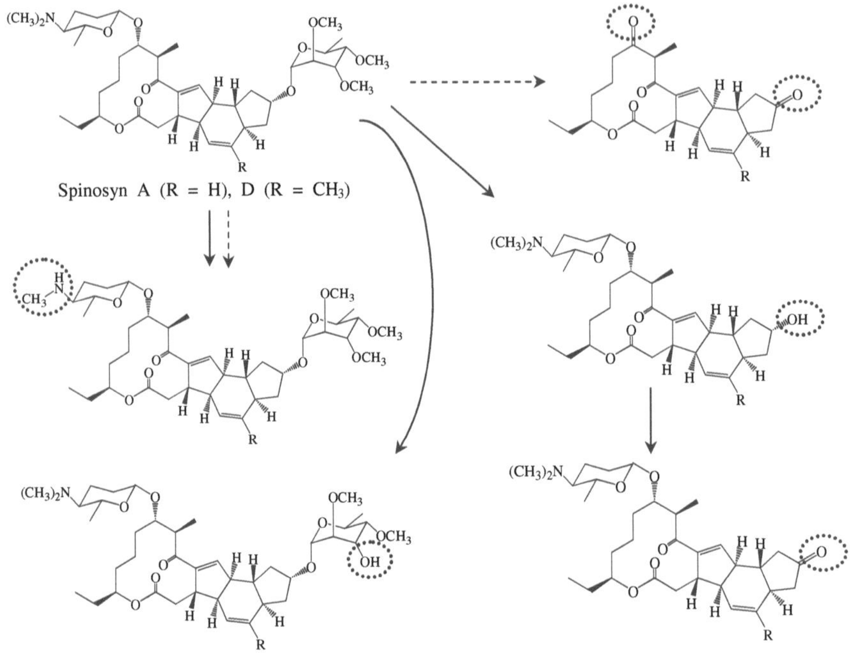

Fig. 11. Aerobic and anaerobic aquatic metabolism of spinosyn A and D. *Dashed lines*, aerobic; *Solid lines*, anaerobic.

al. 1997; EU-Ex 1999b). Tepraloxydim (150) showed moderate aerobic aquatic degradation; the main pathway was hydrolysis of the oxime ether to the corresponding imine derivative (PMRA 2004a). Fluridone (152) was gradually partitioned to sediment with a slow degradation to the acid derivative (Muir and Grift 1982). Detection of ring-hydroxylated derivatives showed involvement of oxidative ring destruction to the carboxyl group. Spinosad (153) is a mixture of spinosyn A and D and gave different metabolites depending on system aerobicity (Cleveland et al. 2002; CDPR 2004). *N*-Demethylation at the forosamine moiety was common to both isomers irrespective of aerobicity (Fig. 11). Forosamine and rhamnose moieties were liberated via aerobic ether cleavage to give the 9,17-diketone isomers with formation of bound residues and mineralization, whereas under anaerobic conditions ether cleavage to liberate the rhamnose moiety followed by oxidation to 9-keto derivative and *O*-demthylation at the rhamnose ring were observed.

VI. Fate of Pesticides in Microcosms and Mesocosms

Many variable factors in natural aquatic environments make the distribution and degradation profiles of a pesticide more complicated than expected from the simple laboratory water–sediment system (Miyamoto et al. 1990).

To assess ecotoxicological impacts by a pesticide and its degradates more realistically, larger-scale indoor and outdoor systems have been investigated by many researchers, and either similarities or differences in its fate among the laboratory and larger-scale studies have been found (Table 17) (Miyamoto et al. 1985; Leeuwangh et al. 1994; Johnson et al. 1994; Traub-Eberhard et al. 1994; Brock et al. 1995; Rönnefahrt et al. 1997; Caquet et al. 2000; Boxall et al. 2001). Møhlenberg et al. (2001) reviewed 91 mesocosm studies for 31 pesticides and evaluated the relationships between ecotoxicological effects and system characteristics. To simulate ditch, stream, and paddy field environments more realistically, corresponding indoor microcosm systems have been proposed (Schäfers and Hassink 2000; Dierksmeier et al. 2002; Florip et al. 2003). Sunlight photodegradation is one of the factors differing most from an ordinary laboratory water–sediment system. The contribution of photolysis, however, is likely to be limited to a shallow water column in the neighborhood of an air–water interface (Maguire and Hale 1980; Reeves 1999; Cleveland et al. 2002; Hein et al. 2003) because of either less transmission of light into a deeper column of natural water resulting from the presence of macrophytes and suspended matter or rapid adsorption onto it (Miller and Zepp 1979a,b; Crosby 1994; Getsinger et al. 2000; Petty et al. 2001).

A. Differences from Laboratory Water–sediment Systems

pH of Water Column. Under illumination by natural sunlight or artificial light, photosynthesis in macrophytes and algae usually occurs. In the case of aquatic angiosperms, utilization of HCO_3^- as the carbon source is known to result in the production of 1 mole OH^- for each mole CO_2 assimilated. Prins et al. (1980) measured the pH change at the upper and lower sides of leaves using a microelectrode and found that pH jumps by at least 1–2 to an alkaline condition, which occurs within 30 min after illumination with extent depending on light intensity. Increase of pH has been demonstrated in the enclosures of pond microcosms where the pH value in the dark is around 7.5 but sunlight exposure increases pH up to 9.2–9.5 (Graham et al. 1999a; Rand et al. 2000; Ensz et al. 2003). The diurnal pH change was also reported for the ditch microcosm treated with *lambda*-cyhalothrin (39) (Leistra et al. 2003). The pH value fluctuated from 7 to 8 at night to maximum values around 10 in the middle of the day, which was considered to originate from the photosynthesis activity of macrophytes and phytoplanktons. Woodruff et al. (1999a,b) reported a similar pH change at a water–sediment interface where algal biofilm was formed by using the fluvarium channel including natural river sediments under flow conditions. The pH value just above the interface was approximately 3 points higher than below and changed diurnally with light : dark cycle by a pH value of 1, which indicated the importance of algal photosynthesis at the

Table 17. Comparison of fate and relevant controlling factors among laboratory water–sediment system, outdoor microcosms, and outdoor mesocosms.

	Laboratory water–sediment system	Indoor model ecosystem	Outdoor microcosm	Outdoor mesocosm
Water chemistry				
Volume	0.1–1 L	1–100 L	0.1–10 m^3	>10–50 m^3
pH and salinity	Mostly constant		Diurnal change, weather- and season dependent	
Suspended matters	Negligible		Moderate to significant	
Biota	Negligible		Dependent on the experimental design	
Sediment system				
Depth	<2–3 cm	5–10 cm	20–50 cm	
Biofilm	Negligible		Moderate to significant	
Bioturbation	Negligible		Moderate to significant	
Application	Less realistic		Good simulation (spray drift, runoff)	
System stability	Shorter period (week–month)		Longer period (month–year)	
System complexity	Simple	Moderate	Complex	
Results				
Reactions	Mainly dark		Various including photolysis	
Rate(s)	For partial processes		Good to simulate realistic ones	
Distribution	Easy to monitor		Sometimes difficult to monitor	
Mass balance	Good	Moderate	Difficult	
Metabolites	Easy to identify		Sometimes difficult to monitor	

sediment–water interface. These pH fluctuations in outdoor microcosms and mesocosms are considered to affect the degradation profiles of a pesticide having a hydrolytically labile structure. In microcosm and pond mesocosm studies of cyfluthrin [racemic mixtures of (38)] conducted under the same conditions, greater degradation in the microcosm having the higher pH value was reported (Johnson et al. 1994). In the case of chlorpyriphos (155), the pH value in the pond microcosms was 8–9.5 and alkaline hydrolysis was considered to be one of the most important degradation pathways (Giddings et al. 1997).

Surface Microlayer. At the air–water interface of a natural water body, various types of organic matter with different origins are known to form a surface microlayer (Norkrans 1980), which is conveniently defined as <1 mm thick with high densities of particles and microorganisms (GESAMP 1995) and is considered to affect gaseous exchanges and transport processes from the water column to the atmosphere and vice versa. The fugacity model incorporating the surface microlayer was found to describe well the fate of several pesticides in microcosm studies, and the lesser capacity of the microlayer for hydrophobic substances than octanol was estimated (Southwood et al. 1999). There are several techniques to collect the microlayer, but the components and thickness are highly dependent on the sampling method (Hardy 1982). Not only natural surface-active substances such as amino acids, proteins, fatty acids, and lipids but also anthropogenic materials including pesticides are known to be enriched in this layer, as decribed by the enrichment factor (EF), defined as $EF = [S]_{sm}/[S]_{sb}$ (S, concentration of a chemical; "sm" and "sb," surface microlayer and subsurface water, respectively). This factor was known to possibly change diurnally from the monitoring of a lake surface microlayer (Södergren 1984). High EF values of $>10\text{–}10^4$ were reported for polychlorinated contaminants (Napolitano and Richmond 1995) and some pesticides (Eisenreich et al. 1978; Gever et al. 1996). In the case of pond microcosm studies of fenitrothion (13) and esfenvalerate (41), the EF values were calculated to be 100–150 immediately after a spray application (<1 hr), but such an enrichment gradually disappeared with time (Maguire and Hale 1980; Samsøe-Petersen et al. 2001). The higher EF value of $10^2\text{–}10^4$ was observed continuously for up to 2 wk when cypermethrin [racemic mixtures of (35)] was sprayed on the experimental pond, and a mixed community of vascular plants and filamentous algae might inhibit efficient diffusion of the pesticide to the subsurface water (Crossland 1982).

Organic matter detected in a surface microlayer comes from various origins (Meyers and Kawka 1982). The algal input results in fatty acid distribution with C_{12}–C_{18} straight chains, while C_{22}–C_{28} straight-chain acids would be dominant in input from terrestrial plants. Phytoplankton gives C_{17} alkanes as well as $C_{27,29,31}$, and the microbial input provides branched lipids. Chemical components detected in the surface microlayer have been investigated extensively by instrumental analyses of extracts such as infrared spectroscopy (Gucinski et al. 1981). Although the EF value for each chemical class was dependent on the origin of water, hydrocarbons, fatty acids, and phospholipids were dominantly enriched in several U.S. stream waters (Napolitano and Richmond 1995). GC-MS analysis of hydrocarbons showed the existence of C_{13}–C_{32} aliphatic straight-chain alkanes, and the greater population of odd-numbered alkanes indicated the importance of leaf origin. Organic carbon detected in a surface microlayer of a mesohumic lake in Norway was found to be mostly in the dissolved form, and size-exclusion analysis showed that acetic and hexadecanoic acids were the dominant species (Knulst et al. 1997). Chlorophyll *a* was also enriched in the microlayer as a result of the dominance of green algae together with phospholipids whose main acids contained $C_{16:0}$, $C_{16:1}$, and $C_{18:1}$ chains. Knulst et al. (1998) analyzed macromolecular fractions (using XAD-8/XAD-4 resins) from the lake water by ^{13}C-NMR and found enrichment of humic and fulvic acids together with hydrophilic neutral fractions in the surface microlayer. Analysis of the wind-generated foam from the surface microlayer of a U.S. lake showed enrichment of particulate matter consisting of organic and inorganic phsopshorus, nitrogen, and carbon, that is, primarily proteinaceous and carbonaceous matter, together with heavy metal ions (Eisenreich et al. 1978). The site-specific abundance of particulate matter was also reported for a lake microlayer (Meyers and Owen 1980). Particulate organic carbon tended to decrease from the river mouth to the open lake due to sedimentation. A similar trend was observed for fatty acids, whose dominant chain was $C_{16:0}$ and $C_{18:1}$. The composition of fatty acids collected from paddy fields was also reported (Gever et al. 1996). Various types of C_{12}–C_{18} fatty acids were detected together with their methyl esters, but represented only 3.2% of the microlayer weight; the remainder consisted of polypeptide, polysaccharides, and polyphenols. Particulate organic matter is enriched in the microlayer with an EF value of 6–10, and the lipids detected are known to consist of neutral fats, waxes, phospholipids, and glycolipids (Hunter and Liss 1981). The higher even-to-odd carbon number (C_{12}–C_{16}) in the chain with a small fraction of branched structures implied planktonic lipids as their origin. Carbohydrates were also the dominant portion of the dissolved organic carbon in the microlayer. Kattner et al. (1985) reported the proportion of individual carbon and nitrogen containing dissolved substances in the seawater microlayer. The typical composition was carbohydrates (46%), proteins (41%), amino acids (7%), and fatty acids (5%) for carbon, and NH_4^+ (38%), protein (24%), NO_3^-+ NO_2^- (8%), and amino acids (4%).

Many chemicals acting as a photosensitizer or photochemically generating active oxygen species such as the HO· radical are also concentrated in this microlayer, and hence either direct or indirect photolysis would become an important degradation pathway (Crosby 1994; GESAMP 1995). The aqueous photodegradation of phloroglucinol was investigated in the presence and absence of the marine surface microlayer (Lin and Carlson 1991). Acceleration of photolysis was observed by addition of the microlayer components, with the extent being proportional to the added amount. Sodium azide retarded the degradation, which may imply the involvement of singlet oxygen in the reaction. Gever et al. (1996) reported the accelerated degradation of thiobencarb (79) by photolysis in the presence of microlayer components collected from rice paddies.

Stratification in Water Column. The laboratory water–sediment study as prescribed in various guidelines is a small-scale system with a water depth usually less than several centimeters, and thus the water phase is considered to quickly become homogeneous after application of a pesticide. Sharom and Solomon (1981) confirmed the homogeneity of a water phase immediately after application of permethrin (34) by comparing its concentration at two different depths. However, it would take more time for a pesticide to be evenly distributed in the deeper water phase. When atrazine (110) was applied to the top of a water–sediment system with depth of 22 cm (water) and 2 cm (sediment), a concentration gradient among water depths of 1, 8, and 15 cm was observed (Mersie et al. 2000). The concentration difference remained for up to 112 d at 5° and 24°C under static conditions, but that between the depth of 1 and 8 cm disappeared after 7 d at 24°C, indicating the importance of a diffusion process. This stratification was also reported for model ecosystems. The lower concentration of carbofuran (161) just above the bottom sediment than at the surface continued up to 15 d when it was applied to the water surface of an outdoor trench system (Garg and Agnihotri 1985).

Brock et al. (1992) examined the stratification in an indoor glass aquarium when a formulation of chlorpyrifos (155) was applied, together with added macrophytes. In the absence of macrophytes, the concentration of (155) at a water depth of 5 cm was larger by a factor of 50 than that at 25 and 45 cm immediately after application, but this difference disappeared by 24 hr postapplication. In contrast, the introduction of macrophytes made the diffusion of (155) to the deeper layer difficult, and the concentration at 45 cm reached only one-half to one-third of that at 5 and 25 cm, even after 8 d. Much less distribution of (155) to the bottom sediment was concomitantly observed, showing the interception by macrophytes for its distribution. In natural environments, water movement caused by either the diurnal change of water temperature or wind would facilitate a vertical mixing of water column, and thus the stratification observed in laboratory systems would disappear within a shorter period. Crum and Brock (1994) investigated the concentration gradient of (155) at three different depths also in an outdoor ditch microcosm and confirmed that homogeneity of the water column is realized within 1 d, even in the presence of macrophytes. In the case of linuron (86) applied to outdoor experimental ditches, however, clear stratification was not detected (Crum et al. 1998). Similar effects by macrophytes, more stratification and reduced partition to sediment, were reported in outdoor pond microcosms treated with a few pesticides (Helweg et al. 2003). The period of disappearing stratification in outdoor microcosms and mesocosms when a pesticide is applied to a water surface is considered to be within 1–2d (Muir et al. 1985; Knuth and Heinis 1992, 1995; Samsøe-Petersen et al. 2001), but when the system has a deeper water column, the concentration gradient disappeared slowly (Day et al. 1987).

Bioturbation in Sediment. There are many kinds of biota inhabiting bottom sediments, such as chironomids, oligochetes, and microorganisms. Especially by the activity of macrofauna, a sediment layer is considered to be partially mixed, that is, bioturbation, and as a result particle transport and solution-phase movement are enhanced (Warren et al. 2003). The concentration of dissolved molecular oxygen also increases with bioturbation, which may encourage activity of aerobes, leading to different types of biodegradation proceeding in the sediment. Transport of a particle-bound chemical by bioturbation has been proposed by Thiobodeaux et al. (2001) to occur within the upper decimeter of bottom sediment to a water–sediment interface, and chemical desorption at the interface would be followed by transport through the benthic boundary layer to the overlying water. Bioturbation is likely to cause greater distribution of a chemical to the sediment. An intact sediment core collected from a lake using a glass tube and an artificially prepared water–sediment system from the sieved sediment were subjected to an aerobic aquatic metabolism study of parathion (11) (Houx and Dekker 1987). Approximately twice the amount of (11) was distributed in the sediment phase of the intact core, showing the effect of bioturbation. Goedkoop and Peterson (2003) reported that higher chironomid density increases the concentration of lindane (1) both in porewater and overlying water with its concomitant decrease in the sediment. The burrowing and feeding activity of chironomid larvae could account for these changes of distribution. Furthermore, the effect of bioturbation on mass transfer has been examined for intact cores collected from a salt marsh and pond using *p*-chlorophenol and ^3H_2O as markers (Pritchard et al. 1987). Their concentration in the sediment decreased with depth under sterile conditions, and only 1/100–1/20 amounts of the markers were detected at 8-cm depth compared with those in the top 1-cm layer, whereas such gradients were not found under nonsterile conditions where chironomids were active and each marker was evenly distributed in the sediment layers. This vertical mixing of the bottom sediment was also proposed in lake and littoral enclosures by detection of pesticides in the lower layers of the bottom sediment (Veith and Lee 1971; Knuth and Heinis 1992).

The change of distribution in a water–sediment system caused by bioturbation has been theoretically analyzed by several researchers. Karickhoff and Morris (1985) have conveniently expressed the burrowing activity as a burial rate of glass beads in sediment and estimated the flux of a few chemicals from sediment to overlying water. Another approach using a compartment model was taken by O'Neill et al. (1989) to describe the depth-dependent distribution of fenthion (16) in sediment. A chemical was assumed to diffuse from overlying water to porewater, being in adsorption/desorption equilibrium with sediment. Fenthion (16) reached the depth of 2.5 cm under sterile conditions, while deeper transport of up to 5.5 cm was observed in the nonsterile sediment, from which approximately 70-fold-higher diffusion rates of (16) in the upper layers of sediments were estimated in the presence of sediment dwellers. Schaffner et al. (1997) measured the time-dependent distribution of aromatic compounds applied to the top of the sediment in both sediment and porewater at each depth and found that the diffusive loss from the top sediment layer is enhanced by bioturbation. Distribution profiles of carbendazim (methyl benzimidazol-2-ylcarbamate) were analyzed by assuming bioturbation as a dispersion of particles in sediment (Koelmans et al. 2000).

B. Degradation Profiles of Pesticides

The degradation profiles in these larger-scale systems are listed in Table 18. After spraying on the surface water of ponds and lakes, triclopyr (154) was gradually partitioned to the bottom sediment and biota with β-oxidation to form the corresponding phenol (Getsinger et al. 2000; Petty et al. 2001). The *O*-methylated phenol was only detected in biota, and the absence of dihydroxylated derivative of (154) by substitution of chlorines showed the insignificant contribution of photolysis. Although detailed degradation profiles for methyl parathion (12) (Crossland and Bennett 1984) and fenthion (16) (O'Neill et al. 1989) are not available, hydrolysis of the P-Oaryl linkage to the corresponding phenol with reduction of the nitro group have been reported for fenitrothion (13) (Maguire and Hale 1980), chloropyrifos (155) (Kale et al. 1999; Nhan et al. 2002) and its methyl derivative (156) (Setzo and Sundaram 1982). In the case of the shallow outdoor aquaria including macrophytes (Weinberger et al. 1982), (13) was partly photodegraded via oxidative desulfuration to its oxon, thiono-thiolo rearrangement to the *S*-isomer and oxidation of the aryl methyl group to the carboxyl, which resulted in less sediment residue than the dark control. The successive *S*-oxidation to form sulfoxide and sulfone was observed for phorate (157) in littoral mesocosms (Dieter et al. 1995), as similarly reported for laboratory studies of fenthion (16), oxydemeton-methyl (20), and tribufos (27). The contribution of microbial degradation was confirmed for glyphosate (28) in an outdoor aquarium containing sediment (Zaranyika and Nyandoro 1993).

DDT (6) was rapidly partitioned to sediment and biota in an indoor model ecosystem with reductive dechlorination and dehydrochlorination (Wandiga et al. 2002). When the system was aerated, the rate of volatilization loss was enhanced. Similar reductive reactions were reported for methoxychlor (7) in a lake enclosure, and the forced mixing of the water

Table 18. Degradation profiles of pesticides in microcosm systems.

No.	Pesticide/structure	System information[a] DT_{50} (w/s)[b]	Route of degradation[c]	Reference
154	Triclopyr 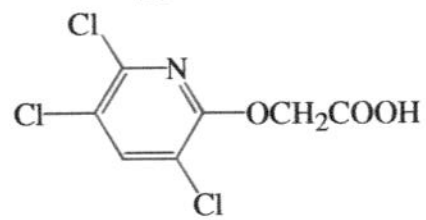	F, O, N.R., Lake Minnetonka, 28 d 3.7–4.7 d / 5.0–5.8 d	O2, C1 (O)	Getsinger et al. (2000)
		F, O, N.R., 3 U.S. ponds, 42 d 5.9–7.5 d / 2.8–4.6 d	O2, C1 (O)	Petty et al. (2001)
12	Methyl parathion 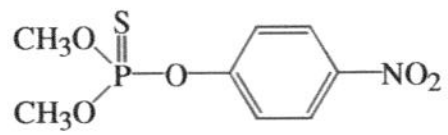	A, O, 10°–17°C, UK pond, 40 d 8.8–17.1 d / N.R.	N.R.	Crossland and Bennett (1984)
13	Fenitrothion 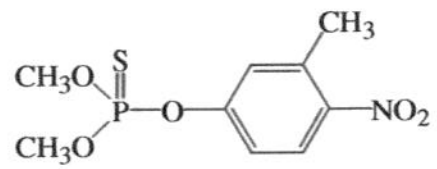	F, O, 14°C, 2 Canadian ponds, 50 hr 9.8 hr / 14.2 hr	R4, H7	Maguire and Hale (1980)
		F, O, 19°–23°C, Canadian lake, 28 d 1 d / N.R.	O1, O7, H7, M5	Weinberger et al. (1982)
16	Fenthion 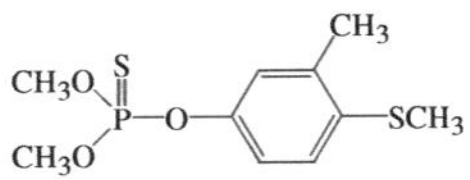	A, I, 20°C, salt marsh model ecosystem, 8 d 1.4–1.5 d / N.R.	N.R.	O'Neill et al. (1989)
155	Chlorpyrifos 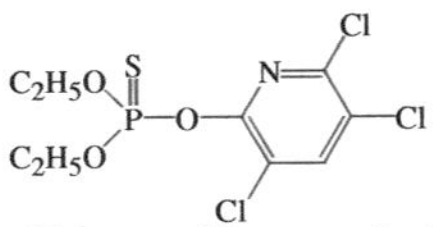	A, I, r.t., marine ecosystem in aquarium, 60 d N.R. / N.R.	H7	Kale et al. (1999)
		A, O, 30°–33°C, brackish water in aquaria, 30 d <2 d / <2 d	H7, B	Nhan et al. (2002)
156	Chlorpyrifos-methyl 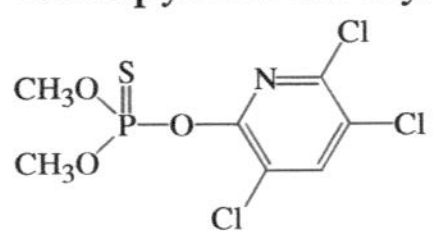	A, I, 15°C, soil–lake water, 90 d N.R. / 4 d	H7	Setzo and Sundaram (1982)

Table 18. *Continued*

No.	Pesticide/structure	System information[a] DT_{50} (w/s)[b]	Route of degradation[c]	Reference
157	Phorate 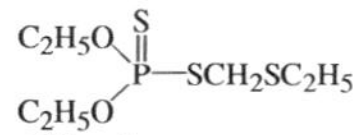	F, O, N.R., South Dakoda littoral, 28 d N.R. / N.R.	O6	Dieter et al. (1995)
28	Glyphosate 	A, O, N.R., Mukuvisi River, 72 d 9.8 d / 15.4 d	N.R.	Zaranyika and Nyandoro (1993)
6	DDT 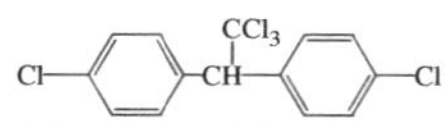	A, I, 23°–32°C, marine model ecosystem, 28 d 3–4 hr / N.R.	R1, R2	Wandiga et al. (2002)
7	Methoxychlor 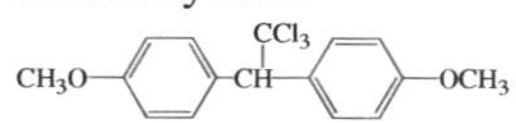	F, O, 10°–25°C, lake enclosure, 4 mon 6–13 d / N.R.	R1, R2	Solomon et al. (1986)
158	PCNB 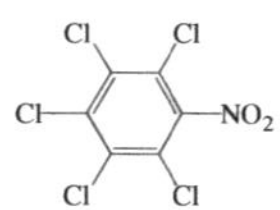	A, O, N.R., experimental pond, 6 mon 1.8 d / N.R.	R4, H8 (C—N)	Schauerte et al. (1982)
159	Bromoxynil (ester) 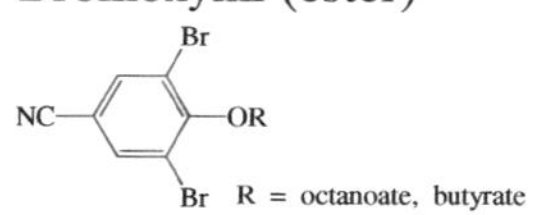	F, O, N.R., Prairie wetland, 120 d 0.3–1 d / N.R.	R1, H3	Muir et al. (1991)
36	Deltamethrin 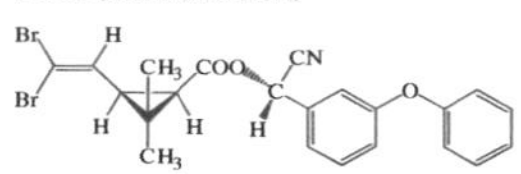	F, O, N.R., Canadian prairie pond, ~1 mon <1 d / N.R.	H3, M1 (benzyl-C)	Muir et al. (1992)
		F, O, 9°C, pond mesocosm, ~1 mon 1 hr / N.R.	H3, M1 (benzyl-C)	Maguire et al. (1989)
		F, O, 19°–24°C, U.S. pond, 7 d 8–18 hr / N.R.	H3, M1 (benzyl-C), M1 (*cis-trans*)	Erstfeld (1999)

Table 18. *Continued*

No.	Pesticide/structure	System information[a] DT_{50} (w/s)[b]	Route of degradation[c]	Reference
160	Tralomethrin 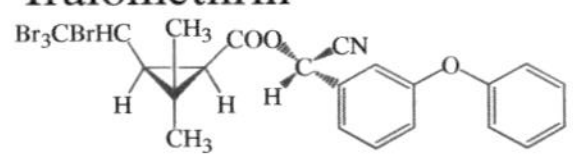	F, O, 19°–24°C, U.S. pond, 7 d 6.9 hr / N.R.	R1, H3, M1 (benzyl-C), M1 (*cis-trans*)	Erstfeld (1999)
39	*Lambda*-Cyhalothrin 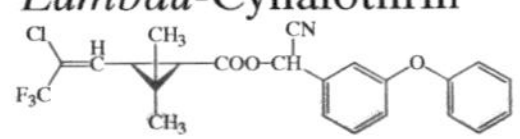	A, O, N.R., soil–water system in tank, N.R. N.R. / N.R.	H3	Bewick et al. (1984)
		A, I, 12°–17°C, U.K. pond, 96 hr <3 hr / <3 hr	H3	Hand et al. (2001)
62	Naproanilide 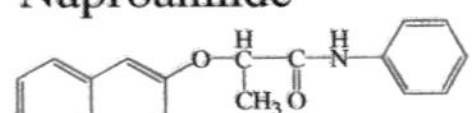	A, I, N.R., rice paddy model ecosystem, 23 d N.R. / N.R.	O5, H4, C1 (O)	Wang et al. (1992)
53	Alachlor 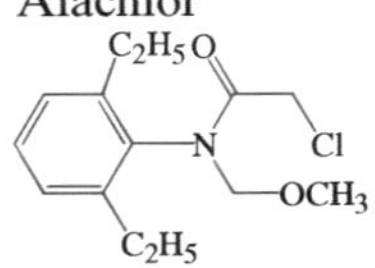	F, O, 26°C, Kansas pond sediment, 85 d 21 d / N.R.	C3 (glutathione) to sulfonic and oxalic acids	Graham et al. (1999b)
		F, O, N.R., aquatic model ecosystem, 40 d 3.3–3.4 d / 5.0–5.3 d	N.R.	Ramesh and Maheswari (2004)
55	Metolachlor 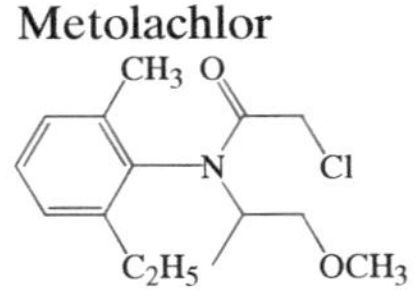	F, O, 26°C), Kansas pond, 85 d 33–46 d / N.R.	C3 (glutathione) to sulfonic and oxalic acids	Graham et al. (1999b)

Table 18. *Continued*

No.	Pesticide/structure	System information[a] DT_{50} (w/s)[b]	Route of degradation[c]	Reference
78	Carbaryl 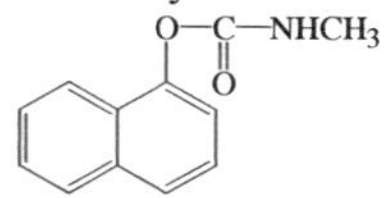	A, I, 23°C, aquatic model ecosystem, 20 d N.R. / N.R.	H5	Kanazawa et al. (1975)
161	Carbofuran 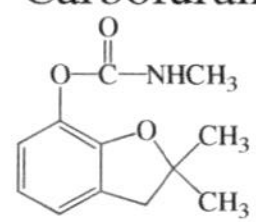	F, I, N.R., rice paddy ecosystem, 93 d N.R. / N.R.	O1 (3-furan), H5	Jinhe et al. (1989)
162	Carbosulfan 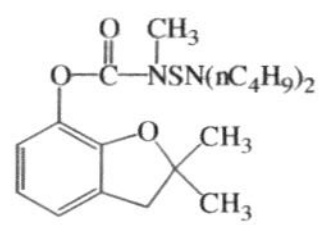	A, I, N.R., rice paddy model ecosystem, N.R. N.R. / N.R.	H8 (N—S)	Tejada et al. (1997)
163	Cyanatryn 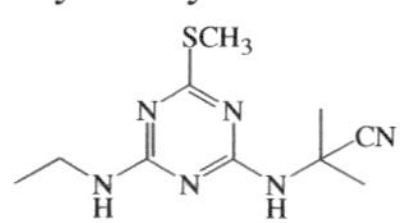	A, O, N.R., pond model ecosystem, 12 wk ~2 wk / N.R.	O3, H2, H8 (C—S)	Roberts (1974)
126	Trifluralin 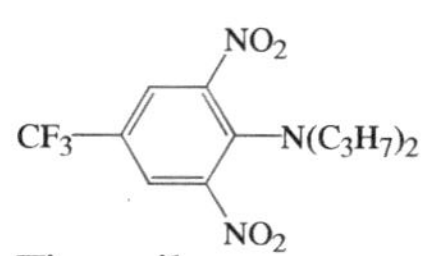	A, I, N.R., soil aquatic model ecosystem, 71 d 2.3 d / 39.9 d	R4, O3, M4 (benzimidazole)	Isensee et al. (1979)
148	Fipronil 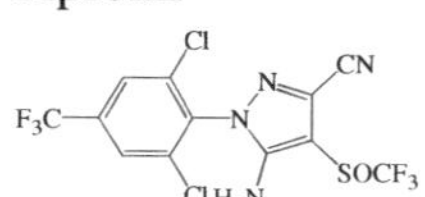	A, O, N.R., estuarine, 28 d 2–8.5 d / N.R.	O6, R5, M2 (C—S)	Walse et al. (2004)

Table 18. *Continued*

No.	Pesticide/structure	System information[a] DT_{50} (w/s)[b]	Route of degradation[c]	
124	Carfentrazone-ethyl 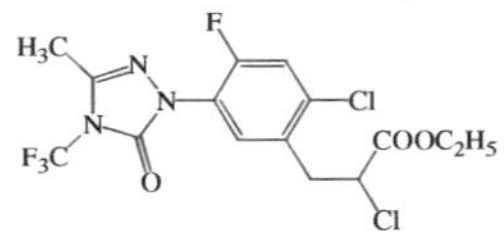	F, O, N.R., 3 rice fields, 1 wk–1 mon 6.5–11.1 hr / 38–174 hr	O1, R1, R2, H3	Ngim and Crosby (2001)
152	Fluridone 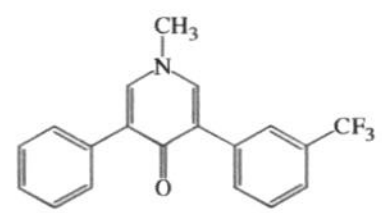	F, O, N.R., two small ponds, 14 wk 2–3.5 d / 7–17 wk	*o*- or *p*-O5, M3 (ph)	Muir and Grift (1982)
164	Epoxiconazole 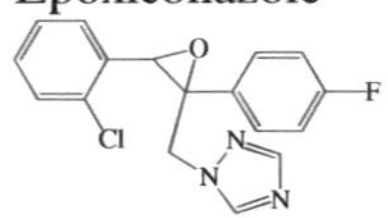	F, O, N.R., paddy field ecosystem, 40–100 d 11–20 d / 20–>97 d	photolysis	Lin et al. (2001)
165	Metribuzin 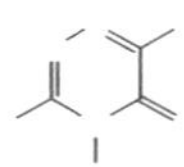	F, O, N.R., 0.1-ha microcosm, 30 d 5 d / N.R.	photolysis	Fairchild and Sappington (2002)
166	Azocyclotin 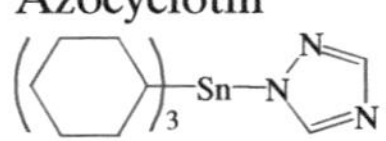	A, I, 20°C, pond, 9 mon 68 d (system)	H8 (Sn—N, Sn—C),	Kördel and Stein (1997)

N.R., information not reported.
[a]A (active ingredient) or F (formulation), O (outdoor) or I (indoor), water temperature (r.t., room temperature), site and duration of the experiment.
[b]Time of 50% disappearance of a pesticide–sediment system(s).
[c]Reaction number listed in Table 2. Terms in parentheses mean a position of reaction or product.

column greatly enhanced its adsorption to sediment (Solomom et al. 1986). In the case of PCNB (158), nitro reduction was observed in the sediment of an outdoor experimental pond together with substitution of the nitro group with methoxy (Schauerte et al. 1982). When mixtures of bromoxynil octanoate and butyrate (1:1) (159) were sprayed on experimental ponds, the butyrate ester disappeared more rapidly (Muir et al. 1991). Greater amounts of the octanoate ester in the suspended solids and bottom sediment showed its favorable partition due to its higher hydrophobicity, which resulted in less hydrolysis in the microcosm. The main product was the corresponding phenol and the photoproduct formed via monodebromination was also detected both in the water column and in bottom sediment.

Many freshwater aquatic field studies have been conducted for pyrethroids (Hill et al. 1994). Their common profiles are rapid dissipation from the water column by adsorption to suspended and bottom sediments followed by hydrolysis when sprayed to a water surface. Deltamethrin (36) more quickly dissipated from the surface microlayer than the subsurface water column, but its dissipation was slightly reduced when injected below the water surface due to suppression of volatilization (Maguire et al. 1989; Muir et al. 1985, 1992). A very low level of (36) was detected only in the bottom sediment due to hydrolysis and epimerization together with adsorption to suspended solids and biota. In the case of tralomethrin (160), rapid conversion to (36) was first detected in a pond study (Erstfeld 1999). The *trans*-isomer of (36) was detected in higher amounts from (160) than (36), which indicated involvement of a photodegradation process. *Lambda*-cyhalothrin (39) similarly produced the hydrolysis product, the corresponding chrysanthemic acid, and PBald in an indoor aquatic microcosm that included macrophytes (Bewick et al. 1984; Hand et al. 2001). Involvement of abiotic hydrolysis was unlikely considering pH of the overlying water, and less partitioning to sediment (approximately 5%) was observed than in the usual water–sediment study, implying that the macrophytes partly affected the distribution and degradation profiles of (39).

The behavior of naproanilide (62) was not significantly affected by the coexistence of several aquatic biota in a rice paddy model ecosystem (Wang et al. 1992). Concentration dependency on the behavior of alachlor (53) was not observed when an EC formulation was applied to the model ecosystem (Ramesh and Maheswari 2004). In tank mesocosms set in a shallow pond, the fate of (53) and metolachlor (55) was examined by focusing on the behavior of their glutathione conjugate metabolites, ethanesulfonic and oxanilic acids (Graham et al. 1999b). Formation of the former metabolite in water was similar between the two herbicides whereas that of the oxanilic acid was greatly enhanced for (53). This difference was considered to stem from the bulkiness of the *N*-alkyl substituent in (55), which hinders further conversion of the cysteine conjugate. Carbaryl (78) partitioned mainly to the bottom sediment, with its portion escaping from the system by volatilization and the main degradation route being cleavage of the carbamate linkage to form 1-naphthol (Kanazawa et al. 1975). Soil and water applications were examined for carbofuran (161) in the rice–fish ecosystem; uptake to plants and fish was greater in the water application (Jinhe et al. 1989). Rapid conversion of carbosulfan (162) to (161) was reported in a model rice paddy ecosystem (Tejada et al. 1997).

The typical degradation patterns in triazine herbicides were observed for cyanatryn (163) applied to a model ecosystem (Roberts 1974). Substitution of the methylthio group with OH proceeded with the successive hydrolysis of the nitrile group in the *N*-alkyl side chain. Partition and degradation profiles of trifluralin (126) have been extensively investigated in an ecosystem chamber including sediment and biota (Isensee et al. 1979). Rapid dissipation of (126) from the water phase was accompanied by adsorption to bottom sediment and formation of the benzimidazole derivative being most likely produced via photolysis. The effect of photolysis has been also reported for fipronil (148) (Walse et al. 2004). The dissipation of (148) in an estuarine mesocosm showed the hockey-stick profile. During the first rapid decline due to adsorption to sediment, photo-induced desulfinylation proceeded in the water column and the resultant degradate was partitioned to every compartment. The sulfone derivative by *S*-oxidation was detected mainly in the water, while the sulfide formed via reduction was detected only in bottom sediment. The contribution of indirect photolysis was reported for carfentrazone-ethyl (124) applied to flooded rice fields (Ngim and Crosby 2001). (124) underwent rapid ester hydrolysis in the side chain followed by reductive dechlorination and dehydrochlorination being observed in the usual water–sediment study. The oxidation product from the resulting acid was also detected and small UV absorption by (124) and its acid at 300 nm indicated the importance of indirect photolysis involving reaction with OH· radicals. Ring hydroxylation proceeded either microbiologically or photochemically for fluridone (152), but the latter process was found dominant in a field pond study (Muir and Grift 1982). The further oxidative ring opening reported for the usual water–sediment study could not be detected in the field study. The contribution of photolysis was proposed for epoxiconazole (164) and metribuzin (165), but the chemical identification of products is not available (Lin et al. 2001; Fairchild and Sappington 2002). In the case of azocyclotin (166) applied as a particle-bound form to simulate runoff events, rapid hydrolysis via cleavage of the C-N(triazole) bond was observed (Kördel and Stein 1997).

Summary

Many experimental reports on the fate of pesticides in either laboratory or outdoor water–sediment systems have been obtained from both research and regulatory aspects that show some trends in distribution and degradation for each chemical class of pesticides. Adsorption, diffusion, hydrolysis, and biodegradation processes are important in controlling the behavior of pesticides in these water–sediment systems. Through these investigations, the contribution of suspended particles and dissolved organic matter has become more accepted in relation to these processes. Not only the physicochemical properties and degradability of a pesticide but also the characteristics of the many phases composing a water–sediment system determine the actual pesticide behavior, and therefore we should appropriately design an experimental system by considering the real situation of the natural aqueous environment to be examined. Many factors controlling experimental results in a laboratory system such as water–sediment ratio, depth of water and sediment phases, and mixing of water column have been clarified; however, there are still many issues to be examined. For example, a pesticide is always used as a formulation, but its effects on pesticide behavior in a water–sediment system have not been extensively examined. When its behavior in a natural aquatic system is considered, the effect and importance of photolysis are necessary to examine as an individual degradation process, but photolysis has been only briefly discussed in outdoor microcosm and mesocosm studies. Many studies discuss the distribution and degradation pathways of a pesticide, but its transport between water and sediment phases has scarcely been investigated because of its complexity, especially for a pesticide that is moderately or easily degraded in a water–sediment system. This form of investigation would be very useful when metabolites or degradates having more toxicological impact on aquatic species and sediment dwellers are found. From this point of view, the behavior of a pesticide and its metabolite(s) in an interstitial sediment porewater should become another critical point to be examined in the future. Other issues to be investigated further are the relevant processes in the neighborhood of interfaces. In an air–water interface, the effect of a surface microlayer has been examined mainly through microcosm and mesocosm studies, but the contribution of interfaces to either volatilization or photodegradation should be examined in more detail to precisely estimate dissipation profiles of a pesticide in the real aquatic environment. Furthermore, the enrichment of a pesticide in this interface should be investigated in relation to an emergence of chironomids. Recently, many kinetic approaches have been attempted to more effectively use experimental data in prediction of the fate of a pesticide by the aid of a simulation model. Most existing rate data usually represent apparent dissipation rates but not degradation rates, and therefore separation of the degradation rate from dissipation by considering adsorption–desorption and transport processes would be of immense value.

References

Aajoud A, Ravanel P, Tissut M (2003) Fipronil metabolism and dissipation in a simplified aquatic ecosystem. J Agric Food Chem 51:1347–1352.

Adriaanse PI (2004) Water-sediment studies for step 3 FOCUS surface water: guidance of FOCUS degradation kinetics. In: Proceedings of the 3rd European Modelling Workshop, Catania, Italy. http://www.pfmodels.org/emw3.html.
Adriaanse PI, Vink JPM, Brouwer WWM, Leistra M, Tas JW, Linders JBHJ, Pol JW (2002) Estimating transformation rates of pesticides, to be used in the TOXSWA model, from water-sediment studies. Alterra rapport 023. Alterra Green World Research, Wageningen, The Netherlands.
Aelion CM, Mathur PP (2001) Atrazine biodegradation to deisopropylatrazine and deethylatrazine in coastal sediments of different land uses. Environ Toxicol Chem 20:2411–2419.
Agnihotri NP, Jain HK (1985) Persistence of thiophanate-methyl residues in soil, water, sediment and plant. Pesticides 19:30a–30d.
Agnihotri NP, Jain HK (1987) Persistence of flucythrinate and fluvalinate in soil, water and sediment. Pesticides 21:36–38.
Agnihotri NP, Jain HK, Gajbhiye VT (1986) Persistence of some synthetic pyrethroid insecticides in soil, water and sediment. Part 1. J Entomol Res 10:147–151.
Agnihotri NP, Jain HK, Gajbhiye VT, Srivastava KP (1989) Persistence of some synthetic pyrethroids and organophosphorus insecticides in soil, water and sediment. Part II. J Entomol Res 13:131–136.
Agriculture Canada (1987) Environmental chemistry and fate guidelines for registration of pesticides in Canada. T-1-255.
Alexander M (1981) Biodegradation of chemicals of environmental concern. Science 211:132–138.
Allan IJ, House WA, Parker A, Carter JE (2004) Transport and distribution of lindane and simazine in a riverine environment: measurements in bed sediments and modeling. Pestic Manag Sci 60:417–433.
Allen R, Arnold DJ (1990) The comparative fate of (^{14}C)-amitraz in different sediment/water types. In: Proceedings, BCPC Conferences: Pest and Diseases. British Crop Protection Council, Brighton, U.K., pp 1023–1028.
Anderson C, Brumhard B, Ditgens K, Reiner H (1999) Metabolism of fenhexamid (KBR 2738) in plants, animals and the environment. Pflanzenschutz-Nachrichten Bayer 52:227–251.
APVMA (2001a) Public release summary on evaluation of the new active fenhexamid in the product Teldor 500 SC Fungicide. Australian Pesticides and Veterinary Medicines Authority, Canberra, Australia.
APVMA (2001b) Public release summary on evaluation of the new active thiamethoxam in the product Cruiser 350FS Insecticide Seed Treatment. Australian Pesticides and Veterinary Medicines Authority, Canberra, Australia.
APVMA (2002a) Public release summary on evaluation of the new active butafenacil in the product Logran B-Powder Herbicide & Touchdown B-Powder Herbicide. Australian Pesticides and Veterinary Medicines Authority, Canberra, Australia.
APVMA (2002b) Public release summary on evaluation of the new active fludioxinil in the product Maxim 100FS Fungicide Seed Treatment. Australian Pesticides and Veterinary Medicines Authority, Canberra, Australia.

APVMA (2002c) Public release summary on evaluation of the new active trifloxysulfuron sodium in the product Envoke Herbicide. Australian Pesticides and Veterinary Medicines Authority, Canberra, Australia.
APVMA (2003) Public release summary on evaluation of the new active pyraclostrobin in the product Cabrio Fungicide. Australian Pesticides and Veterinary Medicines Authority, Canberra, Australia.
Armbrust KL (1999) Photochemical processes influencing pesticide degradation in rice paddies. J Pestic Sci 24:69–73.
Arnold DJ, Leake CR, Somerville L (1986) Clofentezine: degradation and mobility in soil/sediments. In: Proceedings, BCPC Conference: Pests and Diseases. British Crop Protection Council, Brighton, U.K., pp 851–855.
Attaway HH III, Camper ND, Paynter MJB (1982a) Anaerobic microbial degradation of diuron by pond sediment. Pestic Biochem Physiol 17:96–101.
Attaway HH III, Paynter MJB, Camper ND (1982b) Degradation of selected phenylurea herbicides by anaerobic pond sediment. J Environ Sci Health B17:683–699.
Babczinski P (2002) Environmental behavior of spinodiclofen (BAJ 2740; Envidor®). Pflanzenschutz-Nachrichten Bayer 55:197–209.
Bailey GW, Thruston AD Jr, Pope JD Jr, Cochrane DR (1970) The degradation kinetics of an ester of silvex and the persistence of silvex in water and sediment. Weed Sci 18:413–419.
BBA (1990) Degradability and fate of plant protectants in the water/sediment system. In: Guidelines for Examination of Plant Protectants in the Registration Process, Part IV 5-1. Federal Biological Institute for Agriculture and Forestry, Federal Republic of Germany, Braunschweig.
Beigel C, Beulke S, Boesten J, Aden K, Dust M, Dyson J, Fomsgaard I, Jones R, Karlsson S, van der Linden T, Soria OM, Richter O, Soulas G (2004) FOCUS work group on degradation kinetics. 2. Special considerations for metabolites in soil. In: Proceedings of the 3rd European Modelling Workshop, Catania, Italy. http://www.pfmodels.org/emw3.html.
Bennett (1990) Evaluation of the fate of pesticides in water and sediment. In: Hutson DH, Roberts TR (eds) Progress in Pesticide Biochemistry and Toxicology, vol 7. Wiley, New York, pp 149–173.
Berger BM, Wolfe NL (1996) Hydrolysis and biodegradation of sulfonylurea herbicides in aqueous buffers and anaerobic water-sediment systems: assessing fate pathways using molecular descriptors. Environ Toxicol Chem 15:1500–1507.
Beulke S, Brown CD (2001) Evaluation of methods to derive pesticide degradation parameters for regulatory modeling. Biol Fertil Soils 33:558–564.
Beulke S, Boesten J, Aden K, Beigel C, Dust M, Dyson J, Fomsgaard I, Jones R, Karlsson S, van der Linden T, Soria OM, Richter O, Soulas G (2004) FOCUS work group on degradation kinetics. 1. Overview, general data issues and parent kinetics. In: Proceedings of the 3rd European Modelling Workshop, Catania, Italy. http://www.pfmodels.org/emw3.html.
Bewick DW, Hill JR, Hamer M, Bharti H (1984) PP321: behaviour in terrestrial and aquatic ecosystems. In: Proceedings, BCPC Conference: Pests and Diseases, British Crop Protection Council, Brighton, U.K., pp 343–347.
Boesten JJTI, Jones RL, Aden K, Beigel CY, Beulke S, Dust M, Dyson JS, Fomsgaard IS, Karisson S, van der Linden AMA, Richter O, Magrans JO, Soulas G (2003) Calculation of degradation kinetics for use in regulatory exposure

assessment. In: Del Re AAM, Capri E, Padovani L, Trevisan M (eds) Pesticides in Air, Plant, Soil and Water System. Proceedings, 12th Symposium on Pesticide Chemistry, Piacenza, Italy, pp 467–471.

Bohn HL (1971) Redox potentials. Soil Sci 112:39–45.

Bollag JM, Liu SY (1990) Biological transformation processes of pesticides. In: Cheng HH (ed) Pesticides in the Soil Environment: Processes, Impacts, and Modeling. SSSA Book Series 2. Soil Science Society of America, Madison, WI, pp 169–211.

Bollag JM, McGahen LL, Minard RD, Liu SY (1986) Bioconcentration of alachlor in an anaerobic stream sediment. Chemosphere 15:153–162.

Bondarenko S, Gan J (2004) Degradation and sorption of selected organophosphate and carbamate insecticides in urban stream sediments. Environ Toxicol Chem 23:1809–1814.

Borchers H, Stupp HP (2004) Environmental fate of fluoxastrobin (HEC5725). Pflanzenschutz-Nachrichten Bayer 57:337–354.

Boschin G, D'Agostina A, Arnoldi A, Marotta E, Zanaradini E, Negri M, Valle A, Sorlini C (2003) Biodegradation of chlorsulfuron and metsulfuron-methyl by *Aspergillus niger* in laboratory conditions. J Environ Sci Health B38:737–746.

Boxall A, Brown C, Barrett K (2001) Higher tier laboratory aquatic toxicity testing. Cranfield Centre for EcoChemistry research report No. JF 4317E for DETR, Silsoe, U.K.

Brassard P (1997) Measurement of Eh and pH in aquatic sediments. In: Mudroch A, Azcue JM, Mudroch P (eds) Manual of Physico-Chemical Analysis of Aquatic Sediments. Lewis, Boca Raton, pp 47–67.

Brock TCM, Crum SJH, van Wijngaarden R, Budde BJ, Tijink J, Zuppelli A, Leeuwangh P (1992) Fate and effects of the insecticide Dursban® 4E in indoor *Elodea*-dominated and macrophyte-free freshwater model ecosystems: I. Fate and primary effects of the active ingredient chlorpyrifos. Arch Environ Contam Toxicol 23:69–84.

Brock TCM, Bos AR, Crum SJH, Gylstra R (1995) The model ecosystem approach in ecotoxicology as illustrated with a study on the fate and effects of an insecticide in stagnant freshwater microcosms. In: Hock B, Niessner Jr R (eds) Immunochemical Detection of Pesticides and Their Metabolites in the Water Cycle. VCH, Weinheim, pp 167–185.

Bromilow RH, Evans AA, Nichllos PH (2003) The influence of lipophilicity and formulation on the distribution of pesticides in laboratory-scale sediment/water systems. Pest Manag Sci 59:238–244.

Capel PD, Eisenreich SJ (1989) Sorption of organochlorines by lake sediment porewater colloids. In: Suffet IH, MacCarthy PM (eds) Aquatic Humic Substances: Influence on Fate and Treatment of Pollutants. Advances in Chemistry Series 219. American Chemical Society, Washington, DC, pp 185–207.

Caquet T, Lagadic L, Sheffield SR (2000) Mesocosms in ecotoxicology (1). Outdoor aquatic systems. Rev Environ Contam Toxicol 165:1–38.

Carlton RR, Allen R (1994) The use of a compartment model for evaluating the fate of pesticides in sediment/water system. In: Proceedings, BCPC Conference: Pests and Diseases. British Crop Protection Council, Brighton, U.K., pp 1349–1354.

CDPR (2001) Environmental fate of fipronil. In: Fate Reviews. California Department of Pesticide Regulation, Environmental Monitoring Branch, Sacramento, CA.

CDPR (2004) Environmental fate of spinosad. In: Fate Reviews. California Department of Pesticide Regulation, Environmental Monitoring Branch, Sacramento, CA.

Chesters G, Simsiman GV, Levy J, Alhajjar BJ, Fathulla RN, Harkin JM (1989) Environmental fate of alachlor and metolachlor. Rev Environ Contam Toxicol 110:1–74.

Chi J, Huang GL (2002) Modelling air-water exchange process of pentachlorophenol in the aquatic environment. Water Qual Res J Can 37:445–458.

Chiou CT, Peters LJ, Freed VH (1979) A physical concept of soil-water equilibria for nonionic organic compounds. Science 206:831–832.

Cleveland CB, Bormett GA, Saunders DG, Powers FL, McGibbon AS, Reeves GL, Rutherford L, Balcer JL (2002) Environmental fate of spinosad. 1. Dissipation and degradation in aqueous systems. J Agric Food Chem 50:3244–3256.

Cotham WE Jr, Bidleman TF (1989) Degradation of malathion, endosulfan and fenvalerate in seawater and seawater/sediment microcosms. J Agric Food Chem 37:824–828.

Cranwell PA (1976) Organic geochemistry of lake sediments. In: Nriagu JO (ed) Environmental Biogeochemistry, vol 1. Carbon, Nitrogen, Phosphorus, Sulfur and Selenium Cycles. ANN Arbor Science, Ann Arbor, MI, pp 75–88.

Crawford JJ, Sims GK, Simmons FW, Wax LM, Freedman DL (2002) Dissipation of the herbicide [^{14}C]dimethanamid under anaerobic conditions in flooded soil microcosms. J Agric Food Chem 50:1483–1491.

Cripe CR, Pritchard PH (1990) Aquatic test systems for studying the fate of xenobiotic compounds. Aquatic Toxicology Risk Assessment, vol 13. ASTM Special Technical Publication 1096. American Society for Testing and Materials, Washington, DC, pp 29–47.

Cripe CR, O'Neill EJ, Woods ME, Gilliam WT, Pritchard PH (1989) Fate of fenthion in salt-marsh environments: I. Factors affecting biotic and abiotic degradation rates in water and sediment. Environ Toxicol Chem 8:747–758.

Crosby DG (1994) Photochemical aspects of bioavailability. In: Hamelink JL (ed) Bioavailability Proceedings Pellston Workshop, 13th, Session 5. Dynamic Environmental Factors. Lewis, Boca Raton, pp 109–118.

Crossland NO (1982) Aquatic toxicology of cypermethrin. II. Fate and biological effects in pond experiments. Aquatic Toxicol 2:205–222.

Crossland NO, Bennett D (1984) Fate and biological effects of methyl parathion in outdoor ponds and laboratory aquaria. Ecotoxicol Environ Saf 8:471–181.

Crossland NO, Bennett D, Wolff CJM, Swannell RPJ (1986) Evaluation of models used to assess the fate of chemicals in aquatic systems. Pestic Sci 17:297–304.

Crum SJH, Brock TCM (1994) Fate of chlorpyrifos in indoor microcosms and outdoor experimental ditches. In: Hill IR, Heimbach F, Leeuwangh P, Matthiessen P (eds) Freshwater Field Tests for Hazard Assessment of Chemicals. Lewis, Boca Raton, pp 315–322.

Crum SJH, Aslderink GH, Brock TCM (1998) Fate of the herbicide linuron in outdoor experimental ditches. Chemosphere 36:2175–2190.

Daniels WM, House WA, Zhmud BV, Rae JE, Parker A (1998) Diffusive movement of simazine and lindane in river-bed sediments. Pestic Sci 54:212–222.

Daniels WM, House WA, Rae JE, Parlor A (2000) The distribution of micro-organic contaminants in river bed-sediment cores. Sci Total Environ 253:81–92.

Day KE, Kaushik NK, Solomon KR (1987) Impact of fenvalerate on enclosed freshwater planktonic communities and on in situ rates of filtration of zooplankton. Can J Fish Aquat Sci 44:1714–1728.
Deane G, Chroneer Z, Lick W (1999) Diffusion and sorption of hexachlorobenzene in sediments and saturated soils. J Environ Eng 125:689–696.
DeVitre RR, Sulzberger B, Buffle J (1994) Transformation of iron at redox boundaries. In: Buffle J, DeVitre RR (eds) Chemical and Biological Regulation of Aquatic Systems. Lewis, Boca Raton, pp 91–137.
Dierksmeier G, Martinez K, Ricardo C, Garcia M, Orta L, Moreno P, Suárez B (2002) Behavior of pesticides in a water/sediment system under laboratory and field conditions. Environ Technol 23:1303–1307.
Dieter CD, Duffy WG, Flake LD (1995) Environmental fate of phorate and its metabolites in northern prairie wetlands. J Freshwat Ecol 10:103–109.
Ding JY, Wu SC (1993) Laboratory studies of the effects of dissolved organic material on the adsorption of organochlorine pesticides by sediments and transport in rivers. Water Sci Technol 28:199–208.
Droppo IG, Ongley ED (1992) The state of suspended sediment in the freshwater fluvial environment: a method of analysis. Water Res 26:65–72.
Droppo IG, Leppard GG, Flannigan DT, Liss SN (1997) The freshwater floc: a functional relationship of water and organic and inorganic floc constituents affecting suspended sediment properties. Water Air Soil Pollut 99:43–53.
Dunnivant FM, Schwarzenbach RP, Macalady DL (1992) Reduction of substituted nitrobenzenes in aqueous solutions containing natural organic matter. Environ Sci Technol 26:2133–2141.
Dutch Guideline (1995) Data on the degradation in water/sediment systems. In: The Registration Form and Guidelines for the Submission of Applications for Registration of Pesticides. Application form A, Part G.2.1.2.
Dyson JS, Chapman PF, Snell RJ, Tarr JB (1999) Pesticide metabolites in soil: estimation of formation and degradation rates. In: Del Re, Attlio Amerigo Maria (eds) Human and Environmental Exposure to Xenobiotics. Proceedings, 11th Symposium on Pesticide Chemistry. Goliardica Pavese, Pavia, Italy, pp 165–170.
Eadie BJ, Morehead NR, Landrum PF (1990) Three-phase partitioning of hydrophobic organic compounds in Great Lakes waters. Chemosphere 20:161–178.
EC (2002) Guidance document on aquatic ecotoxicology. Sanco/3268/2001 rev 4 (final). European Commission, Health & Consumer Protection Directorate – General, Brussels.
Eisenreich SJ, Elzerman AW, Armstrong DE (1978) Enrichment of micronutrients, heavy metals and chlorinated hydrocarbons in wind-generated lake foam. Environ Sci Technol 12:413–417.
Elmarakby SA, Supplee D, Cook R (2001) Degradation of [^{14}C]carfentrazone-ethyl under aerobic aquatic conditions. J Agric Food Chem 49:5285–5293.
Ensz AP, Knapp CW, Graham DW (2003) Influence of autochthonous dissolved organic carbon and nutrient limitation on alachlor biotransformation in aerobic aquatic systems. Environ Sci Technol 37:4157–4162.
EPA (1994) Reregistration Eligibility Decision (RED) Metalaxyl. EPA 738-R-94-017. U.S. Environmental Protection Agency, Office of Prevention, Pesticides and Toxic Substances, Washington, DC.

EPA (1995a) Reregistration Eligibility Decision (RED) Asulam. EPA 738-R-95-024. U.S. Environmental Protection Agency, Office of Prevention, Pesticides and Toxic Substances, Washington, DC.

EPA (1995b) Reregistration Eligibility Decision (RED) Linuron. EPA 738-R-95-003. U.S. Environmental Protection Agency, Office of Prevention, Pesticides and Toxic Substances, Washington, DC.

EPA (1996a) Reregistration Eligibility Decision (RED) Amitrole. List A, Case 0095. U.S. Environmental Protection Agency, Office of Pesticide Programs, Special Review and Reregistration Division, Washington, DC.

EPA (1996b) Reregistration Eligibility Decision (RED) Desmedipham. EPA 738-R-96-006. U.S. Environmental Protection Agency, Office of Prevention, Pesticides and Toxic Substances, Washington, DC.

EPA (1997a) Reregistration Eligibility Decision (RED) Diflubenzuron. EPA 738-R-97-008. U.S. Environmental Protection Agency, Office of Prevention, Pesticides and Toxic Substances, Washington, DC.

EPA (1997b) Reregistration Eligibility Decision (RED) Thiobencarb. EPA 738-R-97-013. U.S. Environmental Protection Agency, Office of Prevention, Pesticides and Toxic Substances, Washington, DC.

EPA (1998a) Reregistration Eligibility Decision (RED) Bromoxynil. EPA 738-R-98-013. U.S. Environmental Protection Agency, Office of Prevention, Pesticides and Toxic Substances, Washington, DC.

EPA (1998b) Reregistration Eligibility Decision (RED) Dichlobenil. EPA 738-R-98-003. U.S. Environmental Protection Agency, Office of Prevention, Pesticides and Toxic Substances, Washington, DC.

EPA (1999a) Pesticide Reregistration Eligibility Decision (RED). Captan. EPA R-738-R-99-015. U.S. Environmental Protection Agency, Office of Prevention, Pesticides and Toxic Substances, Washington, DC.

EPA (1999b) Pesticide Reregistration Status (REDs, IREDS and TREDs). Documents for Oxydemeton-methyl. Revised Environmental Fate and Effects Assessment. U.S. Environmental Protection Agency, Office of Pesticide Program, Washington, DC, part 1.

EPA (2000a) Pesticide Reregistration Status (REDs, IREDS and TREDs). Documents for Malathion. Environmental Fate and Effects. U.S. Environmental Protection Agency, Office of Pesticide Program, Washington, DC.

EPA (2000b) Interim Reregistration Eligibility Decision (IRED) Profenofos. EPA 738-R-00-006. U.S. Environmental Protection Agency, Office of Prevention, Pesticides and Toxic Substances, Washington, DC.

EPA (2000c) Interim Reregistration Eligibility Decision (IRED) Tribufos. EPA 738-R-00-022. U.S. Environmental Protection Agency, Office of Prevention, Pesticides and Toxic Substances, Washington, DC.

EPA (2001a) Interim Reregistration Eligibility Decision (IRED) Acephate. EPA 738-R-01-013. U.S. Environmental Protection Agency, Office of Prevention, Pesticides and Toxic Substances, Washington, DC.

EPA (2001b) Reregistration Eligibility Decision (RED) Propargite. Case no. 0243. U.S. Environmental Protection Agency, Office of Prevention, Pesticides and Toxic Substances, Washington, DC.

EPA (2002a) Interim Reregistration Eligibility Decision (IRED) for Methamidophos. Case no. 0043. U.S. Environmental Protection Agency, Office of Prevention, Pesticides and Toxic Substances, Washington, DC.

EPA (2002b) Interim Reregistration Eligibility Decision (IRED) Naled. EPA 738-R-02-008. U.S. Environmental Protection Agency, Office of Prevention, Pesticides and Toxic Substances, Washington, DC.
EPA (2003) Interim Reregistration Eligibility Decision (IRED) for Methyl Parathion. Case no. 0153. U.S. Environmental Protection Agency, Office of Prevention, Pesticides and Toxic Substances, Washington, DC.
EPA (2004) Pesticide Fact Sheet, Dinotefuran. U.S. Environmental Protection Agency, Office of Prevention, Pesticides and Toxic Substances, Washington, DC.
Erstfeld KM (1999) Environmental fate of pyrethroids during spray drift and field runoff treatments in aquatic microcosms. Chemosphere 39:1737–1769.
Erzgräber B, Görlitz G, Gottesbüren B, Hosang J, Schäfer H, Resselen H, Aden K, Kloskowski R, Michalski B (2002) Recommendations for the calculation of the degradation behaviour of metabolites. Nachrichtenbl Deutsch Pflanzenschutzd 54:S25–S30.
EU-Ex (1999a) Plant protection products: existing active substances decisions and review reports. Technical review reports. Fluroxypyr. 6848/VI/98-final. European Commission, Health & Consumer Protection Directorate-General, E1 – Plant Health, Brussels.
EU-Ex (1999b) Plant protection products: existing active substances decisions and review reports. Technical review reports. Spiroxamine. 7584/VI/97-final. European Commission, Health & Consumer Protection Directorate-General, E1 – Plant Health, Brussels.
EU-Ex (2000a) Plant protection products: existing active substances decisions and review reports. Technical review reports. Bentazone. 7585/VI/97-final. European Commission, Health & Consumer Protection Directorate-General, E1 – Plant Health, Brussels.
EU-Ex (2000b) Plant protection products: existing active substances decisions and review reports. Technical review reports. Esfenvalerate. 6846/VI/97-final. European Commission, Health & Consumer Protection Directorate-General, E1 – Plant Health, Brussels.
EU-Ex (2000c) Plant protection products: existing active substances decisions and review reports. Technical review reports. Metsulfuron-methyl. 7593/VI/97-final. European Commission, Health & Consumer Protection Directorate-General, E1 – Plant Health, Brussels.
EU-Ex (2000d) Plant protection products: existing active substances decisions and review reports. Technical review reports. Triasulfuron. 7589/VI/97-final. European Commission, Health & Consumer Protection Directorate-General, E1 – Plant Health, Brussels.
EU-Ex (2001a) Plant protection products: existing active substances decisions and review reports. Technical review reports. 2,4-D. 7599 VI/97-final. European Commission, Health & Consumer Protection Directorate-General, E1 – Plant Health, Brussels.
EU-Ex (2001b) Plant protection products: existing active substances decisions and review reports. Technical review reports. *Lambda*-Cyhalothrin. 7572/VI/97-final. European Commission, Health & Consumer Protection Directorate-General, E1 – Plant Health, Brussels.
EU-Ex (2001c) Plant protection products: existing active substances decisions and review reports. Technical review reports. Pyridate. 7576/VI/97-final. European

Commission, Health & Consumer Protection Directorate-General, E1 – Plant Health, Brussels.

EU-Ex (2001d) Plant protection products: existing active substances decisions and review reports. Technical review reports. Thiabendazole. 7603/VI/97-final. European Commission, Health & Consumer Protection Directorate-General, E1 – Plant Health, Brussels.

EU-Ex (2001e) Plant protection products: existing active substances decisions and review reports. Technical review reports. Thifensulfuron-methyl. 7577/VI/97-final. European Commission, Health & Consumer Protection Directorate-General, E1 – Plant Health, Brussels.

EU-Ex (2002a) Plant protection products: existing active substances decisions and review reports. Technical review reports. *Beta*-Cyfluthrin. 6841/VI/97-final. European Commission, Health & Consumer Protection Directorate-General, E1 – Plant Health, Brussels.

EU-Ex (2002b) Plant protection products: existing active substances decisions and review reports. Technical review reports. Cyazofamid. Sanco/10379/2002-final. European Commission, Health & Consumer Protection Directorate-General, E1 – Plant Health, Brussels.

EU-Ex (2002c) Plant protection products: existing active substances decisions and review reports. Technical review reports. Cyhalofop-butyl. 6500/VI/99-final. European Commission, Health & Consumer Protection Directorate-General, E1 – Plant Health, Brussels.

EU-Ex (2002d) Plant protection products: existing active substances decisions and review reports. Technical review reports. 2,4-DB. 7601/VI/97-final. European Commission, Health & Consumer Protection Directorate-General, E1 – Plant Health, Brussels.

EU-Ex (2002e) Plant protection products: existing active substances decisions and review reports. Technical review reports. Deltamethrin. 6504/VI/99-final. European Commission, Health & Consumer Protection Directorate-General, E1 – Plant Health, Brussels.

EU-Ex (2002f) Plant protection products: existing active substances decisions and review reports. Technical review reports. Ethofumesate. 6503/VI/99-final. European Commission, Health & Consumer Protection Directorate-General, E1 – Plant Health, Brussels.

EU-Ex (2002g) Plant protection products: existing active substances decisions and review reports. Technical review reports. Glyphosate. 6511/VI/99-final. European Commission, Health & Consumer Protection Directorate-General, E1 – Plant Health, Brussels.

EU-Ex (2002h) Plant protection products: existing active substances decisions and review reports. Technical review reports. Iprodione. 5036/VI/98-final. European Commission, Health & Consumer Protection Directorate-General, E1 – Plant Health, Brussels.

EU-Ex (2002i) Plant protection products: existing active substances decisions and review reports. Technical review reports. Isoproturon. Sanco/3045/99-final. European Commission, Health & Consumer Protection Directorate-General, E1 – Plant Health, Brussels.

EU-Ex (2002j) Plant protection products: existing active substances decisions and review reports. Technical review reports. Linuron. 7595/VI/97-final. European

Commission, Health & Consumer Protection Directorate-General, E1 – Plant Health, Brussels.

EU-Ex (2003a) Plant protection products: existing active substances decisions and review reports. Technical review reports. Chlorpropham. Sanco/3041/99-final. European Commission, Health & Consumer Protection Directorate-General, E1 – Plant Health, Brussels.

EU-Ex (2003b) Plant protection products: existing active substances decisions and review reports. Technical review reports. Mecoprop-P. Sanco/3065/99-final. European Commission, Health & Consumer Protection Directorate-General, E1 – Plant Health, Brussels.

EU-Ex (2003c) Plant protection products: existing active substances decisions and review reports. Technical review reports. Pendimethalin. 7477/VI/98-final. European Commission, Health & Consumer Protection Directorate-General, E1 – Plant Health, Brussels.

EU-Ex (2003d) Plant protection products: existing active substances decisions and review reports. Technical review reports. Propiconazole. Sanco/3049/99-final. European Commission, Health & Consumer Protection Directorate-General, E1 – Plant Health, Brussels.

EU-Ex (2003e) Plant protection products: existing active substances decisions and review reports. Technical review reports. Propineb. Sanco/7574/97-final. European Commission, Health & Consumer Protection Directorate-General, E1 – Plant Health, Brussels.

EU-Ex (2003f) Plant protection products: existing active substances decisions and review reports. Technical review reports. Propyzamide. 6502/VI/99-final. European Commission, Health & Consumer Protection Directorate-General, E1 – Plant Health, Brussels.

EU-Ex (2004a) Plant protection products: existing active substances decisions and review reports. Technical review reports. *Alpha*-Cypermethrin. Sanco/4335/2000-final. European Commission, Health & Consumer Protection Directorate-General, E1 – Plant Health, Brussels.

EU-Ex (2004b) Plant protection products: existing active substances decisions and review reports. Technical review reports. Benalaxyl. Sanco/4351/2000-final. European Commission, Health & Consumer Protection Directorate-General, E1 – Plant Health, Brussels.

EU-Ex (2004c) Plant protection products: existing active substances decisions and review reports. Technical review reports. Bromoxynil. Sanco/4347/2000-final. European Commission, Health & Consumer Protection Directorate-General, E1 – Plant Health, Brussels.

EU-Ex (2004d) Plant protection products: existing active substances decisions and review reports. Technical review reports. Desmedipham. Sanco/4061/2001-final. European Commission, Health & Consumer Protection Directorate-General, E1 – Plant Health, Brussels.

EU-Ex (2004e) Plant protection products: existing active substances decisions and review reports. Technical review reports. Ioxynil. Sanco/4349/2000-final. European Commission, Health & Consumer Protection Directorate-General, E1 – Plant Health, Brussels.

EU-Ex (2004f) Plant protection products: existing active substances decisions and review reports. Technical review reports. Phenmedipham. Sanco/4060/2001-final.

European Commission, Health & Consumer Protection Directorate-General, E1 – Plant Health, Brussels.
EU-New (1998a) Plant protection products: new active substances decisions and review reports. Technical review reports. Azoxystrobin. 7581/VI/97-final. European Commission, Health & Consumer Protection Directorate-General, E1 – Plant Health, Brussels.
EU-New (1998b) Plant protection products: new active substances decisions and review reports. Technical review reports. Kresoxim-methyl. 7583/VI/97-final. European Commission, Health & Consumer Protection Directorate-General, E1 – Plant Health, Brussels.
EU-New (2001a) Plant protection products: new active substances decisions and review reports. Technical review reports. Cyclanilide. 7463/VI/98-final. European Commission, Health & Consumer Protection Directorate-General, E1 – Plant Health, Brussels.
EU-New (2001b) Plant protection products: new active substances decisions and review reports. Technical review reports. Flupyrsulfuron-methyl. 5050/VI/97-final. European Commission, Health & Consumer Protection Directorate-General, E1 – Plant Health, Brussels.
EU-New (2002a) Plant protection products: new active substances decisions and review reports. Technical review reports. Acibenzolar-s-methyl. 6506/VI/99-final. European Commission, Health & Consumer Protection Directorate-General, E1 – Plant Health, Brussels.
EU-New (2002b) Plant protection products: new active substances decisions and review reports. Technical review reports. Cinidon-ethyl. 6498/VI/99-final. European Commission, Health & Consumer Protection Directorate-General, E1 – Plant Health, Brussels.
EU-New (2002c) Plant protection products: new active substances decisions and review reports. Technical review reports. Ethoxysulfuron. 7461/VI/98-final. European Commission, Health & Consumer Protection Directorate-General, E1 – Plant Health, Brussels.
EU-New (2002d) Plant protection products: new active substances decisions and review reports. Technical review reports. Famoxadone. 6505/VI/99-final. European Commission, Health & Consumer Protection Directorate-General, E1 – Plant Health, Brussels.
EU-New (2002e) Plant protection products: new active substances decisions and review reports. Technical review reports. Florasulam. Sanco/1406/2001-rev 6. European Commission, Health & Consumer Protection Directorate-General, E1 – Plant Health, Brussels.
EU-New (2002f) Plant protection products: new active substances decisions and review reports. Technical review reports. Flumioxazin. 7471/VI/98-final. European Commission, Health & Consumer Protection Directorate-General, E1 – Plant Health, Brussels.
EU-New (2002g) Plant protection products: new active substances decisions and review reports. Technical review reports. Foramsulfuron. Sanco/10324/2002-final. European Commission, Health & Consumer Protection Directorate-General, E1 – Plant Health, Brussels.
EU-New (2002h) Plant protection products: new active substances decisions and review reports. Technical review reports. Imazamox. Sanco/4325/2000-final.

European Commission, Health & Consumer Protection Directorate-General, E1 – Plant Health, Brussels.
EU-New (2002i) Plant protection products: new active substances decisions and review reports. Technical review reports. Iprovalicarb. Sanco/2034/2000-final. European Commission, Health & Consumer Protection Directorate-General, E1 – Plant Health, Brussels.
EU-New (2002j) Plant protection products: new active substances decisions and review reports. Technical review reports. Metalaxyl-M. Sanco/3037/99-final. European Commission, Health & Consumer Protection Directorate-General, E1 – Plant Health, Brussels.
EU-New (2002k) Plant protection products: new active substances decisions and review reports. Technical review reports. Oxadiargyl. Sanco/3053/99-final. European Commission, Health & Consumer Protection Directorate-General, E1 – Plant Health, Brussels.
EU-New (2002l) Plant protection products: new active substances decisions and review reports. Technical review reports. Oxasulfuron. Sanco/4323/2000-final. European Commission, Health & Consumer Protection Directorate-General, E1 – Plant Health, Brussels.
EU-New (2002m) Plant protection products: new active substances decisions and review reports. Technical review reports. Picolinafen. Sanco/1418/2001-final. European Commission, Health & Consumer Protection Directorate-General, E1 – Plant Health, Brussels.
EU-New (2002n) Plant protection products: new active substances decisions and review reports. Technical review reports. Prosulfuron. Sanco/3055/99-final. European Commission, Health & Consumer Protection Directorate-General, E1 – Plant Health, Brussels.
EU-New (2002o) Plant protection products: new active substances decisions and review reports. Technical review reports. Pymetrozine. 7455/VI/98-final. European Commission, Health & Consumer Protection Directorate-General, E1 – Plant Health, Brussels.
EU-New (2002p) Plant protection products: new active substances decisions and review reports. Technical review reports. Pyraflufen-ethyl. Sanco/3039/99-final. European Commission, Health & Consumer Protection Directorate-General, E1 – Plant Health, Brussels.
EU-New (2002q) Plant protection products: new active substances decisions and review reports. Technical review reports. Sulfosulfuron. 7459/VI/98-final. European Commission, Health & Consumer Protection Directorate-General, E1 – Plant Health, Brussels.
EU-New (2003a) Plant protection products: new active substances decisions and review reports. Technical review reports. Carfentrazone-ethyl. 7473/VI/99-final. European Commission, Health & Consumer Protection Directorate-General, E1 – Plant Health, Brussels.
EU-New (2003b) Plant protection products: new active substances decisions and review reports. Technical review reports. Dimethanamid-P. Sanco/1402/2001-final. European Commission, Health & Consumer Protection Directorate-General, E1 – Plant Health, Brussels.
EU-New (2003c) Plant protection products: new active substances decisions and review reports. Technical review reports. Fenamidone. Sanco/1404/2001-final.

European Commission, Health & Consumer Protection Directorate-General, E1 – Plant Health, Brussels.
EU-New (2003d) Plant protection products: new active substances decisions and review reports. Technical review reports. Flufenacet. 7469/VI/98-final. European Commission, Health & Consumer Protection Directorate-General, E1 – Plant Health, Brussels.
EU-New (2003e) Plant protection products: new active substances decisions and review reports. Technical review reports. Flurtamone. Sanco/10162/2003-final. European Commission, Health & Consumer Protection Directorate-General, E1 – Plant Health, Brussels.
EU-New (2003f) Plant protection products: new active substances decisions and review reports. Technical review reports. Iodosulfuron. Sanco/10166/2003-final. European Commission, Health & Consumer Protection Directorate-General, E1 – Plant Health, Brussels.
EU-New (2003g) Plant protection products: new active substances decisions and review reports. Technical review reports. Isoxaflutole. Sanco/3136/99-final. European Commission, Health & Consumer Protection Directorate-General, E1 – Plant Health, Brussels.
EU-New (2003h) Plant protection products: new active substances decisions and review reports. Technical review reports. Mesotrione. Sanco/1416/2001-final. European Commission, Health & Consumer Protection Directorate-General, E1 – Plant Health, Brussels.
EU-New (2003i) Plant protection products: new active substances decisions and review reports. Technical review reports. Picoxystrobin. Sanco/10196/2003-final. European Commission, Health & Consumer Protection Directorate-General, E1 – Plant Health, Brussels.
EU-New (2003j) Plant protection products: new active substances decisions and review reports. Technical review reports. Propoxycarbazone. Sanco/4067/2001-rev, final. European Commission, Health & Consumer Protection Directorate-General, E1 – Plant Health, Brussels.
EU-New (2003k) Plant protection products: new active substances decisions and review reports. Technical review reports. Silthiofam. Sanco/1424/2001-final. European Commission, Health & Consumer Protection Directorate-General, E1 – Plant Health, Brussels.
EU-New (2003l) Plant protection products: new active substances decisions and review reports. Technical review reports. Trifloxystrobin. Sanco/4339/2000-final. European Commission, Health & Consumer Protection Directorate-General, E1 – Plant Health, Brussels.
Fairchild JF, Sappington LC (2002) Fate and effects of the triazinone herbicide metribuzin in experimental pond mesocosms. Arch Environ Contam Toxicol 43:198–202.
FAO (1999a) JMPR: Evaluations of Pesticide Residues in Food. Bitertanol (144). Food and Agriculture Organization of the United Nations, New York, pp 3–95.
FAO (1999b) JMPR: Evaluations of Pesticide Residues in Food. Pyriproxyfen (200). Food and Agriculture Organization of the United Nations, New York, pp 673–728.
Florip P, Pancaldid D, Braschi I, Gessa C (2003) Behavior of four herbicides in a paddy field: simulation by a laboratory microcosm. In: Del Re AAM, Capri E,

Padovani L, Trevisan M (eds) Pesticides in Air, Plant, Soil and Water Systems. Proceedings, 12th Symposium on Pesticide Chemistry, Piacenza, Italy, pp 223–233.

Formica SJ, Baron JA, Thibodeaux LJ, Valsaraj KT (1988) PCB transport into lake sediments, computational model and laboratory simulation. Environ Sci Technol 22:1435–1440.

Gallé T, Van Lagen B, Kurtenbach A, Bierl R (2004) An FTIR-DRIFT study on river sediment particle structure: Implications for biofilm dynamics and pollutant binding. Environ Sci Technol 38:4496–4502.

Gambrell RP, Reddy CN, Collard V, Green G, Patrick Jr WH (1981) Behavior of DDT, kepone and permethrin in sediment-water systems under different oxidation-reduction and pH conditions. EPA-600/3-81-038. U.S. Environmental Protection Agency, Washington, DC.

Gambrell RP, Taylor BA, Reddy KS, Patrick Jr WH (1984) Fate of selected toxic compounds under controlled redox potential and pH conditions in soil and sediment-water systems. EPA-600/3-83-018. U.S. Environmental Protection Agency, Washington, DC.

Gao JP, Maguhn J, Spitzauer P, Kettrup A (1997) Distribution of pesticides in the sediment of the small Teufelsweiher pond (southern Germany). Water Res 31:2811–2819.

Gao JP, Maguhn J, Spitzauer P, Kettrup A (1998a) Sorption of pesticides in the sediment of the Teufelsweiher pond (southern Germany) I: Equilibrium assessments, effect of organic carbon content and pH. Water Res 32:1662–1672.

Gao JP, Maguhn J, Spitzauer P, Kettrup A (1998b) Sorption of pesticides in the sediment of the Teufelsweiher pond (southern Germany). II: Competitive adsorption, desorption of aged residues and effect of dissolved organic carbon. Water Res 32:2089–2094.

Garg PK, Agnihotri NP (1985) Persistence of carbofuran and bendiocarb in water and sediment. Ann Agric Res 6:183–186.

GESAMP (1995) The sea-surface microlayer and its role in global change. GESAMP Reports and Studies, No. 59. WHO, Geneva. http://gesamp.imo.org/publicat.htm.

Getsinger KD, Petty DG, Madsen JD, Skogerboe JG, Houtman BA, Haller WT, Fox AM (2000) Aquatic dissipation of the herbicide triclopyr in lake Minnetonka, Minnesota. Pestic Manag Sci 56:388–400.

Gever JR, Maburg SA, Crosby DG (1996) Rice field surface micrlayers: collection, composition and pesticide enrichment. Environ Toxicol Chem 15:1676–1682.

Giddings JM, Biever RC, Racke KD (1997) Fate of chlorpyrifos in outdoor pond microcosms and effects on growth and survival of bluegill sunfish. Environ Toxicol Chem 16:2353–2362.

Gilliom RJ (2001) Pesticides in the hydrologic system – what do we know and what's next? Hydrol Process 15:3197–3201.

Gobas FAPC, Maclean LG (2003) Sediment-water distribution of organic contaminants in aquatic ecosystems: the role of organic carbon mineralization. Environ Sci Technol 37:735–741.

Goedkoop W, Peterson M (2003) The fate, distribution and toxicity of lindane in tests with *Chironomus riparius*: effects of bioturbation and sediment organic matter content. Environ Toxicol Chem 22:67–76.

Golab T, Althaus WA, Wooten HL (1979) Fate of [^{14}C] trifluralin in soil. J Agric Food Chem 27:163–179.

Graebing PW, Chib JS, Hubert TD, Gingerich WH (2004) Metabolism of niclosamide in sediment and water systems. J Agric Food Chem 52:5924–5932.

Graetz DA, Chesters G, Daniel TC, Newland LW, Lee GB (1970) Parathion degradation in lake sediments. J Water Pollut Control Fed 42:R76–R94.

Graham DW, Miles D, DeNoyelles F, Smith VH (1999a) Development of small outdoor microcosms for studying contaminant transformation rates and mechanisms under various water column conditions. Environ Toxicol Chem 18: 1124–1132.

Graham WH, Graham DW, Denoyelles F Jr, Smith VH, Larive CK, Thurman EM (1999b) Metolachlor and alachlor breakdown product formation patterns in aquatic field mesocosms. Environ Sci Technol 33:4471–4476.

Groenendijk P, van der Kolk JWH, Travis KZ (1994) Prediction of exposure concentrations in surface waters. In: Hill IR, Heinbach F, Leeuwangh P, Matthiessen P (eds) Freshwater Field Tests for Hazard Assessment of Chemicals. Lewis, Boca Raton, pp 105–125.

Gschwend PM, Wu SC (1985) On the constancy of sediment-water partition coefficients of hydrophobic organic pollutants. Environ Sci Technol 19:90–96.

Gucinski H, Goupil DW, Baier RE (1981) Sampling and composition of the surface microlayer. In: Eisenreich SJ (ed) Atmospheric Pollutants in Natural Waters. Ann Arbor Press, Ann Arbor, Mi, pp 165–180.

Gupta S, Gajbhiye VT (2005) Dissipation of β-cyfluthrin in water as affected by sediment, pH and temperature. Bull Environ Contam Toxicol 74:40–47.

Guth JA (1981) Experimental approaches to studying the fate of pesticides in soil. In: Hutson DH, Roberts TR (eds) Progress in Pesticide Biochemistry, vol 1. Wiley, New York, pp 85–114.

Häggblom MM (1992) Microbial breakdown of halogenated aromatic pesticides and related compounds. FEMS Microbiol Rev 103:29–72.

Halarnkar PP, Leimkuehler WM, Green DL, Marlow VA (1997) Degradation of [^{14}C]tebupirimphos under anaerobic aquatic conditions. J Agric Food Chem 45:2349–2353.

Hand LH, Kuet SF, Lane MCG, Maund SJ, Warinton JS, Hill IR (2001) Influences of aquatic plants on the fate of the pyrethroid insecticide lambda-cyhalothrin in aquatic environments. Environ Toxicol Chem 20:1740–1745.

Handoo JK (1986) The nature of sediments in some wetlands. Acta Hydrochim Hydrobiol 14:485–493.

Hardy JT (1982) The sea surface microlayer: biology, chemistry and anthropogenic enrichment. Prog Oceanogr 11:307–328.

Hastings ME, Poppe LJ, Hathaway JC (2001) Surficial sediment database. In: USGS East-coast sediment analysis: procedures, database and georeferenced displays. Open-file report 00-358. U.S. Geological Survey, Washington, DC.

Headley JV, Gandrass J, Kuballa J, Peru KM, Gong Y (1998) Rates of sorption and partitioning of contaminants in river biofilm. Environ Sci Technol 32:3968–3973.

Hein W, Fent G, Kubiak R (2003) A new bath test system to investigate the fate of ^{14}C-labeled pesticides in water/sediment systems under outdoor conditions. In: Del Re AAM, Capri E, Padovani L, Trevisan M (eds) Pesticides in Air, Plant,

Soil and Water Systems. Proceedings, 12th Symposium on Pesticide Chemistry, Piacenza, Italy, pp 63–72.
Hellpointner E (2001) Environmental behavior of fentrazamide. Pflanzenschutz-Nachrichten Bayer 54:75–86.
Hellpointner E, Borchers H (2004) Behavior of prothioconazole (JAU 6476) in the environment. Pflanzenschuts-Nachrichten Bayer 57:163–180.
Helweg C, Mogensen BB, Søtensen PB, Madsen T, Bossi R, Rasmussen D, Petersen S (2003) Fate of pesticides in surface waters, laboratory and field experiments. Pesticide Research Nr. 68. Danish Environmental Protection Agency, Denmark.
Henneböle J (1999) Environmental behaviour of iprovalicarb (SZX 0722). Pflanzenschutz-Nachrichten Bayer 52:107–114.
Higashi RM, Crosby DG (1987) A physical-chemical microcosm for ricefield environmental fate studies. In: Greenhalgh R, Roberts TR (eds) Proceedings, Pesticide Science and Biotechnology, 6th IUPAC Congress of Pesticide Chemistry. Blackwell, Oxford, pp 445–448.
Hill IR, Shaw JL, Maund SJ (1994) Fate of chlorpyrifos in indoor microcosms and outdoor experimental ditches. In: Hill IR, Heimbach F, Leeuwangh P, Matthiessen P (eds) Freshwater Field Tests for Hazard Assessment of Chemicals. Lewis, Boca Raton, pp 249–271.
Hoagland RE, Zablotowicz RM (2000) The role of plant and microbial hydrolytic enzymes in pesticide metabolism. In: Hall JC, Hoagland RE, Zablotowicz RM (eds) Pesticide Biotransformation in Plants and Microorganisms: Similarities and Divergences. ACS Symposium Series 777. American Chemical Society, Washington, DC, pp 58–88.
Houx NWH, Dekker A (1987) A test system for the determining of the fate of pesticides in surface water. J Environ Anal Chem 29:37–59.
Huang W, Peng P, Yu Z, Fu J (2003) Effects of organic matter heterogeneity on sorption and desorption of organic contaminants by soils and sediments. Appl Geochem 18:955–972.
Huber G, Gémes E (1981) Decomposition of urea herbicide linuron [3-(3,4-dichlorophenyl)-1-methoxy-1-methylurea] in the water of lake balaton. J Ind Chem 9:113–124.
Hulbert MH, Bennett RH, Baerwald RJ, Long RL, Curry KJ, Curry AL, Abril MT (2002) Observation of the sediment-water interface: marine and freshwater environments. Mar Georesour Geotechnol 20:255–274.
Hunter KA, Liss PS (1981) Organic sea surface films. In: Ouursma EK, Dawson R (eds) Marine Organic Chemistry: Evolution, Composition, Interactions and Chemistry of Organic Matter in Seawater. Elsevier Oceanography Series 31. Elsevier, Amsterdam, pp 259–298.
Ide A, Niki Y, Sakamoto F, Watanabe I, Watanabe H (1972) Decomposition of pentachlorophenol in paddy soil. Agric Biol Chem 36:1937–1944.
Inao K, Kitamura Y (1999) Pesticide paddy field model (PADDY) for predicting pesticide concentrations in water and soil in paddy fields. Pestic Sci 55:38–46.
Isensee AR, Kearney PC, Jones GE (1979) Modeling aquatic ecosystems for metabolic studies. In: Khan MAQ, Lech JJ, Menn JJ (eds) Pesticide and Xenobiotic Metabolism in Aquatic Organisms. ACS Symposium Series 99. American Chemical Society, Washington, DC, pp 195–216.
Jackson R, Douglas M (1999) An aquatic risk assessment for cyhalofop-butyl: a new herbicide for control of barnyard grass in rice. In: Del Re, Attlio Amerigo Maria

(eds) Human and Environmental Exposure to Xenobiotics. Proceedings, 11th Symposium on Pesticide Chemistry. Goliardica Pavese, Pavia, Italy, pp 345–354.

Jinhe S, Jianying G, Ziyuan C (1989) Fate of carbofuran in model rice/fish ecosystems. Pestic Sci 26:147–157.

JMAFF (2000) Studies of fate in flooded aerobic soil. Guidelines for preparation of study results submitted when applying for registration of agricultural chemicals. Notification No. 12-Nousan-8147, 2-5-1. Director-General, Agricultural Protection Bureau, Ministry of Agriculture, Forestry and Fisheries, Japan.

Johnson PC, Kennedy JH, Morris RG, Hambleton FE, Graney RL (1994) Fate and effects of cyfluthrin (pyrethroid insecticide) in pond mesocosms and concrete microcosms. In: Graney RL, Kennedy JH, Rodgers JH (eds) Aquatic Mesocosm Studies in Ecological Risk Assessment. Lewis, Boca Raton, pp 337–371.

Jones TW, Kemp WM, Stevenson JC, Means JC (1982) Degradation of atrazine in estuarine water/sediment systems and soils. J Environ Qual 11:632–638.

Kaiser JP, Feng Y, Bollag JM (1996) Microbiol metabolism of pyridine, quinoline, acridine and their derivatives under aerobic and anaerobic conditions. Microbiol Rev 60:483–498.

Kale SP, Carvalho FP, Raghu K, Sherkhane PD, Pandit GG, Mohan Rao A, Mukherjee PK, Murthy NB (1999) Studies on degradation of ^{14}C-chlorpyrifos in the marine environment. Chemosphere 39:969–976.

Kalsch W, Knacker T, Robertz M, Studinger G, Franke C (1998) Partitioning and mineralization of [^{14}C] lindane in a laboratory sediment-water system. Environ Toxicol Chem 17:662–669.

Kanazawa J, Isensee AR, Kearney PC (1975) Distribution of carbaryl and 3,5-xylyl methylcarbamate in an aquatic model ecosystem. J Agric Food Chem 23:760–763.

Karickhoff SW, Morris KR (1985) Impact of tubificid oligocheates on pollutant transport in bottom sediments. Environ Sci Technol 19:51–56.

Karickhoff SW, Brown DS, Scott TA (1979) Sorption of hydrophobic pollutants on natural sediments. Water Res 13:241–248.

Karns JS, Muldoon MT, Mulbry WW, Derbyshire MK, Kearney PC (1987) Use of microorganisms and microbial systems in the degradation of pesticides. In: LeBaron HM, Mumma RO, Honeycutt RC, Duesing JH (eds) Biotechnology in Agricultural Chemistry. ACS Symposium Series 334. American Chemical Society, Washington, DC, pp 156–170.

Katagi T (2002) Abiotic hydrolysis of pesticides in the aquatic environment. Rev Environ Contam Toxicol 175:79–261.

Katagi T (2004) Photodegradation of pesticides on plant and soil surfaces. Rev Environ Contam Toxicol 182:1–195.

Kattner G, Nagel K, Eberlein K, Hammer KD (1985) Components of natural surface and subsurface waters. Oceanol Acta 8:175–183.

Kaufman DD (1974) Degradation of pesticides by soil microorganisms. In: Dinauer RC (ed) Pesticides in Soil and Water. Soil Science Society of America, Madison, WI, pp 133–202.

Kazumi J, Capone DG (1995) Microbial aldicarb transformation in aquifer, lake and salt marsh sediments. Appl Environ Microbiol 61:2820–2829.

Kewu Y, Hanhong M, Fengahun A (1997) Fate of pesticide *N′*-(2,4-dimethylphenyl)-*N*-methylformamide hydrochloride in simulated aquatic ecosystem. J Environ Sci 9:372–375.

Kile DE, Chiou CT, Zhou H, Li H, Xu O (1995) Partition of nonpolar organic pollutants from water to soil and sediment organic matters. Environ Sci Technol 29:1401–1406.
Kim J, Liu KH, Kang SH, Koo SJ, Kim JH (2003) Degradation of the sulfonylurea herbicide LGC-42153 in flooded soil. Pest Manag Sci 59:1260–1264.
Klaus U, Mohamed S, Volk M, Spiteller M (1998) Interaction of aquatic humic substances with anilazine and its derivatives: the nature of the bound residues. Chemosphere 37:341–361.
Klausen J, Tröber SP, Haderlein SB, Schwarzenbach RP (1995) Reduction of substituted nitrobenzenes by Fe (II) in aqueous mineral suspensions. Environ Sci Technol 29:2396–2404.
Klein O, Reiner H, Scholz K (1997) Metabolism and residues of spinoxamine (KWG 4168) in plants, animals and the environment. Pflanzenschutz-Nachrichten Bayer 50:71–98.
Klupinski TP, Chin YP (2003) Abiotic degradation of trifluralin by Fe (II): kinetics and transformation pathways. Environ Sci Technol 37:1311–1318.
Knulst JC, Backlund P, Hessen DO, Riise G, Södergren A (1997) Response of surface microlayers to artificial acid precipitation in a meso-humic lake in Norway. Water Res 31:2177–2186.
Knulst JC, Boerschke RC, Loemo S (1998) Differences in organic surface microlayers from an artificially acidified and control lake, elucidated by XAD-8/XAD-4 tandem separation and solid state ^{13}C NMR spectroscopy. Environ Sci Technol 32:8–12.
Knuth ML, Heinis LJ (1992) Dissipation and persistence of chlorpyrifos within littoral enclosures. J Agric Food Chem 40:1257–1263.
Knuth ML, Heinis LJ (1995) Distribution and persistence of diflubenzuron within littoral enclosure mesocosms. J Agric Food Chem 43:1087–1097.
Kodaka R, Sugano T, Katagi T, Takimoto Y (2002) Aerobic aquatic metabolism of fenitrothion and its oxon analog in water-sediment systems. J Pestic Sci 27: 235–241.
Kodaka R, Sugano T, Katagi T, Takimoto Y (2003) Comparative metabolism of oragnophosphorus pesticides in water-sediment systems. J Pestic Sci 28:175–182.
Kodaka R, Sugano T, Katagi T, Takimoto Y (2004) Aerobic aquatic soil metabolism of pesticides in water- and sediment-spiked systems. J Pestic Sci 29:332–338.
Koelmans AA, Hubert E, Koopman HW, Portielje R, Crum SJH (2000) Modeling the vertical distribution of carbendazim in sediments. Environ Toxicol Chem 19:793–800.
Kollig HP, Parrish RS, Holm HW (1987) An estimate of the variability in biotransformation kinetics of xenobiotics in natural waters by aufwuchs communities. Chemosphere 16:49–60.
Kördel W, Stein B (1997) Fate of the organotin pesticide azocyclotin in aquatic microcosms. Chemosphere 35:191–207.
Krieger A, Weinberger P, Greenhalgh R (1989) A mathematical model used to predict the modifying effects of adjuvants on fenitrothion insecticide in lake and estuarine water systems. In: Chow PNP (ed) Adjuvants and Agrochemicals, vol 2. CRC Press, Boca Raton, pp 157–167.
Krohn J (2001) Behavior of triacloprid in the environment. Pflanzenschutz-Nachrichten Bayer 54:281–290.
Krohn J, Hellpointner E (2002) Environmental fate of imidacloprid. Pflanzenschutz-Nachrichten Bayer 55(special edition):3–26.

Kurogochi S, Köster J (1998) Metabolism and residues of carpropamid (KTU 3616) in plants, animals and the environment. Pflanzenschutz-Nachrichten Bayer 51:219–244.
Lal R (1982) Accumulation, metabolism and effects of organophosphorus insecticides on microorganisms. Adv Appl Microbiol 28:149–200.
Lalah JO, Schramm KW, Henkelmann B, Lenoir D, Behechti A, Günther K, Kettrup A (2003) The dissipation, distribution and fate of a branched ^{14}C-nonylphenol isomer in lake water/sediment systems. Environ Pollut 122:195–203.
Laskowski DA (2002) Physical and chemical properties of pyrethroids. Rev Environ Contam Toxicol 174:49–170.
Lawrence JR, Kopf G, Headley JV, Neru TR (2001) Sorption and metabolism of selected herbicides in river biofilm communities. Can J Microbiol 47:634–641.
Leake CR (2003) Fenamidone: Metabolism in plants. Pflanzenschutz-Nachrichten Bayer 56:533–546.
Lee RF, Ryan C (1979) Microbial degradation of organochlorine compounds in estuarine waters and sediments. EPA-600/9-79-102. In: Proceedings, Workshop: Microbial Degaradation Pollution Marine Environment, 1978. U.S. Environmental Protection Agency, Washington, DC, pp 443–450.
Lee S, Gan J, Kim JS, Kabashima JN, Crowley DE (2004) Microbial transformation of pyrethroid insecticides in aqueous and sediment phases. Environ Toxicol Chem 23:1–6.
Leeuwangh P, Brock TCM, Kersting K (1994) An evaluation of four types of freshwater model ecosystem for assessing the hazard of pesticides. Hum Exp Toxicol 13:888–899.
Lehmann RG, Miller JR, Cleveland CB (1993) Fate of fluroxypyr in water. Weed Res 33:197–204.
Leistra M, Zwears AJ, Warinton JS, Crum SJH, Hand LH, Beltman WHJ, Maund SJ (2003) Fate of the insecticide *lambda*-cyhalothrin in ditch enclosures differing in vegetation density. Pest Manag Sci 60:75–84.
Lewis DL, Holm HW (1981) Rates of transformation of methyl parathion and diethyl phthalate by aufwuchs microorganisms. Appl Environ Microbiol 42:698–703.
Lin HT, Wong SS, Li GC (2001) Dissipation of epoxiconazole in the paddy field under subtropical conditions of Taiwan. J Environ Health Sci B36:409–420.
Lin K, Carlson DJ (1991) Photo-induced degradation of tracer phenols added to marine surface microlayers. Mar Chem 33:9–22.
Linders J, Adriaanse P, Allen R, Capri E, Gouy V, Hollis J, Jarvis N, Klein M, Lolos P, Maier WM, Maund S, Pais C, Russel M, Smeets L, Teixeira JL, Vizantinopoulos S, Yon D (2003) FOCUS Surface Water Scenarios in the EU Evaluation Process Under 91/414/EEC. Sanco/4802/2001-rev 2 final. European Commission, Brussels.
Liu W, Gan JJ, Lee S, Kabashima JN (2004) Phase distribution of synthetic pyrethroids in runoff and stream water. Environ Toxicol Chem 23:7–11.
Löffler D, Römbke J, Meller M, Ternes TA (2005) Environmental fate of pharmaceuticals in water/sediment systems. Environ Sci Technol 39:5209–5218.
Loor-Vela SX, Simmons JJC, Simmons FW, Raskin L (2003) Dissipation of [^{14}C]acetochlor herbicide under anaerobic aquatic conditions in flooded soil microcosms. J Agric Food Chem 51:6767–6773.

Lotse EG, Graetz DA, Chesters G, Lee GB, Newland LW (1968) Lindane adsorption by lake sediments. Environ Sci Technol 2:353–357.
Lutnicka H, Bogacka T, Wolska L (1999) Degradation of pyrethroids in an aquatic ecosystem model. Water Res 33:3441–3446.
Macalady DL, Wolfe NL (1984) Abiotic hydrolysis of sorbed pesticides. In: Krueger RF, Seiber JN (eds) Treatment and Disposal of Pesticide Wastes. ACS Symposium Series 259. American Chemical Society, Washington DC, pp 221–224.
Macalady DL, Wolfe NL (1985) Effects of sediment sorption and abiotic hydrolyses. 1. Organophosphorothioate esters. Agric Food Chem 33:167–173.
MacRae IC (1989) Microbial metabolism of pesticides and structurally related compounds. Rev Environ Contam Toxicol 109:1–87.
Maguire RJ, Hale EJ (1980) Fenitrothion sprayed on a pond: kinetics of its distribution and transformation in water and sediment. J Agric Food Chem 28:372–378.
Maguire RJ, Carey JH, Hart JH, Tkacz RJ, Lee HB (1989) Persistence and fate of deltamethrin sprayed on a pond. J Agric Food Chem 37:1153–1159.
Malekani K, Hellpointner E (2002) The environmental profile of propoxycarbazone-sodium (BAY MKH 6561). Pflanzenschutz-Nachrichten Bayer 55: 101–110.
Maloney SE, Maule A, Smith ARW (1988) Microbial transformation of the pyrethroid insecticides: permethrin, deltamethrin, fastac, fenvalerate and fluvalinate. Appl Environ Microbiol 54:2874–2876.
Marshall WK, Roberts JR (1977) Simulation modeling of the distribution of pesticides in ponds. In: Fentrothion: The Long-Term Effects of Its Use in Forest Ecosystems. NRCC No.16073. National Research Council of Canada (NRCC), Associate Committee on Scientific Criteria for Environmental Quality, Ottawa, pp 253–278.
Martin JD, Crawford CG, Larson SJ (2003) Pesticides in streams: summary statistics. Preliminary results from cycle I of the national water quality assessment program (NAWQA), 1992–2001. U.S. Geological Survey, NAWQA Pesticide National Synthesis Project. http://ca.water.usgs.gov/pnsp.
Maund SJ, Hamer MJ, Lane MCG, Farrelly E, Rapley JH, Goggin VM, Gentle WE (2002) Partitioning, bioavailability and toxicity of the pyrethroid insecticide cypermethrin in sediments. Environ Toxicol Chem 21:9–15.
Mayer P, Vaes WHJ, Wijnker F, Legierse KCHM, Karaaj RH, Tolls J, Hermens JLM (2000) Sensing dissolved sediment porewater contaminants of persistent and bioaccumulative pollutants using disposable solid-phase microextraction fibers. Environ Sci Technol 34:5177–5183.
Mersie W, Liu J, Seybold C, Tierney D (1998a) Extractability and degradation of atrazine in a submerged sediment. Weed Sci 46:480–486.
Mersie W, Seybold C, Tierney D, McNamee C (1998b) Effect of temperature, disturbance and incubation time on release and degradation of atrazine in water columns over two types of sediments. Chemosphere 36:1867–1881.
Mersie W, McNamee C, Seybold CA, Tierney DP (2000) Diffusion and degradation of atrazine in a water/sediment system. Environ Toxicol Chem 19:2008–2014.
Mersie W, McNamee C, Seybold C, Wu J, Tierney D (2004) Degradation of metolachlor in bare and vegetated soils and in simulated water-sediment systems. Environ Toxicol Chem 23:2627–2632.

Meyers PA, Kawka OE (1982) Fractionation of hydrophobic organic materials in surface microlyaers. J Great Lakes Res 8:288–298.
Meyers PA, Owen RM (1980) Sources of fatty acids in lake Michigan surface microlayers and subsurface waters. Geophys Res Lett 7:885–888.
Meyers P, Schelske CL (2000) An inexpensive, optical (infrared) detector to locate the sediment/water interface in lakes with unconsolidated sediments. J Paleolimnol 23:201–205.
Meylan W, Howard PH, Boethling RS (1992) Molecular topology/fragment contribution method for predicting soil sorption coefficients. Environ Sci Technol 26:1560–1567.
Miller GC, Zepp RG (1979a) Effects of suspended sediments on photolysis rates of dissolved pollutants. Water Res 13:453–459.
Miller GC, Zepp RG (1979b) Photoreactivity of aquatic pollutants sorbed on suspended desiments. Environ Sci Technol 13:860–863.
Miyamoto J, Mikami N, Takimoto Y (1990) The fate of pesticides in aquatic ecosystems. In: Hutson DH, Roberts TR (eds) Progress in Pesticide Biochemistry and Toxocology. Wiley, New York, pp 123–147.
Miyamoto J, Klein W, Takimoto Y, Roberts TR (1985) Critical evaluation of model ecosystems. Pure Appl Chem 57:1523–1536.
Miyazaki S, Sikka HC, Lynch RS (1975) Metabolism of dichlobenil by microorganisms in the aquatic environment. J Agric Food Chem 23:365–368.
Mortland MM, Halloran LJ (1976) Polymerization of aromatic molecules on smectite. Soil Sci Soc Am J 40:367–370.
Møhlenberg F, Petersen S, Gustavsson K, Lauridsen T, Friberg N (2001) Mesocosm experiments in the approval procedure for pesticides. A literature study on effects of mesocosm characteristics and validity of extrapolation methods to protect sensitive species. Pesticide Research No. 56. Danish Environmental Protection Agency, Denmark.
Muir DCG (1991) Dissipation and transformation in water and sediment. In: Grover R, Cessna AJ (eds) Environmental Chemistry of Herbicides, vol 2. CRC Press, Boca Raton, pp 1–89.
Muir DCG, Grift NP (1982) Fate of fluridone in sediment and water in laboratory and field experiments. J Agric Food Chem 30:238–244.
Muir DCG, Yarechewski AL (1982a) Degradation of niclosamide (2′,5-dichloro-4′-nitrosalicylamide) in sediment and water systems. J Agric Food Chem 30:1028–1031.
Muir DCG, Yarechewski AL (1982b) Degradation of terbutryn in sediments and water under various redox conditions. J Environ Sci Health B17:363–380.
Muir DCG, Yarechewski AL (1984) Degradation of methoxychlor in sediments under various redox conditions. J Environ Sci Health B19:271–295.
Muir DCG, Rawn GP, Grift NP (1985) Fate of pyrethroid insecticide deltamethrin in small ponds: a mass balance study. J Agric Food Chem 33:603–609.
Muir DCG, Kenny DF, Grift NP, Robinson RD, Titman RD, Murkin HR (1991) Fate and acute toxicity of bromoxynil esters in an experimental prairie wetland. Environ Toxicol Chem 10:395–406.
Muir DCG, Yarechewski AL, Neal BR (1992) Influence of surface films on the fate of deltamethrin following aerial application to prairie ponds. Environ Toxicol Chem 11:581–591.

Nakamura Y, Ishikawa K, Kuwatsuka S (1977) Degradation of benthiocarb in soils as affected by soil conditions. J Pestic Sci 2:7–16.
Napolitano GE, Richmond JE (1995) Enrichment of biogenic lipids, hydrocarbons and PCBs in stream-surface foams. Environ Toxicol Chem 14:197–201.
Ngim KK, Crosby DG (2001) Fate and kinetics of carfentrazone-ethyl herbicide in California, USA, flooded rice fields. Environ Toxicol Chem 20:485–490.
Nhan DD, Carvalho FP, Nam BQ (2002) Fate of ^{14}C-chlorpyrifos in the tropical estuarine environment. Environ Technol 23:1229–1234.
Norkrans B (1980) Surface microlayers in aquatic environments. Adv Microb Ecol 4:51–85.
OECD (2000) OECD Guideline for the testing of chemicals: adsorption–desorption using a batch equilibrium method. No. 106. OECD Publishing, Brussels.
OECD (2002) OECD Guideline for the testing of chemicals: aerobic and anaerobic transformation in aquatic sediment systems. No. 308. OECD Publishing, Brussels.
Ohyama H, Kuwatsuka S (1978) Degradation of bifenox, a diphenyl ether herbicide, methyl 5-(2,4-dichlorophenyl)-2-nitrobenzoate in soils. J Pestic Sci 3: 401–410.
Oliver BG, Cosgrove EG, Carey JH (1979) Effect of suspended sediments on the photolysis of organics in water. Environ Sci Technol 13:1075–1077.
O'Neill EJ, Cripe CR, Mueller LH, Connolly JP, Pritchard PH (1989) Fate of fenthion in salt-marsh environments: II. Transport and biodegradation in microcosms. Environ Toxicol Chem 8:759–768.
Ongley E (1996) Sediment measurements. In: Water quality monitoring: a practical guide to the design and implementation of freshwater quality studies and monitoring programmes. UNEP/WHO, chap 13. http://www.who.int/docstore/water_sanitation_health/wqmonitor/begin.htm.
Othman MAS, El-Nager M, Abdel-All A, Ibrahim S, Khamis A (1986) Analysis of fenvalerate and flucythrinate in aquatic ecosystems in relation to their toxicity. Alex Sci Exch 7:245–261.
Ou LT (2000) Pesticide biodegradation. In: Lederberg J (ed) Encyclopedia of Microbiology, 2nd Ed, vol 3. Academic Press, New York, pp 594–606.
Oyamada M, Kuwatsuka S (1979) Degradation of CNP, a diphenyl ether herbicide, in flooded soil under oxidative and reductive conditions. J Pestic Sci 4:157–163.
Oyamada M, Igarashi K, Kuwatsuka S (1980) Degradation of the herbicide naproanilide, 1-(2-naphthoxy)-propionanilide, in flooded soils under oxidative and reductive conditions. J Pestic Sci 5:495–501.
Paris DF, Steen WC, Baughman GL, Barnett JT Jr (1981) Second-order model to predict microbial degradation of organic compounds in natural waters. Appl Environ Microbiol 41:603–609.
Park RA, Firlie B, Camacho R, Sappington K, Coombs M, Mauriello D (1995) AQUATOX, a general fate and effects model for aquatic ecosystems. In: Toxic Substances in Water Environments: Assessment and Control. Proceedings of the Water Environment Federation Specialty Conference, Water Environment Federation, Alexandria, VA, part 3, pp 7–17.
Patrick WH, Gambrell RP, Faulkner SP (1996) Redox measurements of soils. In: Bigham JM (ed) Methods of Soil Analysis. Part 3, Chemical Methods. SSSA Book Series No. 5. Soil Science Society of America, Madison, WI, pp 1255–1273.

Peijnenburg WJGM, 'T Hart MJ, Den Hollander HA, Van De Meent D, Verboom HH, Wolfe NL (1992a) Reductive transformation of halogenated aromatic hydrocarbons in anaerobic water-sediment systems: kinetics, mechanisms and products. Environ Toxicol Chem 11:289–300.
Peijnenburg WJGM, 'T Hart MJ, Den Hollander HA, Van De Meent D, Verboom HH, Wolfe NL (1992b) QSARs for predicting reductive transformation rate constants of halogenated aromatic hydrocarbons in anoxic sediment systems. Environ Toxicol Chem 11:301–314.
Percival JB, Lindsay PJ (1997) Measurement of physical properties of sediments. In: Mudroch A, Azcue JM, Mudroch P (eds) Manual of Physico-Chemical Analysis of Aquatic Sediments. Lewis, Boca Raton, pp 7–45.
Petty DG, Skogerboe JG, Getsinger KD, Foster DR, Houtman BA, Fairchild JF, Anderson LW (2001) The aquatic fate of troclopyr in whole-pond treatment. Pest Manag Sci 57:764–775.
Pignatello JJ, Xing B (1996) Mechanisms of slow sorption of organic chemicals to natural particles. Environ Sci Technol 30:1–11.
Plane JMC, Zika RG, Zepp RG, Burns LA (1985) Photochemical modeling applied to natural waters. In: Zika RG, Cooper WJ (eds) Photochemistry of Environmental Aquatic Systems. ACS Symposium Series 327. American Chemical Society, Washington, DC, pp 250–267.
PMRA (1996) Proposed Regulatory Decision Document. PRDD 96-01. Tebufenozide. Pesticide Management Regulatory Agency, Health Canada, Ottawa.
PMRA (2001a) Proposed Regulatory Decision Document. PRDD 2001-05. Trinexapac-ethyl. Pesticide Management Regulatory Agency, Health Canada, Ottawa.
PMRA (2001b) Regulatory Note. REG 2001-08. Cloransulam-methyl. Pesticide Management Regulatory Agency, Health Canada, Ottawa.
PMRA (2001c) Regulatory Note. REG 2001-09. Zoxamide Zoxium® 80W Fungicide, Gravel® 75DF Fungicide. Pesticide Management Regulatory Agency, Health Canada, Ottawa.
PMRA (2002) Regulatory Note. REG 2002-05. Acetamiprid. Assail Brand 70WP Insecticide, Chipco Brand Tristar 70WSP Insecticide, Pristine Brand RTU Insecticide. Pesticide Management Regulatory Agency, Health Canada, Ottawa.
PMRA (2003a) Regulatory Note. REG 2003-03. Fenbuconazole. Pesticide Management Regulatory Agency, Health Canada, Ottawa.
PMRA (2003b) Regulatory Note. REG 2003-12. Fluazinam. Pesticide Management Regulatory Agency, Health Canada, Ottawa.
PMRA (2004a) Proposed Regulatory Decision Document. PRDD 2004-01. Tepraloxydim. Equinox EC, Dash HC. Pesticide Management Regulatory Agency, Health Canada, Ottawa.
PMRA (2004b) Regulatory Note. REG 2004-02. Boscalid. BAS 510. Pesticide Management Regulatory Agency, Health Canada, Ottawa.
PMRA (2004c) Regulatory Note. REG 2004-08. Methoxyfenozide. Pesticide Management Regulatory Agency, Health Canada, Ottawa.
Prins HBA, Snel JFH, Helder RJ, Zanstra PE (1980) Photosynthetic HCO_3^- utilization and OH^- excretion in aquatic angiosperms. Plant Physiol 66:818–822.
Pritchard PH, Bourquin AW, Frederickson HL, Maziarz T (1979) System design factors affecting environmental fate studies in microcosms. EPA-600/9-79-012.

Proceedings, Workshop: Microbial Degradation Pollution Marine Environment, 1978. U.S. Environmental Protection Agency, Washington, DC, pp 251–272.
Pritchard PH, Monti CA, O'Neill EJ, Connolly JP, Ahearn DG (1986) Movement of kepone® (chlordane) across an undisturbed sediment-water interface in laboratory systems. Environ Toxicol Chem 5:647–657.
Pritchard PH, O'Neill EJ, Spain CM, Ahearn DG (1987) Physical and biological parameters that determine the fate of *p*-chlorophenol in laboratory test systems. Appl Environ Microbiol 53:1833–1838.
Radosevich M, Tuovinen OH (2004) Microbial degradation of atrazine in soils, sediments and surface water. In: Gan JJ, Zhu PC, Aust SD, Lemley AT (eds) Pesticide Decontamination and Detoxification. ACS Symposium Series 863. American Chemical Society, Washington, DC, pp 129–139.
Ramanand K, Nagarajan A, Suflita JM (1993) Reductive dechlorination of the nitrogen heterocyclic herbicide picloram. Appl Environ Micobiol 59:2251–2256.
Ramesh A, Maheswari ST (2004) Dissipation of alachlor in cotton plant, soil and water and its bioaccumulation in fish. Chemosphere 54:647–652.
Rand GM, Clark JR, Holmes CM (2000) Use of outdoor freshwater pond microcosms. I. Microcosm design and fate of pyridaben. Environ Toxicol Chem 19:387–395.
Reeves GL (1999) An experimental design for outdoor water/sediment studies using the ^{14}C-labeled pesticides. In: Del Re, Attilio Amerigo Maria (eds) Human and Exvironmental Exposure to Xenobiotics, 11th Symposium on Pesticide Chemistry. Goliardica Pavese, Pavia, Italy, pp 181–187.
Rice PJ, Anderson TA, Coats JR (2004) Effect of sediment on the fate of metolachlor and atrazine in surface water. Environ Toxicol Chem 23:1145–1155.
Roberts TR (1974) The fate of WL63611 in a static aquatic system. In: Beran F (ed) Proceedings, European Weed Research Council, 4th International Symposium, Wasserunkraeuter Aquatic Weeds. Bundesanst Pflanzen, Vienna, pp 232–239.
Rodgers JH Jr, Dickson KL, Saleh FY, Staoles CA (1983) Use of microcosms to study transport, transformation and fate of organics in aquatic systems. Environ Toxicol Chem 2:155–167.
Rönnefahrt I, Traub-Eberhard U, Kördel W, Stein B (1997) Comparison of the fate of isoproturon in small- and large-scale water/sediment systems. Chemosphere 35:181–189.
Rosenberg A (1984) 2,3,6-Trichlorophenylacetic acid (Fenac) degradation in aqueous and soil systems. Bull Environ Contam Toxicol 32:383–390.
Salvestrini S, Coppola E, Capasso S (2004) Determination of the microscopic rate constants for the hydrolysis of diuron in soil/water mixture. Chemosphere 55:333–337.
Samsøe-Petersen L, Gustavson K, Madsen T, Mogensen BB, Lassen P, Skjemov K, Christoffersen K, Jørgensen E (2001) Fate and effects of esfenvalerate in agricultural ponds. Environ Toxicol Chem 20:1570–1578.
Scifres CJ, Allen TJ, Leinweber CL, Pearson KH (1973) Dissipation and phytotoxicity of dicamba residues in water. J Environ Qual 2:306–309.
Schäfers C, Hassink J (2000) Experimental simulation of fate in rivers: a mesocosm for studies with radiolabeled substances. In: Comejo J, Jamet P (eds) Pesticide/Soil Interactions. Institut National de la Recherche Agronomique, Paris, pp 389–395.

Schaffner LC, Dickhut RM, Mitra S, Lay PW, Brouwer-Riel C (1997) Effects of physical chemistry and bioturbation by estuarine macrofauna on the transport of hydrophobic organic contaminants. Environ Sci Technol 31:3120–3125.
Schauerte W, Lay JP, Klein W, Korte F (1982) Long-term fate of organochlorine xenobiotics in aquatic ecosystems: Distribution, residual behaviour and metabolism of hexachlorobenzene, pentachloronitrobenzene and 4-chloroaniline in small experimental ponds. Ecotoxicol Environ Saf 6:560–569.
Scholz K, Fritz R, Anderson C, Spiteller M (1988) Degradation of pesticides in an aquatic model ecosystem. Proceedings, BCPC Conference—Pests and Diseases. British Crop Protection Council, Brighton, UK, pp 149–158.
SETAC (1995) Lynch MR (ed) Procedures for assessing the environmental fate and ecotoxicity of pesticides. Society of Environmental Toxicology and Chemistry (SETAC–Europe), Brussels.
Sethunathan N (1973) Microbial degradation of insecticides in flooded soil and in anaerobic cultures. Residue Rev 47:143–165.
Sethunathan N, MacRae IC (1969) Persistence and biodegradation of diazinon in submerged soils. J Agric Food Chem 17:221–225.
Setzo SY, Sundaram KMS (1982) Behaviour and degradation of chlorpyrifos-methyl in two aquatic models. J Agric Food Chem 30:1032–1035.
Setzo SY, MacCarthy HR, Oloffs PC, Shepherd RF (1979) The fate of acephate and carbaryl in water. J Environ Sci Health B14:635–654.
Seybold CA, Mersie W, McName C, Tierney D (1999) Release of atrazine (^{14}C) from two undisturbed submerged sediments over a two-year period. J Agric Food Chem 47:2156–2162.
SFERM (2004) San Francisco Estuary Regional Monitoring Program Data. San Francisco Estuaary Institute, Oakland, CA. http://www.sfei.org/rmp/data.htm.
Sharak Genthner BR, Price II AW, Pritchard PH (1989) Anaerobic degradation of chloroaromatic compounds in aquatic sediments under a variety of enrichment conditions. Appl Environ Microbiol 55:1466–1472.
Sharom MS, Solomon KR (1981) Adsorption-desorption, degradation, and distribution of permethrin in aqueous systems. J Agric Food Chem 29:1122–1125.
Sijm RTHM, Hallen M, Schrap SM (1997) Influence of storage on sediment characteristics and of drying sediment on sorption coefficients of organic contaminants. Bull Environ Contam Toxicol 58:961–968.
Singles SK, Dean GM, Kirkpatrick DM, Mayo BC, Langford-Pollard AD, Barefoot AC, Bramble FQ Jr (1999) Fate and behavior of flupyrsulfuron-methyl in soil and aquatic systems. Pestic Sci 55:288–300.
Site AD (2001) Factors affecting sorption of organic compounds in natural sorbent/water systems and sorption coefficients for selected pollutants. A review. J Phys Chem Ref Data 30:187–439.
Smalling KL, Aelion CM (2004) Distribution of atrazine into three chemical fractions: impact of sediment depth and organic carbon content. Environ Toxicol Chem 23:1164–1171.
Smith AM, Mao J, Doane RA (1995) Metabolic fate of [^{14}C]acrolein under aerobic and anaerobic aquatic conditions. J Agric Food Chem 43:2497–2503.
Södergren A (1984) Small-scale temporal changes in the biological and chemical composition of surface microlayers in a eutrophic lake. Verh Int Verein Limnol 22:765–771.

Solomon KR, Yoo JY, Lean NK, Day KE, Stephenson GL (1986) Methoxychlor distribution, dissipation and effects in freshwater limnocorrals. Environ Toxicol Chem 5:577–586.
Sørensen SR, Bending GD, Jacobsen CS, Walker A, Aamand J (2003) Microbial degradation of isoproturon and related phenylurea herbicides in and below agricultural fields. FEMS Microbiol Ecol 45:1–11.
Southwood JM, Muir DCG, Mackay D (1999) Modelling agrochemical dissipation in surface microlayers following aerial depositions. Chemosphere 38:121–141.
Spain JC (1995) Biodegradation of nitroaromatic compounds. Annu Rev Microbiol 49:523–555.
Spiteller M, Klaus U, Pfeifer T (2002) Analysis of the binding of amitrole and anilazine to aquatic and terrestrial refractory organic substances. In: Frimmel FH (ed) Refractory Organic Substances in the Environment. Wiley-VCH, Weinheim, pp 446–474.
Stamper DM, Tuovinen OH (1998) Biodegradation of the acetanilide herbicides alachlor, metolachlor and propachlor. Crit Rev Microbiol 24:1–22.
Strek HJ (1998) Fate of chlorsulfuron in the environment. 1. Laboratory evaluations. Pestic Sci 53:29–51.
Stupp HP, Fahl U (2003) Environmental fate of clothianidin (TI-435; Poncho®). Pflanzenschutz-Nachrichten Bayer 56:59–74.
Suedel BC, Rodgers JH Jr (1991) Variability of bottom sediment characteristics of the continental United States. Water Resour Bull 27:101–109.
Susarla S, El Hefnawy MM, Masunaga S, Yamashita N, Yonezawa Y, Salem Rizk MM, Urushigawa Y (1997) Anaerobic biotransformation of organochlorine pesticides in Manzala Lake Eygypt. Toxicol Environ Chem 62:149–160.
Tarr JB, Dyson JS, Chapman PF, Dark RJ (2002) Modeling pesticide transformation in soil and aquatic environments: Development of a common approach. In: Phelps W, Winton K, Effland WR (eds) Pesticide Environmental Fate. ACS Symposium Series 813. American Chemical Society, Washington, DC, pp 103–123.
Tejada AW, Varca LM, Calumpang SMF, Bajet CM, Medina MJB (1997) Fate of pesticides in a model rice paddy ecosystem. In: Proceedings of International Symposium on the Use of Nuclear and Related Techniques for Studying Environmental Behavior of Crop Protection Chemicals. International Atomic Energy Agency, Vienna, pp 265–278.
The ARS Pesticide Properties Database. United States Department of Agriculture, Agricultural Research Service, Washington DC. http://www.ars.usda.gov/Services/docs.htm?docid=6433.
Thibodeaux LJ, Valsaraj KT, Reible DD (2001) Bioturbation-driven transport of hydrophobic organic contaminants from bed sediment. Environ Eng Sci 18:215–223.
Thomas RG (1990) Volatilization from water. In: Lyman WJ, Reehl WF, Rosenblatt DH (eds) Handbook of Chemical Property Estimation Methods. American Chemical Society, Washington DC, chap 15.
Thurman EM (1985) In: Thurman EM (ed) Organic Geochemistry of Natural Waters. Nijhoff/Junk, Dordrecht, Netherlands.
Tor JM, Xu C, Stucki JM, Wander MM, Sims GK (2000) Trifluralin degradation under microbiologically induced nitrate and Fe (III) reducing conditions. Environ Sci Technol 34:3148–3152.

Traub-Eberhard U, Schäfer H, Debus R (1994) New experimental approach to aquatic microcosm systems. Chemosphere 28:501–510.
Tuominen L, Kairesalo T, Hartikainen H (1994) Comparison of methods for inhibiting bacterial activity in sediment. Appl Environ Microbiol 60:3454–3457.
USEPA (1982) Aerobic aquatic metabolism studies. In: Pesticide assessment guidelines. Subdivision N, Chemistry, Environmental fate: 162-4. U.S. Environmental Protection Agency, Washington, DC.
USEPA (1997) The incidence and severity of sediment contamination in surface waters of the United States. National Sediment Quality Survey, vol. EPA 823-R-97-006. U.S. Environmental Protection Agency, Washington DC, vol 1.
Valsaraj KT, Sojitra I (1997) Transport of hydrophobic organic compounds by colloids through porous media. 3. Diffusion from sediment porewater to overlying water in laboratory microcosms. Colloids Surfaces A Phys Eng Asp 121:125–133.
van der Kolk JWH, Crum SJH (1993) Laboratory micro-ecosystems as physical models for predicting the fate of pesticides in aquatic systems. Sci Total Environ Suppl:1429–1437.
Van Rees KCJ, Sudicky EA, Rao PSC, Reddy KR (1991) Evaluation of laboratory techniques for measuring diffusion coefficients in sediments. Environ Sci Technol 25:1605–1611.
Van Veld PA, Spain JC (1983) Degradation of selected xenobiotic compounds in three types of aquatic test systems. Chemosphere 12:1291–1305.
Veith GD, Lee GF (1971) Water chemistry of toxaphene: role of lake sediments. Environ Sci Technol 5:230–234.
Vink JPM, van der Zee SEATM (1997) Effect of oxygen status on pesticide transformation and sorptions in undisturbed soil and lake sediments. Environ Toxicol Chem 16:608–616.
Voice TC, Weber WJ Jr (1983) Sorption of hydrophobic compounds by sediments, soils and suspended solids: I. Theory and background. Water Res 17:1433–1441.
Walker WW (1984) Development of a fate/toxicity screening test. EPA/600/4-84/074. U.S. Environmental Protection Agency, Washington, DC.
Walker WW, Cripe CR, Pritchard PH, Bourquin AW (1988) Biological and abiotic degradation of xenobiotic compounds in in-vitro estuarine water and sediment/water systems. Chemosphere 17:2225–2270.
Wandiga SO, Ongeri DMK, Mbuvi L, Lalah JO, Jumba IO (2002) Accumulation, distribution and metabolism of ^{14}C-1,1,1-trichloro-2,2-bis(*p*-chlorophenyl)ethane (*p,p'*-DDT) residues in a model tropical marine ecosystem. Environ Technol 23:1285–1292.
Wang YS, Hwang KL, Hsieh YN, Chen YL (1992) Fate of the herbicide naproanilide in a rice paddy model ecosystem. J Pestic Sci 17:161–167.
Walse SS, Pennington PL, Scott GI, Ferry JL (2004) The fate of fipronil in modular estuarine mesocosms. J Environ Monit 6:58–64.
Warren N, Allan IJ, Carter JE, House WA, Parker A (2003) Pesticides and other micro-organic contaminants in freshwater sedimentary environments: a review. Appl Geochem 18:159–194.
Weinberger P, Greenhalgh R, Moody RP, Boulton B (1982) Fate of fenitrothion in aquatic microcosms and the role of aquatic plants. Environ Sci Technol 16: 470–473.
Wijayaratne RD, Means JC (1984) Affinity of hydrophobic pollutants for natural estuarine colloids in aquatic environments. Environ Sci Technol 18:121–123.

Willis GH, Wander RC, Southwick LM (1974) Degradation of trifluralin in soil suspensions as related to redox potential. J Environ Qual 3:262–265.
Wolfe NL, Macalady DL (1992) New perspectives in aquatic redox chemistry: abiotic transformations of pollutants in groundwater and sediments. J Contam Hydrol 9:17–34.
Wolfe NL, Zepp RG, Paris DF, Baughman GL, Hollis RC (1977) Methoxychlor and DDT degradation in water: Rates and products. Environ Sci Technol 11:1077–1081.
Wolfe NL, Kitchens BE, Macalady DL, Grundl TJ (1986) Physical and chemical factors that influence the anaerobic degradation of methyl parathion in sediment systems. Environ Toxicol Chem 5:1019–1026.
Wolfe NL, Mingelgrin U, Miller GC (1990) Abiotic transformations in water, sediments and soil. In: Cheng HH (ed) Pesticides in the Soil Environment: Processes, Impacts, and Modeling. SSSA Book Series 2. Soil Science Society of America, Madison, WI, pp 103–168.
Wolt JD, Nelson HP Jr, Cleveland CB, van Wesenbeeck IJ (2001) Biodegradation kinetics for pesticide exposure assessment. Rev Environ Contam Toxicol 169:123–164.
Wood M, Harold J, Johnson A, Hance R (1991) The potential for strazine degradation in aquifer sediments. BCPC Monograph. British Crop Protection Council, Brighton. Pestic Soil Water 47:175–182.
Woodruff SL, House WA, Callow ME, Leadbeater BSC (1999a) The effects of biofilms on chemical processes in surficial sediments. Freshw Biol 41:73–89.
Woodruff SL, House WA, Callow ME, Leadbeater BSC (1999b) The effects of a developing biofilm on chemical changes across the sediment-ester interface in a freshwater environment. Int Rev Hydrobiol 84:509–532.
Wu SC, Gschwend PM (1986) Sorption-kinetics of hydrophobic organic compounds to natural sediments and soils. Environ Sci Technol 20:717–725.
Yin Y, Wang Z, Zhang Y, Xu Y, Xu L, Wu W, Chen G (1994) Ecotoxicological study of fenpropathrin on some common Chinese aquatic organisms and evaluation of its effect on the aquatic ecosystem. EPA-600-R-94-138. Fish Physiology, Toxicology and Water Quality Management. U.S. Environmental Protection Agency, Washington, DC, pp 47–60.
Ying GG, Williams B (2000) Laboratory study on the interaction between herbicides and sediments in water systems. Environ Pollut 107:399–405.
Zablotowicz RM, Hoagland RE, Lee H, Alber T, Trevers JT, Hill JC, Locke MA (2000) Transformation of nitroaromatic pesticides and related xenobiotics by microorganisms and plants. In: Hill JC, Hoagland RE, Zablotowicz RM (eds) Pesticide Biotransformation in Plants and Microorganisms. Similarities and Divergences. ACS Symposium Series 777. American Chemical Society, Washington DC, pp 194–216.
Zanaradini E, Arnoldi A, Boschin G, D'Agostina A, Negri M, Sorlini C (2002) Degradation pathways of chlorsulfuron and metsulfuron-methyl by a *Pseudomonas fluorescens* strain. Ann Microbiol 52:25–37.
Zaranyika MF, Nyandoro MG (1993) Degradation of glyphosate in the aquatic environment: an enzymatic kinetic model that takes into account microbial degradation of both free and colloidal (or sediment) particle adsorbed glyphosate. J Agric Food Chem 41:838–842.
Zepp RG, Schlotzhauer PF (1981) Effects of equilibration time on photoreactivity of the pollutant DDE sorbed on natural sediments. Chemosphere 10:453–460.

Zepp RG, Schlotzhauer PF (1983) Influence of algae on photolysis rates of chemicals in water. Environ Sci Technol 17:462–468.
Zepp RG, Wolfe NL (1987) Abiotic transformation of organic chemicals at the particle-water interface. In: Stumm W (ed) Aquatic Surface Chemistry: Chemical Processes at the Particle-Water Interface. Wiley, New York, pp 423–455.
Zhong C, Chen S, Carvalho FP, Zhao X (1998) A mini model ecosystem about the fate of labeled pesticides. Zhongshan Daxue Xuebao Ziran Kexueban 37:93–97.
Zhou JL, Rowland S, Mantoura FC (1995) Partition of synthetic pyrethroid insecticides between dissolved and particulate phases. Water Res 29:1023–1031.

Manuscript received August 31, accepted September 9, 2005.

Index

MIX
Papier aus verantwortungsvollen Quellen
Paper from responsible sources
FSC® C105338

If you have any concerns about our products,
you can contact us on
ProductSafety@springernature.com

In case Publisher is established outside the EU,
the EU authorized representative is:
Springer Nature Customer Service Center GmbH
Europaplatz 3, 69115 Heidelberg, Germany

Printed by Libri Plureos GmbH
in Hamburg, Germany